14-95

An Introdu...
Cardiovasc...

D1643181

An Introduction to Cardiovascular Physiology

J R Levick, MA, DPhil, BM, BCh
Reader in Physiology, St George's Hospital Medical School, London

Butterworths
London Boston Singapore Sydney Toronto Wellington

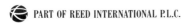 PART OF REED INTERNATIONAL P.L.C.

First published, 1991

© Butterworth & Co (Publishers) Ltd, 1991

British Library Cataloguing in Publication Data

Levick, J. R.
  An introduction to cardiovascular physiology.
  1. Man. Cardiovascular system. Physiology
  I. Title
  612.1

ISBN 0-750-61028-X

Library of Congress Cataloging-in-Publication Data

Levick, J. R. (J. Rodney)

  An introduction to cardiovascular physiology/J.R. Levick.

    p. cm.
  Includes bibliographical references.
  Includes index.
  ISBN 0-750-61028-X :
  1. Cardiovascular system–Physiology.   I. Title.
  [DNLM: 1. Cardiovascular System–physiology.   WG 102 L664i]
  QP101.L47   1990
  612.1–dc20
  DNLM/DLC
  for Library of Congress   90-1811
                                                      CIP

Composition by Genesis Typesetting, Laser Quay, Rochester, Kent
Printed and bound in Great Britain by Alden Press (London & Northampton) Ltd,
London, England

Preface

This is an introductory text designed primarily for students of medicine and physiology. The teaching style is necessarily didactic in many places ('The way it works is like this, . . .') but also, where space permits, I have tried to show how our knowledge of the circulation is derived from experimental observations. The latter not only puts flesh on the didactic bones, but ultimately keep the student (and the writer) in contact with reality. Human data are presented where possible, and their relevance to human disease is emphasized. The occasional anecdotes and doggerel betray a deplorable levity on my part, but will have earned their place if they interest the reader, and doubly so if they help to make a point memorable. The undergraduate will find a useful guide to learning objectives in the coloured box at the beginning of each chapter.

The traditional weighting of subject matter has been re-thought, resulting in a fuller account of microvascular physiology than is usual. This reflects the explosion of microvascular research over the past two decaddes. Even setting aside these advances, it seems self-evident that the culminating, fundamental function of the cardiovascular system – the transfer of nutrients from plasma to the tissue – merits more than the few lines usually accorded to it in introductory texts. Major advances continue apace in other fields too, for example the elucidation of the biochemical events underlying Starling's law of the heart, the discovery of new vasoactive substances produced by endothelium, the exploration of non-adrenergic, non-cholinergic neurotransmission, rapid advances in vascular smooth muscle physiology, and new concepts on how the central nervous control of the circulation is organized.

I would like to thank many friends and colleagues – Tom Bolton, John Gamble, Max Lab, William Large, Janice Marshall, Charles Michel, Mark Noble, Peter Simkin, Laurence Smaje, Mike Spyer and John Widdicombe – for helpful comments on sections of the text. Any mistakes or muddles that remain are, of course, entirely my own; please do not hesitate to point them out to me. Perhaps I should thank the cardiovascular system too, for proving to be even more fascinating than I had realized before writing this book!

Rodney Levick
St. George's Hospital Medical School

Contents

Chapter 1
Overview of the cardiovascular system

The heart and blood vessels form a system for the rapid transport of oxygen, nutrients, waste products and heat around the body. Small primitive organisms lack such a system because their needs can be met by direct diffusion from the environment, and even in man diffusion remains the fundamental transport process between blood and cells. In order to appreciate fully the need for a cardiovascular system we must begin by considering some properties of the diffusion process.

1.1 Diffusion: its virtues and limitations

The 'drunkard's walk' theory. Diffusion is a passive process in that it is not driven by metabolic energy but arises from the innate random thermal motion of molecules in a solution or gas. Although each individual movement of a solute molecule occurs in a random direction (the 'drunkard's walk') this nevertheless produces a net movement of solute in the presence of a concentration

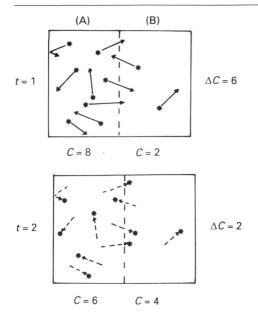

Figure 1.1 Sketch illustrating how random molecular steps result in a net movement of solute down a concentration gradient. At time 1 (upper sketch) there are 8 molecules per unit volume in (A) and 2 in (B). At time 2 (lower sketch) each molecule has moved a unit step in a random direction. Because there was a greater density of molecules in A there was a greater probability of random movement from A to B, resulting in a net 'downhill' flux

gradient. Figure 1.1 illustrates how this happens. Notice that although the net transfer of solute is from compartment A into compartment B there is also a smaller backflux into compartment A. This can be proved by adding a trace of radiolabelled solute to compartment B; some labelled molecules appear in compartment A even though the net diffusion is from A to B.

The importance of diffusion distance The rate at which diffusional transport occurs is critically important because the supply of nutrients must keep up with cellular demand. However, as Albert Einstein showed the time (t) that it takes a randomly jumping particle to move a distance x in one specific

direction increases with the square of distance:

$$t \propto x^2 \tag{1.1}$$

(see footnote to Table 1.1); and as a result diffusional transport is extremely slow over large distances. While diffusion across a short distance such as the neuromuscular

Table 1.1 Time taken for a glucose molecule to diffuse a specified distance in one direction

Distance (x)	Time (t)*	Comparable distance in vivo
0.1 μm	0.000005 s	Neuromuscular gap
1.0 μm	0.0005 s	Capillary wall
10.0 μm	0.05 s	Cell to capillary
1 mm	9.26 min	Skin, artery wall
1 cm	15.4 h	Ventricle wall

* Times are calculated by Einstein's equation $t = x^2/2D$. 'D' is the solute diffusion coefficient. For glucose in water at 37°C, D is 0.9×10^{-5} cm²/s (Einstein, A. (1905) *Theory of Brownian Movement* (trans. and ed. by R. Fürth and A. D. Cowper, 1956), Dover Publications, New York)

gap (0.1 μm) takes only 5 millionths of a second, diffusion across the heart wall (approximately 1 cm) is hopelessly slow, taking over half a day (Table 1.1). Sadly, Nature often proves the validity of Einstein's equation and Figure 1.2 is an example of this: it shows the heart of a patient who suffered a coronary thrombosis (obstruction of the blood supply to the heart wall). The pale area in the wall is muscle which has died from lack of oxygen even though the adjacent cavity (the left ventricle) was fully of richly oxygenated blood; the patient died simply because a distance of a few millimetres reduced diffusional transport to an inadequate rate.

Convection for fast long-distance transport Clearly then, for distances greater than approximately 0.1 mm a faster transport system is needed and this is provided by the cardiovascular system (Figure 1.3). The

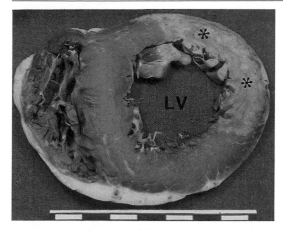

Figure 1.2 Section through the left ventricle of a human heart after a coronary thrombosis. The section is stained to show intracellular enzyme content. The pale area marked by asterisks is an infarct, an area of muscle severely damaged or killed by oxygen lack; the pallor is due to the intracellular enzyme having leaked out of the dying cells. The infarct was caused by a thrombus in a coronary artery, blocking the convectional delivery of oxygen. Diffusion of oxygen from the blood in the adjacent cavity of the left ventricle is unaffected yet only a thin rim of tissue (approximately 1 mm) can survive on this diffusional flux. (Courtesy of Professor M. Davies, St. George's Hospital Medical School, London)

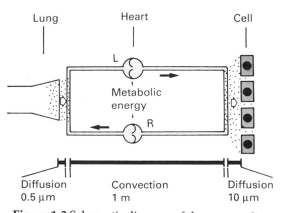

Figure 1.3 Schematic diagram of the mammalian cardiovascular system to illustrate the roles of diffusion and convection in oxygen transport. L, left side of heart. R, right side of heart. The pulmonary and systemic circulations lie in series

cardiovascular system still relies on diffusion for the uptake of molecules at points of close proximity to the environment (e.g. oxygen uptake into lung capillaries) but it then transports them rapidly over large distances by sweeping them along in a stream of pumped fluid. This form of transport is called bulk flow or convective transport. Convective transport requires an energy input and this is provided by a pump, the heart. In man convection takes only 30 s to carry oxygen over a metre or more from the lungs to the smallest blood vessels of the limbs (capillaries). Over the final 10–20 microns separating the capillary from the cells, diffusion is again the main transport process.

1.2 Functions of the cardiovascular system

The first and foremost function is the *rapid convection* of oxygen, glucose, amino acids, fatty acids, vitamins, drugs and water to the tissues and the rapid washout of metabolic waste products like carbon dioxide, urea and creatinine. The cardiovascular system is also part of a *control system* in that it distributes hormones to the tissues and even secretes some hormones itself (e.g. atrial natriuretic peptide; see Chapter 11). In addition, the circulation plays a vital role in *temperature regulation*, for it regulates the delivery of heat from the core of the body to the skin, and a vital role in *reproduction*, as it provides the mechanism for penile erection.

1.3 Circulation of blood

The heart is an intermittent muscular pump, or rather two adjacent pumps, the right and left ventricles (see Figure 1.4). Each pump is filled from a reservoir, the right or left atrium. The right ventricle pumps blood through the lungs to the left side (the pulmonary circulation) and the left

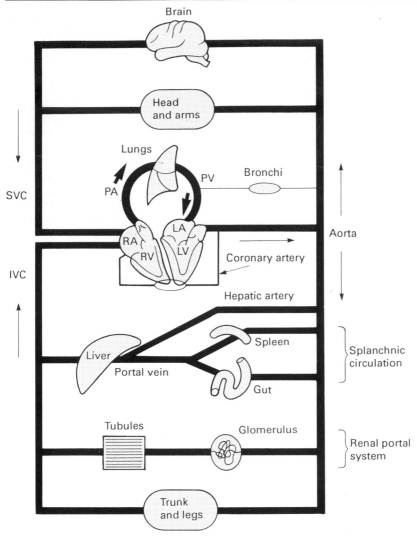

Figure 1.4 General arrangement of the circulation showing right and left sides of the heart in series. Circulations to individual organs are mostly in parallel (e.g. cerebral and coronary circulations) but a few are in series (liver, renal tubules). Note that the bronchial venous blood drains anomolously into the left rather than right atrium. PA, PV, pulmonary artery and vein; RA, LA, right and left atrium (an 'atrium' was a Roman hall); RV, LV, right and left ventricle; SVC, IVC, superior and inferior vena cava

ventricle simultaneously pumps blood through the rest of the body and back to the right side (the systemic circulation). The blood is compelled to follow a circular pathway by one-way valves located in the heart and veins, as was first established by the London physician, William Harvey, in a celebrated book entitled *De Motu Cordis* (*Concerning the Movement of the Heart*) in 1628.

Pulmonary circulation Venous blood enters the right atrium from the two major veins, the superior and inferior venae cavae, then flows through a valve into the right ventricle. The ventricle, which is composed mainly of cardiac muscle, receives the blood while it is in a state of relaxation called diastole (pronounced die-a-stole-ea). Contraction, or 'systole' (pronounced sis-tole-ea), then forces part of the blood out through the pulmonary artery and into the lungs at a low pressure. Gases exchange by diffusion in the lung air sacs (alveoli) raising the blood oxygen content from approximately 150 ml/l (venous blood) to 195 ml/l. The oxygenated blood returns through the pulmonary veins to the left atrium and left ventricle.

Systemic circulation The left ventricle contracts virtually simultaneously with the right and ejects the same volume of blood but at a much higher presure. The blood flows through the aorta and the branching arterial system into fine thin-walled tubes called capillaries. Here the ultimate function of the cardiovascular system is fulfilled as dissolved gases and nutrients diffuse between the capillary blood and the tissue cells. The circulation of the blood is completed by the venous system which conducts blood back to the venae cavae.

1.4 Cardiac output and its distribution

The cardiac output is the volume of blood ejected by one ventricle during one minute, and this depends on both the volume ejected per contraction (the stroke volume) and the number of contractions per minute (heart rate). In a resting 70 kg adult, the stroke volume is 70–80 ml and the heart rate is approximately 65–75 beats/min, so the resting cardiac output is approximately 75 ml × 70 per min or roughly 5 l per min. The output is not fixed, however, and

adapts rapidly to changing internal or external circumstances. In severe exercise for example, when oxygen demand can increase tenfold, the heart responds with a fourfold increase in output, or even more in athletes. These changes imply that special control systems must exist for regulating the heart beat, and these controls are the subject of Chapters 3 and 6.

Distribution of cardiac output The output of the right ventricle passes to the lungs alone. The output of the left ventricle is in general distributed to the peripheral tissues in proportion to their metabolic rate; resting skeletal muscle for example accounts for around 20% of human oxygen consumption and the muscle receives roughly 20% of the cardiac output (Figure 1.5). This egalitarian principle is over-ridden, however, where the particular function of an organ requires a higher blood flow; the kidneys consume only 6% of the body's oxygen yet receive 20% of the cardiac output since this is necessary for their excretory function. As a result some other tissues are relatively ill-supplied and, rather surprisingly, cardiac muscle is one of them. Consequently, it is compelled to extract an unusually high proportion of the oxygen content of the blood, namely 65–75%. The distribution of the cardiac output is not fixed, however, but is actively adjusted to meet varying conditions. A good example of this is provided by heavy exercise, where the proportion of the cardiac output going to skeletal muscle increases to 80% or more, owing to widening of the vessels supplying blood to the muscle (vasodilatation).

1.5 Introducing some hydraulic considerations: pressure and flow

Blood pressure What drives blood along the blood vessels after it has left the heart? The main factor is the gradient of pressure

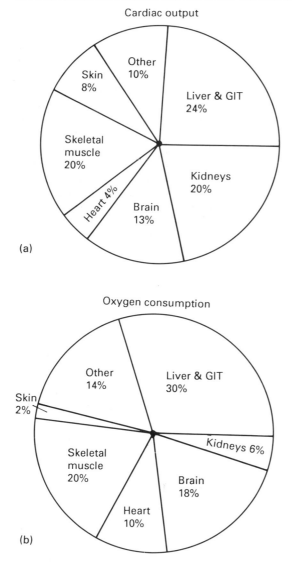

(a)

(b)

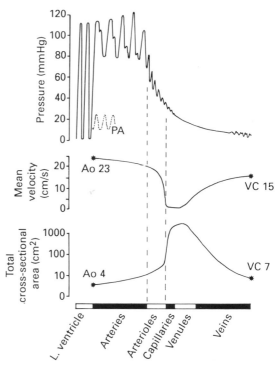

Figure 1.6 The profile of blood pressure and velocity in the systemic circulation of a resting man. The abscissa represents distance along the vessels. Velocity at any level is the cardiac output divided by total cross-sectional area of the vascular bed at that point. Pressure in the pulmonary artery is shown as a dotted line. Ao, human aorta. VC, human vena cava. (From several sources)

Figure 1.5 The distribution of left ventricular output in a resting man (top) compared with the oxygen consumption (bottom) of the various tissues. GIT = gastrointestinal tract. (From Wade, O. L. and Bishop, J. M. (1962) *Cardiac Output and Regional Blood Flow*, Blackwell, Oxford, by permission)

along the vessel. Ventricular ejection raises aortic blood pressure to approximately 120 mmHg above atmospheric pressure while the pressure in the great veins is close to atmospheric pressure, and the pressure difference drives blood from artery to vein. Arterial pressure is pulsatile, however, not steady, because the heart ejects blood intermittently: between successive ejection phases the systemic arterial pressure decays from 120 mmHg to approximately 80 mmHg, while pulmonary pressure decays from 25 mmHg to 10 mmHg (Figure 1.6). The conventional way of writing this is 120/80 mmHg and 25/10 mmHg. The conventional units are mmHg above atmospheric pressure because human blood pressure is measured clinically with a mercury column, taking atmospheric pressure as the reference of zero level (see Appendix, 'Pressure').

Simple 'law of flow' The relation between a pulsatile flow and pulsatile driving pressure is quite complex (Chapter 7), but it is useful at this stage to consider a simpler situation, such as water flowing along a rigid tube under a steady pressure gradient. Under these conditions, flow ($\dot{Q}$) is directly proportional to the pressure difference between the inlet (P_1) and outlet (P_2) of the tube:

$$\dot{Q} \propto P_1 - P_2 \qquad (1.2)$$

Flow is often represented by $\dot{Q}$ because Q stands for quantity of fluid and the dot denotes rate of passage, this being Newton's original calculus notation. It should be noted that flow is by definition a rate (the passage of a volume or mass per unit time) and the common expression 'rate of flow' is really rather muddling and best avoided. By inserting a proportionality factor (K) into the above expression we can change it into an equation describing flow:

$$\dot{Q} = K \,.\, (P_1 - P_2) \qquad (1.3)$$

where K is called the *hydraulic conductance* of the tube. Conductance is the reciprocal of *resistance* (R), so we can also write:

$$\dot{Q} = \frac{(P_1 - P_2)}{R} \qquad (1.4)$$

This expression is a form of Darcy's law of flow and is analogous to Ohm's law for an electrical current ($I = \Delta V/R$). It states that flow is proportional to driving pressure ($P_1 - P_2$) and is inversely proportional to the hydraulic resistance. The total resistance of the systemic circulation in man is around 0.02 mmHg per ml/min while that of the pulmonary circulation is only 0.003 mmHg per ml/min, and the latter low value explains why a very low pressure suffices to drive the cardiac output through the lungs.

The law of flow also helps us to understand how the blood flow to an organ is regulated. Equation 1.3 shows that there are essentially only two ways of altering flow: either the driving pressure must be changed or else the vascular resistance. In normal subjects blood pressure is in fact kept roughly constant, and it is changes in vascular resistance that regulate local blood flow. During salivation, for example, blood flow to the salivary glands can increase 10 times due to a fall in vascular resistance to one-tenth its former value, while the driving pressure (arterial pressure) does not increase at all. Changes in vascular resistance are brought about by contraction or relaxation of the vessels, so we should next consider their structure.

1.6 Structure and functional classification of blood vessels

The aorta and pulmonary artery divide into smaller arteries, which branch progressively to form narrow high-resistance vessels called arterioles (see Figure 1.8). Arterioles branch into innumerable capillaries which then converge to form venules and veins. The characteristic dimensions of these various vessels are set out in Table 1.2.

Structure of the blood vessel wall

With the exception of capillaries all blood vessels have the same basic three-layered plan (see Figure 1.7) consisting of a tunica intima (innermost layer), tunica media (middle layer) and tunica adventitia (outer layer). The *intima* consists of flat endothelial cells resting on a thin layer of connective tissue. The endothelial layer is the main barrier to plasma proteins and also secretes many vasoactive products, but it is mechanically weak. The *media* supplies mechanical strength and contractile power. It consists of spindle-shaped smooth muscle cells arranged circularly and embedded in a matrix of elastin and collagen fibres. Internal and external elastic laminae (sheets)

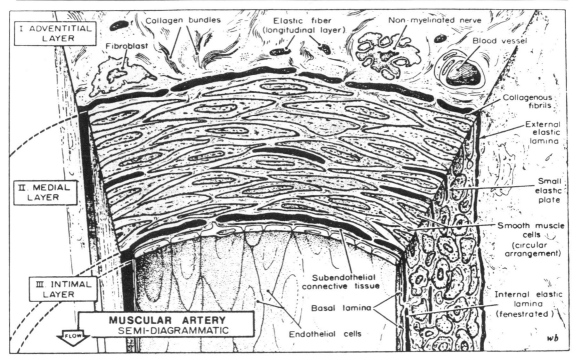

Figure 1.7 Sketch of the structure of a muscular artery. (From Rhodin, J. A. G. 1980, see Further reading, by permission)

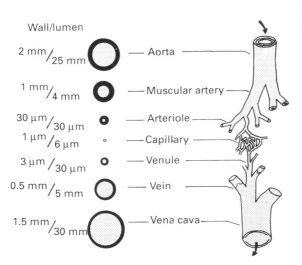

Figure 1.8 The changing thickness of the wall relative to the diameter of the lumen in the various types of blood vessel. The dimensions on the left apply to human vessels. (Sources as for Table 1.3)

mark the boundaries of the media. The *adventitia* is a connective tissue sheath with no distinct outer border which holds the vessel loosely in place. The adventitia of the larger arteries contains small blood vessels, the vasa vasorum (literally 'vessels of vessels'), and in the largest arteries they penetrate into the outer two-thirds of the media too. Their task is to nourish the thick media of large vessels.

Functional classification

The circulation is constructed on the sound economical principle that each vessel must fulfil at least one other function in addition to the conduction of blood. The structure of the vessel is specially adapted to this function and the following functional categories are recognized.

Table 1.2 Average dimensions of blood vessels in a bat wing†

Vessel	Length (mm)	Diameter (µm)	Number	Total cross-section area (µm²)	Volume capacity (%)
Main artery	17.0	53	1	2260	10.1
Arteriole	0.95	7	119	5100	1.1
Exchange vessel*	0.22	4–7	549–1727	6550–78230	4.7
Venule	1.0	21	345	128000	33.7
Small vein	3.4	37	24	27890	25.0
Main vein	16.6	76	1	4880	21.4

† Bat wing vessels are easily observed, unlike most mammalian circulations. The general pattern is similar, however, in other tissues and species. (After Wiedeman, M. P., Tuma, R. F. and Mayrowitz, H. N. (1981) *An Introduction to Microcirculation*. New York, Academic Press)
* Capillaries and postcapillary venules

Elastic arteries (diameter 1–2 cm in man) The pulmonary artery, aorta and major branches, like the iliac arteries, have very distensible walls because their tunica media is particularly rich in elastin, a protein which is six times more extensible than rubber (see Table 1.3). This enables the major arteries to expand and receive the stroke volume during ventricular ejection and to recoil during diastole, thereby converting the intermittent ejection of blood by the heart into a continuous flow through the more distal vessels. Another protein, collagen, forms a meshwork of strong fibrils in the media. Collagen is 100 times stiffer than elastin and its role seems to be to prevent overdistension.

Muscular arteries (diameter 0.1–1.0 cm in man) In medium to small arteries like the popliteal, radial, cerebral and coronary arteries the tunica media is thicker relative to the lumen diameter (see Figure 1.8), and it contains more smooth muscle (see Table 1.3). The muscular arteries act as low-resistance conduits and their thick walls help prevent collapse at sharp bends like the knee joint. They have a rich autonomic nerve supply and can contract but they are not in general important in the regulation of blood flow because their resistance is low. The importance of their contractile ability was, however, vividly demonstrated to me one day as a student in a casualty department, when a motorcycle crash victim was

Table 1.3 The changing composition of the blood vessel wall (%)

	Endothelium	Smooth muscle	Elastic tissue	Connective tissue
Elastic artery	5	25	40	27
Arteriole	10	60	10	20
Capillary	95	–	–	5 (basal lamina)
Venule	20	20	–	60

(After Caro, C. G., Pedley, T. J., Schroter, R. C. and Seed, W. A. (1978) *The Mechanics of the Circulation*. Oxford University Press, Oxford, and Burton, A. C. (1972) *Physiology and Biophysics of the Circulation*. Year Book Medical Publishers, Chicago)

brought in with one leg almost completely severed at the knee. The popliteal artery was torn in half yet the proximal stump was scarcely bleeding – a profound contraction of the media had prevented the patient from bleeding to death. Contraction of muscular arteries can also occur physiologically in cerebral arteries (see Chapter 12) and the limbs of diving animals (see Chapter 14).

Resistance vessels The arterioles have the thickest walls of all vessels relative to their lumen, the ratio of wall thickness to lumen diameter being approximately 1.0 (Figure 1.8). The muscular walls of the larger arterioles are richly innervated by vasoconstrictor nerve fibres but the terminal arterioles or metarterioles (diameter 10–40 µm) are poorly innervated and possess only 1–3 layers of smooth muscle cells. Because of their narrow lumen and limited numbers (Table 1.2) the arterioles form the chief resistance to blood flow in the systemic circulation: the pressure profile in Figure 1.6 confirms this since the pressure drop across the systemic arterioles is much bigger than across any other class of vessel. Because arterioles dominate the resistance to flow through an organ they are able to control the local blood flow and their major role is to match local blood flow to local need. When they dilate (*vasodilatation*) the resistance to flow falls and blood flow increases, while *vasoconstriction* has the reverse effect. Arterioles may thus be regarded as the taps of the circulation, turning local blood flow up or down under the guidance of neural and chemical signals.

The terminal arteriole has one further role: by contracting hard it can prevent blood from flowing through the group of capillaries which arise from it, and can thus regulate the number of functioning capillaries in the tissue. (This job used to be attributed to 'precapillary sphincters' but it now seems that discrete sphincters only occur in a few tissues, such as the mesentery.)

Exchange vessels The capillaries are tiny (diameter 4–7 µm) and short (250–750 µm), and they are so numerous that most cells are no further than 10–20 µm from the nearest capillary. The capillary wall is reduced to a single layer of endothelial cells and the thinness of this layer, approximately 0.5 µm, facilitates the rapid passage of metabolites between blood and tissue. Some exchange also takes place further downstream across slightly larger vessels called postcapillary or pericytic venules (diameter 15–50 µm), which are microscopic venules lacking the complete smooth muscle coat of larger venules. Some gas exchange also occurs across the walls of small arterioles before blood even reaches capillaries, so the functional category of 'exchange vessel' actually embraces both sides of the true capillary network.

Although capillaries are extremely narrow, the capillary bed as a whole offers only a moderate resistance to flow. This is partly because of a special kind of flow in capillaries called bolus flow (see Chapter 7) and partly because the total cross-sectional area of the capillary bed is very large (see Figure 1.6 and Table 1.2). The large cross-sectional area of the capillary bed has the added advantage of slowing the velocity of the blood down to 0.5–1.0 mm/s, rather as a river slows down when its channel broadens. This slowing allows the red cell a period of 1–2 s in the capillary, time enough for it to unload its oxygen and take up carbon dioxide.

The arteriovenous anastomosis In a few tissues, notably the skin and nasal mucosa, there are shunt vessels (diameter 20–135 µm) which pass directly from arterioles to venules and bypass the capillary bed. Their thick muscular walls are innervated by sympathetic nerves and in human skin they help in temperature regulation (see Chapter 12).

Capacitance vessels Venules (diameter 50–200 µm) and veins differ principally in size and number rather than wall structure.

They have thin walls which are easily distended or collapsed, and as a result their blood content can vary enormously. The wall comprises an intima, a thin media composed of collagen and smooth muscle, and an adventitia. In limb veins the intima possesses pairs of semilunar valves (discovered by the gloriously named Hieronymous Fabricius ab Aquapendente in 1603) that prevent any backflow of venous blood; but the large central veins and veins of the head and neck lack functional valves. Venules and small veins are more numerous than arterioles and arteries (Table 1.2) so they offer a low resistance to flow and a pressure difference of just 10–15 mmHg suffices to drive the cardiac output from venule to vena cava.

The main function of the venous system, besides returning blood to the heart, is to act as a variable reservoir of blood holding about two-thirds of the circulating blood volume – their 'capacitance' function (see Figure 1.9). Much of this capacity is located in the numerous venules and small veins of diameter 20 μm to 4 mm (Table 1.2).

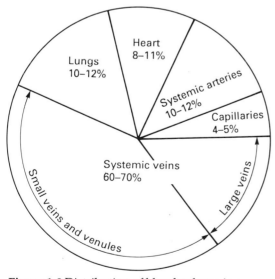

Figure 1.9 Distribution of blood volume in a resting man (5.5 litres). (From Folkow B. and Neil E. (1971) *Circulation*. London, Oxford University Press by permission)

Moreover many veins are innervated by vasoconstrictor nerve fibres, so their volume can be actively controlled; at times of physiological stress they constrict and displace blood into the heart and arterial system.

1.7 Plumbing of the vascular circuits

The systemic circulation is made up of numerous specialized individual circuits supplying the brain, kidneys, gut etc. Usually the blood supply to an organ arises directly from the aorta so that each organ is supplied at full pressure, without any interference by other organs (see Figure 1.4). This form of plumbing is called 'in parallel'. A few organs, however, are connected 'in series' with another organ – that is to say they obtain their blood 'second-hand' from the venous outflow of another organ, an arrangement called a portal system. The biggest portal system is that supplying the liver, which receives approximately 72% of its blood from the intestine and spleen via the portal vein (see Figure 1.4). (The portal vein enters the liver at the 'porta hepatis', or gateway of the liver, and this is how the term 'portal system' arose.) The liver also receives a direct arterial supply via the hepatic artery, so its circuitry is partly in series and partly in parallel. Portal systems have the advantage of transporting a valuable commodity directly from one site to another (e.g. products of digestion from the intestine to the liver) without any dilution of the material in the general circulation. Portal systems also exist in the kidney where effluent blood from the glomerulus supplies the tubules, and in the brain where a portal system carries hormones from the hypothalamus to the anterior pituitary gland. A portal system has one serious weakness, however; the down/stream tissue receives partially deoxygenated blood under a reduced pressure head, and as a result the downstream tissue is very vulnerable to damage during episodes of hypotension (low arterial pressure). Renal tubular damage in particular is a not uncommon complication of severe hypotension.

1.8 Central control of the cardiovascular system

The behaviour of the heart and blood vessels has to be regulated in order to deal with varying environmental and internal stresses. This involves nervous and neuroendocrine reflexes, which are coordinated by the brainstem and higher regions of the brain. One of the most important cardiovascular reflexes, the arterial baroreceptor reflex, safeguards blood flow to the brain by maintaining arterial blood pressure. Baroreceptors in the walls of major arteries sense changes in blood pressure, and reflexly alter the activity of autonomic nerves controlling the heart and blood vessels. This produces changes in cardiac output, peripheral resistance and venous capacitance, and these responses help to restore arterial blood pressure to normal.

Final comment In a system as complex as the cardiovascular system there is a real danger of 'not seeing the wood for the trees'. The above outline should help avoid this. In Chapters 2–11 (cardiac electricity, haemodynamics etc.) we bump into the trees and peep under the bark. In Chapters 12–15 we again stand back to gain the broader view of how the system responds as a whole to physiological and medical challenges.

Further reading

Cliff, W. J. (1976) *Blood Vessels*, Cambridge University Press, Cambridge

Harvey, W. (1628) *The Movement of the Heart and Blood* (trans. by G. Whitteridge) (1976), Blackwell Scientific Publications, Oxford

Henderson, J. R. and Daniel, P. M. (1984) Capillary beds and portal circulations. In *Handbook of Physiology, The Cardiovascular System, Vol. IV, Part 2* (eds E. M. Renkin and C. C. Michel), The American Physiological Society, Maryland, pp. 1035–1046

Neil, E. (1983) Peripheral circulation: historical aspects. In *Handbook of Physiology, Vol. III, Part 1* (eds J. T. Shepherd and F. M. Abboud), American Physiological Society, Maryland, pp. 1–20

Rhodin, J. A. G. (1980) Architecture of the vessel wall. In *Handbook of Physiology, Cardiovascular System, Vol. II* (eds D. F. Bohr, A. P. Somlyo and H. V. Sparks), American Physiological Society, Bethesda, pp. 1–32

Chapter 2
Cardiac cycle

The adult human heart weighs only 300–350 g, yet most of us can reasonably expect it to pump out around 200 million litres of blood over our allotted 'three score years and ten'. In this chapter the mechanical events underlying this remarkable performance are described.

2.1 Gross structure of the heart

The mature heart is built upon a collagenous 'skeleton' in the shape of a fibrotendinous ring (the annulus fibrosus) which is located at the atrioventricular junction (Figure 2.1). The muscular atria and ventricles are attached to either side of this ring, and the ring is perforated by four apertures, each containing a valve. As well as functioning as the mechanical base of the heart, the fibrotendinous ring insulates the ventricles electrically from the atria.

Right atrium and tricuspid valve The right atrium is a thin-walled muscular chamber which receives the venous return from the venae cavae and the coronary sinus (the main vein draining heart muscle; see Figure 2.2a). The wall near the entrance of the superior vena cava also contains the cardiac pacemaker, the 'sparking plug' that initiates each heart beat (see Chapter 3). The right atrium communicates with the right ventricle through the tricuspid valve which, as its name implies has three cusps, although it is sometimes difficult to distinguish all three. It is the large anterior cusp which is mainly responsible for valve closure. Each cusp is a flexible flap of connective tissue, roughly 0.1 mm thick, covered by endothelium. The free margin of the cusp is tethered by tendinous strings, called chordae tendineae, to an inward projection of the ventricle wall, the papillary muscle. The papillary muscle contracts and tenses the chordae tendineae during systole and this helps to prevent the valve from inverting into the atrium during systole.

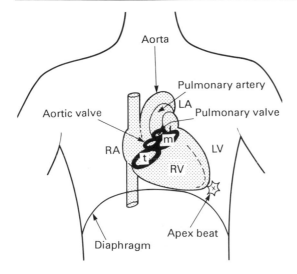

Figure 2.1 The heart lies obliquely across the chest. The fibrotendinous ring (black) forms the base of the heart. It contains the tricuspid (t), mitral (m), aortic and pulmonary valves grouped in an oblique planc beneath the sternum. The apex of the heart is formed by the left ventricle (LV), and the anterior surface is formed by the right ventricle (RV) and right atrium (RA). The inferior surface of the heart and the pericardium (not shown) rest on the central tendon of the diaphragm

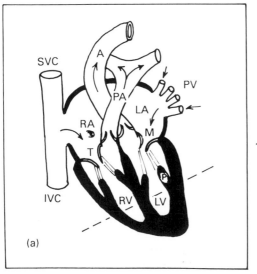

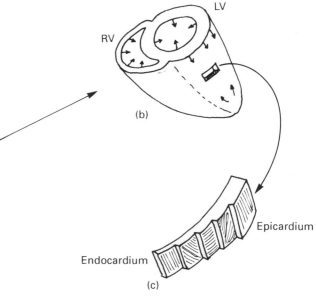

Figure 2.2 Sections through the heart. (a) Schematic diagram of an oblique section. (b) Section across the ventricles to illustrate mode of emptying. (c) Arrangement of muscle fibres in the ventricle wall. RA, LA, right and left atrium. The opening just below the label RA is the coronary sinus. RV, LV, right and left ventricle; T, M, tricuspid and mitral valves; P, papillary muscle with chordae tendineae; A, aorta; PA, PV, pulmonary artery and veins; SVC, IVC, superior and inferior venae cavae

Right ventricle and pulmonary valve The anterior wall of the right ventricle is about 0.5 cm thick in man, and resembles a pocket tacked around the septum (Figure 2.2b). Expulsion of blood is produced chiefly by the free anterior wall approaching the septum, rather like an old-fashioned bellows. The outlet from the ventricle into the pulmonary artery is guarded by the pulmonary valve which, like the aortic valve, consists of three equal sized, baggy cusps.

Left atrium and mitral valve The left atrium receives blood from the pulmonary veins and transmits it into the left ventricle through a bicuspid valve. The large anterior and small posterior cusps are thought to look like a bishop's mitre, hence the name 'mitral valve'. The cusp margins are attached by chordae tendineae to two papillary muscles in the left ventricle.

Left ventricle and aortic valve The chamber of the left ventricle is conical and ejection of blood is produced by a reduction in both diameter and length. The wall is around three times thicker than that of the right ventricle because it has to generate higher pressures. The innermost (endocardial) muscle fibres are orientated longitudinally, running from the base of the heart (the fibrotendinous ring) to the apex (tip of left ventricle); the central fibres run circumferentially; the outermost or epicardial fibres again run longitudinally; and intermediate fibres run obliquely (Figure 2.2c). In other words, the muscle orientation changes progressively across the wall. When the chamber contracts, it twists forwards and the apex taps against the chest wall, producing the *apex beat*. This can be felt in the fifth, left intercostal space, about 10 cm from the midline. The root of the aorta contains a three-cusp valve similar to the pulmonary valve.

The heart is enclosed in a fibrous sac or pericardium, which is lined by a layer of mesothelium and is lubricated by pericardial fluid. The lower surface of the pericardium is fused to the diaphragm, and as the diaphragm descends during inspiration it pulls the heart into a more vertical orientation.

2.2 Mechanical events of the cardiac cycle

The atria and ventricles contract in sequence, resulting in a cycle of pressure and volume changes, and a thorough knowledge of the cycle is needed for the diagnosis of valvular defects. The cardiac cycle has four phases and we will begin, arbitrarily, at a moment when both the atria and ventricles are in diastole (relaxed). The timings below refer to a human cycle of 0.9 s duration (67 beats/min) and the data have been acquired by a combination of echocardiography (Section 2.5), cardiac catheterization (Section 2.5), electrocardiography (see Chapter 4) and cardiometry (Section 6.3).

Ventricular filling

Duration: 0.5 s
Inlet valves (tricuspid and mitral): open
Outlet valves (pulmonary and aortic): closed

Ventricular diastole lasts for nearly two-thirds of the cycle at rest, providing ample time for refilling the chamber. Initially the atria too are in diastole and blood flows passively from the great veins through the open atrioventricular valves into the ventricles. There is an initial phase of rapid filling, lasting about 0.15 s (Figure 2.3), which has a curious feature; even though ventricular volume is increasing, ventricular pressure is falling (see Figure 2.4; also Figure 6.10). The reason is that the ventricle wall is recoiling elastically from the deformation of systole, and is in effect sucking blood into the chamber. As the ventricle reaches its natural volume, filling slows down and further

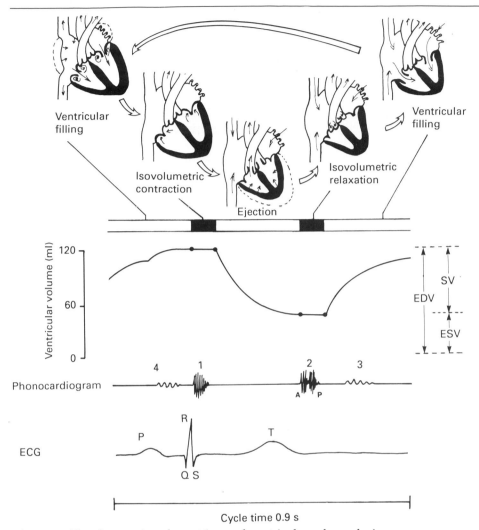

Figure 2.3 The changes in valve setting and ventricular volume during one cardiac cycle lasting 0.9 s. EDV, end-diastolic volume; ESV, end-systolic volume; SV, stroke volume. The ejection fraction is SV/EDV. The heart sounds on the phonocardiogram are numbered 1 to 4 and the second sound is split here into an aortic component (A) and pulmonary component (P). The electrocardiogram waves are described in the text

filling requires distension of the ventricle by the pressure of the venous blood; ventricular pressure now begins to rise. In the final third of the filling phase, the atria contract and force some additional blood into the ventricle. In resting subjects, this atrial boost is quite small and enhances ventricular filling by only 15–20%: indeed, the absence of an atrial boost in patients suffering from atrial fibrillation (ineffective atrial contractions; Section 4.8) makes little difference to resting cardiac output. During exercise, however, when heart rate is high the time available for passive ventricular filling is curtailed (see later), and the atrial boost becomes important.

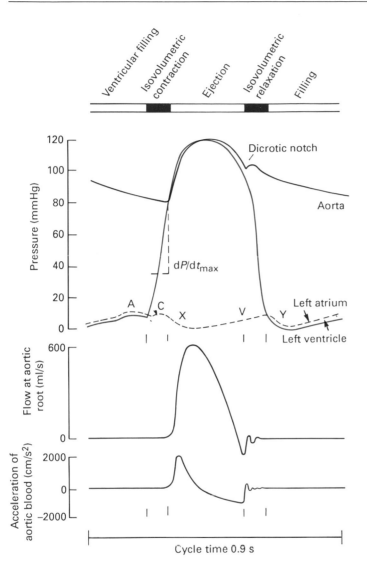

Figure 2.4 Diagram of pressure and outflow on the left side of the human heart, based on data from intracardiac catheters and velocity measurements at the aortic root. $(dP/dt)_{max}$ is the maximum rate of rise of ventricular pressure, a measure of myocardial contractility. The stippled region highlights the pressure gradient which decelerates outflow during the late ejection phase. Note the slight reversal of aortic flow at aortic valve closure. For explanation of the atrial waveform see Section 2.3. The waveforms in the right heart are of similar shape but the pressures are lower (see Table 2.1). (After Noble, M. I. M. (1968) *Circulation Research*, **23**, 663–670)

The volume of blood in a ventricle at the end of the filling phase is called the *end-diastolic volume*, or EDV, and is typically around 120 ml in an adult human. The corresponding *end-diastolic pressure*, or EDP, is a few mmHg. As Table 2.1 shows, the EDP is a little higher in the left ventricle than in the right, the reason being that the left ventricle wall is thicker and therefore less easily distended.

Table 2.1 Mean pressures during the human cardiac cycle in mmHg (various sources)*

	Right	Left
Atrium	3	8
Ventricle –		
end of diastole	4	9
peak of systole	25	120

* adult, resting supine.

Isovolumetric contraction

Duration: 0.05 s
Inlet valves: closed
Outlet valves: closed

As atrial systole begins to wane, ventricular systole commences. It lasts 0.35 s and is divided into a brief isovolumetric phase and a longer ejection phase. As soon as ventricular pressure rises fractionally above atrial pressure, the atrioventricular valves are forced shut by the reversed pressure gradient. Backflow during closure is minimal because the cusps are already approximated by vortices behind them in the late filling phase. The ventricle is now a closed chamber, and the growing wall tension causes a steep rise in the pressure of the trapped blood; indeed the maximum rate of rise of pressure, $(dP/dt)_{max}$, is frequently used as an index of cardiac contractility (see Section 6.5).

Ejection

Duration: 0.3 s
Inlet valves: closed
Outlet valves: open

When ventricular pressure exceeds arterial pressure, the outflow valves are forced open and ejection begins. Three-quarters of the stroke volume are ejected in the first half of the ejection phase (phase of rapid ejection, approximately 0.15 s), and at first blood is ejected faster than it can escape out of the arterial tree. As a result, much of it has to be accommodated by distension of the large elastic arteries, and this drives arterial pressure up to its maximum or 'systolic' level. Vortices behind the cusps of the open aortic valve prevent the cusps from blocking the adjacent openings of the coronary arteries.

As systole weakens and the rate of ejection slows down, the rate at which blood flows away through the arterial system begins to exceed the ejection rate and pressure begins to fall. Active ventricular contraction actually ceases about two-thirds of the way through the ejection phase, but a slow outflow continues for a while owing to the momentum of the blood. As the ventricle begins to relax, ventricular pressure falls below arterial pressure by 2 to 3 mmHg (see Figure 2.4) but the outward momentum of the blood prevents immediate valve closure. The reversed pressure gradient, however, progressively decelerates the outflow, as shown in the lower trace of Figure 2.4, until finally a brief backflow (comprising less than 5% of stroke volume) closes the outflow valve. Valve closure creates a brief pressure rise in the arterial pressure trace called the dicrotic wave. For the rest of the cycle, arterial pressure gradually declines as blood runs to the periphery.

It must be emphasized that the ventricle does not empty completely; the average *ejection fraction* in man is 0.67, corresponding to a *stroke volume* of 70–80 ml in adults. The residual *end-systolic volume* of about

50 ml acts as a reserve which can be utilized to increase stroke volume in exercise.

Isovolumetric relaxation

Duration: 0.08 s
Inlet valves: closed
Outlet valves: closed

With closure of the aortic and pulmonary valves, each ventricle once again becomes a closed chamber. Ventricular pressure falls very rapidly owing to the mechanical recoil of collagen fibres within the myocardium, which were tensed and deformed by the contracting myocytes. When ventricular pressure has fallen just below atrial pressure, the atrioventricular valves open and blood floods in from the atria. The atria have been refilling during ventricular systole, and this leads us to consider next the atrial cycle.

2.3 Atrial cycle and central venous pressure cycle

The cycle of events in the atria produces a cycle of pressure changes in the veins of the thorax and neck (jugular veins) because the veins are in open communication with the atria. A direct record of pressure in an atrium or jugular vein reveals that there are two main pressure waves per cycle, called the A and V waves, and a third smaller wave, the C wave (see dashed line in Figure 2.4).

The *A wave* is an increase in pressure caused by atrial systole, and the 'A' stands for atrial. Atrial systole produces a slight reflux of blood through the valveless venous entrances: this briefly reverses the flow in the venae cavae and raises central venous pressure to its maximum point (3–5 mmHg). The next event, the C wave, occurs earlier in the right atrium than in the neck. In the atrium, it is caused by the tricuspid valve bulging back into the atrium

as it closes. In the internal jugular vein, the C wave is caused partly by expansion of the carotid artery, which lies alongside the vein and presses on it during systole; 'C' stands for 'carotid'. After the C wave there comes a sharp fall in pressure, called the *X descent*, which is caused by atrial relaxation, and venous inflow reaches its peak velocity during this phase (see Figure 7.20; Chapter 7). As the atria fill, atrial pressure begins to rise again, producing the *V wave*; the 'V' refers to the simultaneously occurring ventricular systole. Finally, the atrioventricular valves open and the atria empty passively into the ventricles, producing the sharp *Y descent*.

The cycle of right atrial pressure is mirrored in the internal and external jugular veins of the neck, because the latter are in open communication with the superior vena cava. The pulsating jugular veins are readily visible in a recumbent lean subject and this enables the physician to assess the central venous pressure cycle by simple inspection. What the eye particularly notices in the neck are two sudden collapses of the vein, corresponding to the X and Y descents. Examination of the jugular pulse is a regular clinical procedure because certain cardiac diseases produce characteristic abnormalities in the pulse. Tricuspid incompetence, for example, can produce exaggerated V waves, because blood leaks back through the incompetent valve during ventricular systole.

2.4 Effect of heart rate on phase duration

The timings given earlier for the cardiac cycle refer to a resting subject, but when the heart is beating 180 times per min (which is close to the normal maximum), the duration of the entire cycle is only 0.33 s, and all phases of the cycle have to be shortened. The various phases do not, however, all shorten to an equal degree (see Figure 2.5). Ventricular systole does shorten, but only to

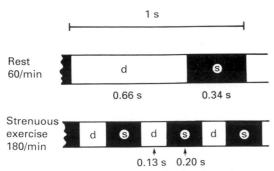

Figure 2.5 Effect of heart rate on the diastolic period available for filling. d, diastole; s, systole. Diastole is curtailed more than systole as heart rate increases

about 0.2 s, and this leaves a mere 0.13 s for refilling during diastole. Passive filling remains important but atrial systole contributes relatively more than at rest. Even with the help of atrial systole, 0.12 s is about the minimum interval that allows an adequate refilling of the human ventricle. Further increase in heart rate, such as the pathological tachycardia which occurs in the Wolf–Parkinson–White syndrome (>250 beats/min), actually causes cardiac output to decline rather than increase, because refilling during diastole becomes inadequate. Diastolic interval is thus the chief factor limiting the maximum useful heart rate.

2.5 Clinical aspects of the human cardiac cycle

The cardiac cycle is assessed in routine clinical practice by examining various physical signs, such as the arterial pulse, the jugular venous pulse and the apex beat. Another important sign, not yet described, is the sound generated by closing valves.

The heart sounds

When a heart valve closes, the cusps balloon back as they suddenly check the momentum of refluxing blood. The sudden tension in the cusps sets up a brief vibration, rather as a sail slaps audibly when suddenly filled by a gust of wind. The vibration is transmitted through the walls of the heart and arteries to the chest wall, where it can be heard through a stethoscope. Provided that the valve is normal, it is only closure that is audible: as with a well-oiled door, opening is silent.

Two heart sounds are normally clearly audible per beat, the first and second heart sounds. They are usually represented as lubb-dupp followed by a pause, roughly in waltz time; the first heart sound (lubb) is the one immediately after the pause. (Lubb-dupp should not be taken too seriously, for it appears that only English-speaking hearts go lubb-dupp, while German ones go doop-teup and Turkish ones rrupp-ta.) The heart sounds can be recorded by a microphone placed on the precordium, and a tracing of the sound is called a phonocardiogram (see Figure 2.3). The first heart sound, a vibration of roughly 100 cycles/s (100 Hertz), is caused by closure of the tricuspid and mitral valves, which close virtually simultaneously. The second sound is of similar frequency and is caused by closure of the aortic and pulmonary valves. The second sound is sometimes audibly 'split', with an initial aortic component and a fractionally delayed pulmonary component; the sounds might then be represented as lubb-terrupp. Splitting of the second sound is common in healthy young people during inspiration, because inspiration increases the filling of the right ventricle; this raises its stroke volume (as described in Section 6.3), which in turn prolongs the right ventricular ejection time and slightly delays pulmonary valve closure.

Two additional sounds besides the first and second sounds can be detected by phonocardiography, but they are of low frequency and difficult for untrained ears to detect. The third heart sound is common in young people and is caused by the rush of blood into the relaxing ventricles during early diastole. The fourth sound occurs just before the first sound and is caused by atrial systole.

Anatomically, the four heart valves lie very close together under the sternum (see Figure 2.1) but fortunately each valve is best heard over a distinct 'auscultation area' some distance away, because the vibration from each valve propagates through the chamber fed by the valve. The mitral valve is best heard in the mid-clavicular line of the 4–5th left intercostal space, the tricuspid valve in the 5th interspace at the left sternal edge, the aortic valve in the 2nd interspace at the right sternal edge, and the pulmonary valve in the 2nd interspace at the left sternal edge.

There are two fundamental classes of valvular abnormality, stenosis and incompetence. *Stenosis* is a narrowing of the valve, and a high pressure-gradient is needed to force blood through a stenosed valve. *Incompetence* is failure of the valve to close tightly, thus allowing a regurgitation of blood. In either case blood passes through the valve in a turbulent jet, setting up a high-frequency vibration which is heard as a 'murmur' through the stethoscope. Given four valves and two pathologies, there are evidently eight basic murmurs, but just one example must suffice here. In mitral valve incompetence there is regurgitation into the left atrium during ventricular systole, producing a murmur that occurs throughout systole (a pansystolic murmur) and sounds loudest over the mitral area. The heart sounds may then be represented, roughly, by a lu-shshshsh-tupp. Hypochondriacs should note, however, that a benign murmur is not uncommon during the ejection phase in the young. It is not caused by a valve lesion but by turbulence in the ventricular outflow tract. This 'benign systolic murmur' is especially marked during pregnancy, strenuous exercise and anaemia (Section 7.2).

Electrocardiography

The electrocardiogram (ECG) is a record of cardiac electrical events obtained from the body surface. The ECG is described fully in Chapter 4, so here we will simply note the relation between the electrical blips of the ECG and the mechanical events of the cycle (see Figure 2.3). The P wave of the ECG is produced by electrical activation of the atria, and its peak coincides with the onset of atrial contraction. The QRS complex is produced by electrical activation of the ventricles, so the QRS complex is followed almost immediately by the onset of ventricular contraction and the first heart sound. The T wave is produced by electrical recharging of the ventricles, and since this marks the onset of diastole it is closely followed by the second heart sound.

Echocardiography

Echocardiography is an invaluable, relatively new tool for the non-invasive assessment of the human cardiac cycle. A beam of ultrasound is directed across the heart from a precordial ultrasonic emitter (a piezo-electric crystal), and reflections of the sound from the walls and valves are collected and used to build up a linear record of their motion. An example of this will be found in Figure 6.20.

Cardiac catheterization

This is a direct and powerful investigative procedure. Under local anaesthesia, a fine catheter is threaded through the antecubital vein in the crook of the elbow and advanced under X-ray guidance through the right atrium into the right ventricle, or even into the pulmonary artery. The aorta and left ventricle can be reached by a catheter introduced through the femoral artery. The intracardiac catheter can then be put to one of the following uses.

Cine-angiography A radio-opaque contrast medium is injected through the catheter and the progress of the medium through the cardiac chambers is followed by X-ray cinematography (cardiac angiography). This displays the movement of the heart wall and reveals any valvular regurgitation.

Radionuclide angiography A recent extention of the angiographic method is to inject a gamma-ray emitting isotope into a central vein and record the gamma emission with a scintillation camera placed over the precordium. Not only can images of the heart in diastole and systole be computed but also, from the fall in counts produced by each ejection, the ventricular ejection fraction can be measured.

Intracardiac pressure measurement Chamber pressures can be recorded by connecting the catheter to an external pressure transducer, or by mounting a miniature transducer in the tip of the catheter. The pressure drop across a closed valve serves as an excellent test of its competence. A pulmonary artery catheter can also be wedged in the pulmonary arterioles, and the recorded 'wedge pressure' is often used as an estimate of pulmonary capillary pressure.

Intracardiac pacing This is a therapeutic application of the cardiac catheter, in which a wire catheter is wedged in the ventricle and used to stimulate each heart beat from an external electrical device, thereby replacing the heart's own pacemaker.

Further reading

Braunwald, E. and Ross, J. (1979) Control of cardiac performance. In *Handbook of Physiology, The Cardiovascular System, Vol. I The Heart* (ed. R. M. Berne), American Physiological Society, Bethesda, pp. 533–579

Caro, C. G., Pedley, T. J., Schroter, R. C. and Seed, W. A. (1978) *The Mechanics of the Circulation*, Oxford University Press, Oxford

Parmley, W. W. and Talbot, L. L. (1979) Heart as a pump. In *Handbook of Physiology, The Cardiovascular System, Vol. I The Heart* (ed. R. M. Berne), American Physiological Society, Bethesda, pp. 429–426

Robinson, T. F., Factor, S. M. and Sonnenblick, E. H. (1986) The heart as a suction pump. *Scientific American* (June), 62–69

Chapter 3
Cardiac excitation and contraction

Overview An isolated heart taken from a cold-blooded animal will continue to beat for a long period if placed in a beaker containing an appropriate solution. This simple experiment proves that cardiac contraction is initiated within the heart itself; unlike skeletal muscle, external nerves are not essential. The heart beat is in fact initiated by a special electrical system in the walls of the heart, and this system is constructed of modified muscle cells. Cardiac muscle cells, or 'myocytes', thus fall into two broad classes; the majority are work cells whose task is contraction, while a minority are specialized cells making up an electrical system whose task is to excite the work cells.

The electrical system consists of (1) a group of cells forming the sino-atrial node or 'pacemaker' which discharges spontaneously at regular intervals, and (2) elongated cells called conduction fibres that transmit the resulting electrical impulse quickly to the ventricular work cells. When the electrical stimulus reaches the ordinary work cell, the latter is excited and fires off an action potential. This leads to a rise in calcium ion concentration within the cell, which in turn activates the contractile proteins of the cell. The purpose of this chapter is to describe the nature of the contractile process and the related electrical events, and in order to do so we must first consider the fine structure of the cardiac myocyte.

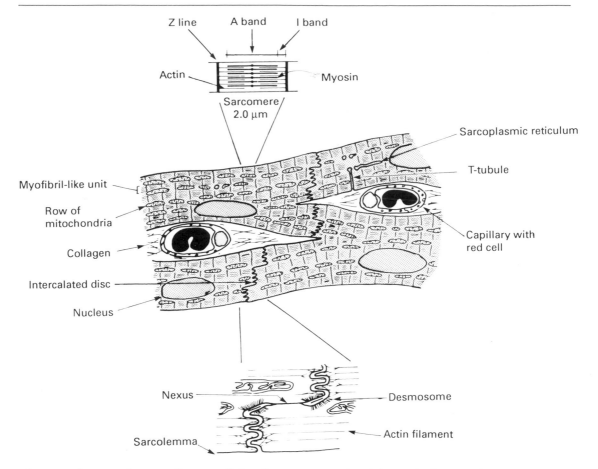

Figure 3.1 Section of myocardium parallel to fibre axis, based on electron microscopic studies. The width of the sarcomere (2 μm) and red cell (7 μm) indicate the scale. The enlargement of part of the intercalated disc at the bottom shows a gap junction or nexus along the horizontal step (interplicate segment) and two desmosomes.

3.1 Ultrastructure of the work cell

Branching cells and their junctions

The human work cell is typically 10–20 μm in diameter and 50–100 μm long, with a single central nucleus. The cell is branched and is attached to adjacent cells in an end-to-end fashion (see Figure 3.1). The end-to-end junction, or intercalated disc, has a characteristic stepped profile in cross-section, and it contains two kinds of smaller, specialized junctions, namely desmosomes and gap junctions. Desmosomes hold the adjacent cells together, probably by means of a proteoglycan glue located in the 25 nm-wide gap between the cell membranes. The gap junction, or nexus, is a region of very close apposition of the adjacent cell membranes and is thought to be the electrically-conductive region through which ionic currents flow from cell to cell. As a result of these electrical connections the myocardium acts as an electrically continuous sheet and this enables excitation to reach every cell.

The myofibril and the sarcomere

The work cell is packed with long branching contractile bundles whose diameter is around 1 μm. They are called myofibril-like units because of their resemblance to the myofibrils of skeletal muscle, or myofibrils for brevity. Each myofibril is composed of smaller units called sarcomeres, which are joined end to end; they are also aligned across the cell, giving the myocyte its characteristic striated appearance under the microscope. The sarcomere is the basic contractile unit and is defined as the material between two Z lines: a Z line is a thin dark-staining partition composed of a protein, α-actinin. The sarcomere is 2.0–2.2 μm long in resting myocytes and contains two kinds of interdigitating filament: a thick filament made of the protein myosin and a thin filament composed chiefly of actin, another protein. The thick filaments, of diameter 11 nm and length 1.6 μm, lie in parallel in a central region of the sarcomere called the A band. (The 'A' stands for 'anisotropic', a reference to its appearance through a polarizing microscope.) The thin filaments, of diameter 6 nm and length 1.05 μm, are rooted in the Z line and form the pale I band (isotropic band); the latter is only approximately 0.25 μm wide because most of the length of the thin filament protrudes into the A band in between the myosin rods. As well as actin, the thin filaments contain the proteins troponin and tropomyosin.

Tubular systems

The surface membrane, or sarcolemma, is invaginated opposite the Z line into a series of fine transverse tubules (T tubules) which run into the cell interior (see Figure 3.2). The T-tubules transmit the electrical stimulus rapidly into the interior of the cell and thus help to activate the numerous myofibrils almost simultaneously. This system is

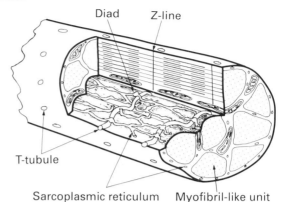

Figure 3.2 Three dimensional reconstruction of the tubular system based on electron microscopy. The transverse tubules form approximately 5% of the cell volume in ventricular cells. The sarcotubular system forms approximately 1% and often lies just under the sarcolemma; there is debate as to whether it is completely sealed off from the extracellular space as in skeletal muscle

well developed in ventricular myocytes but is scanty in atrial cells. The internal ends of the T-tubules are closed, so the extracellular fluid is never in direct contact with intracellular fluid.

Within the cell there is a second, separate system of tubules called the sarcoplasmic reticulum, with few if any surface connections. The sarcoplasmic reticulum is developed from endoplasmic reticulum and it consists of a closed set of anastomosing tubules coursing over the myofibrils. The sarcoplasmic tubules expand into flattened sacs called subsarcolemmal cisternae near the T-tubules, and in section a T-tubule and adjacent cisterna are often seen as a pair, or 'diad'. The cisternae contain a store of calcium ions that can be released to activate the contractile machinery. This brings us to the question of how contraction is produced.

3.2 Mechanism of contraction

Contraction of the myocyte is caused by shortening of its sarcomeres. Direct inspection shows that the I bands shorten, but the

A band does not. This is one of the key observations indicating that contraction is caused by the thin filaments of the I band sliding into the spaces between the thick filaments of the A band – the *sliding filament mechanism*. The filaments are propelled past each other by the repeated making and breaking of crossbridges between the thin and thick filaments. These crossbridges are actually the heads of myosin molecules, which protrude from the side of the thick filament as illustrated in Figure 3.3. At rest the actin sites, with which the myosin heads would react, are blocked by tropomyosin. Contraction is initiated by a sudden rise in the concentration of free intracellular calcium ions (explained later) which bind to troponin C, a component of the troponin complex. This alters the position of the adjacent tropomyosin molecule and thereby exposes a specific myosin-binding site on the actin chain, allowing the myosin head to bind to the actin. Force and movement is produced by a subsequent change in the angle of this crossbridge (i.e. the attached myosin head), after which the head disengages and the process repeats itself at a new actin site. This process occurs at numerous similar sites along the filament, and in this way the thick filament 'rows' itself into the space between the thin filaments. The most important point about the whole process from the physiological point of view is that the number of crossbridges formed, and therefore the force of the contraction, depends directly on the concentration of free calcium ions within the myocyte.

The energy for crossbridge cycling is provided by adenosine triphosphate (ATP), which is broken down during the process into inorganic phosphate and adenosine diphosphate (ADP) by an ATPase site on the myosin head. In order to maintain an adequate supply of ATP, the myocyte possesses an exceptionally high density of mitochondria, which lie in rows between the myofibrils and form 30–35% of the cell volume. ATP is manufactured in mitochondria by oxidative phosphorylation, for which oxygen is obligatory, and this is why

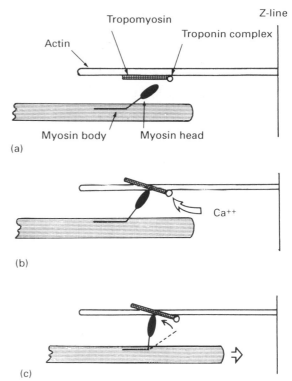

(a)

(b)

(c)

Figure 3.3 Diagram of the contractile proteins at three stages during the crossbridge cycle. (a) Resting state. The actin binding site is blocked by tropomyosin and the myosin heads are disengaged. (b) Calcium ions bind to troponin C of the troponin complex, displacing the tropomyosin. The myosin head cross-links to the exposed actin binding site. (c) The angle of the myosin head changes, 'rowing' the thick filament towards the Z line. The head then disengages and the cycle repeats at a new actin site. Only one myosin molecule is shown here but the thick filament contains approximately 400

cardiac performance is directly dependent on coronary blood flow. Cardiac muscle is almost incapable of anaerobic phosphorylation, unlike skeletal muscle.

The release of calcium from the cisternal store, which activates the contractile machinery, is itself triggered by a change in the electrical potential across the cell membrane, so we must consider next the nature of the electrical potential in a myocyte.

3.3 Resting membrane potential of a work cell

The potential difference between the interior and exterior of a myocyte can be measured by driving a fine microelectrode into the cell (a microelectrode is simply a piece of glass tubing which has been heated, drawn out and filled with a conducting solution). The intracellular electrode is connected to an amplifier and a voltmeter, and the other lead of the voltmeter is connected to an electrode outside the cells. The intracellular potential of the resting myocyte is then found to be -60 mV to -90 mV (i.e. 60–90 mV lower than the extracellular potential). In work cells this 'resting membrane potential' is stable, until external excitation is applied (see Figure 3.4a), but in sino-atrial (SA) node cells and many conduction fibres it is unstable, drifting towards zero with time; this more complex situation is considered in Section 3.7.

Potassium: generator of the resting membrane potential

The resting potential is due primarily to two factors: the high concentration of potassium ions in the intracellular fluid and the high permeability of the cell membrane to potassium ions compared with other ions. The intracellular K^+ concentration is about 35

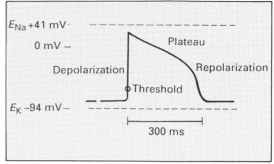

(a) Ventricular cell

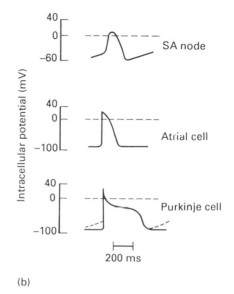

(b)

Figure 3.4 Intracellular potential of a ventricular work cell (a) measured by a glass microelectrode. The stable baseline potential of -80 mV is the resting membrane potential. The depolarization stage of the action potential is so rapid (500 V/s) that it appears vertical. The theoretical potassium equilibrium potential E_K and sodium equilibrium potential E_{Na} are marked. (b) Three records to illustrate the different shapes of action potential at other sites. Note the unstable resting potential in the pacemaker cell (SA node). Some Purkinje fibres also have unstable resting potentials (dotted line)

Table 3.1 Concentration of ions in myocardial cells†

	Intracellular (mM)	Extracellular (mM)	Nernst equilibrium potential (mV)
K^+	140	4	-94
Na^+	10	140	$+41$
Ca^{2+}	0.0001	2	$+133$
Cl^-	30	120	-36

(† After Noble, D. (1979) *The Initiation of the Heartbeat*, Oxford, Clarendon Press)

times higher than the extracellular K^+ concentration (see Table 3.1), so there is a continuous tendency for K^+ to diffuse out of the cell down its concentration gradient.

However, the negative intracellular ions, mainly organic phosphates and charged proteins, cannot accompany the K^+ ions because the cell membrane is impermeable to them (see Figure 3.5). The outward diffusion of a small number of potassium ions therefore creates a very slight separation of charge and leaves the cell interior negative with respect to the exterior. Because the electrical charge on a single ion is very large indeed (see Faraday's constant, Appendix), just one excess negative ion per 10^{15} ion pairs is sufficient to produce the resting potential. A numerical imbalance of this size is of course far too tiny to be detected by chemical methods.

Potassium equilibrium potential

If the negative intracellular potential were big enough, the electrical attraction of the cell interior for the positive potassium ions could fully offset the outward diffusion tendency of the ions, creating a dynamic equilibrium in which there was no further net movement of K^+ out of the cell. The electrical potential at which this would happen is called the potassium equilibrium potential. The equilibrium potential is, by definition, equal in magnitude to the outward-driving effect of the concentration gradient, or chemical potential as it is called: and the latter depends on the ion concentration outside the cell (C_o) relative to that inside (C_i). The exact relation between equilibrium potential of ionic species X (E_x) and the ionic concentration ratio is given by the Nernst equation, namely:

$$E_x = \frac{RT}{zF} \cdot \log_e(C_o/C_i) \qquad (3.1)$$

where z is the ionic valency, R is the gas constant (Appendix), T is absolute temperature and F is Faraday's constant. For a monovalent ion at body temperature (310 K) this works out, in millivolts, as 61 × $\log(C_o/C_i)$: and since the myocyte potassium ratio is 1:35, the potassium equilibrium potential works out to be −94 mV. Ex-

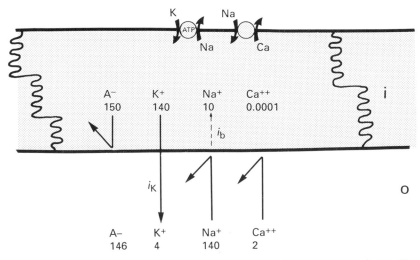

Figure 3.5 Chemical gradients and currents across the resting membrane (i, inside. o, outside). A^- refers to impermeant intracellular anions. Straight arrows show concentration gradients for ions permeating the sarcolemma. Reflected arrows indicate ions unable to penetrate the resting membrane. i_b is the inward background current and i_K the outward background current. Cell membrane pumps are shown at top

perimentally, it is found that the myocyte's resting potential is indeed close to the potassium equilibrium potential (but never quite equal to it – see Figure 3.6 and later). It is also found that when the extracellular potassium concentration is increased, as can happen in certain medical conditions, the myocyte resting potential declines (grows less negative) in proportion to the logarithm of the extracellular potassium concentration, as predicted by the Nernst equation (see Figure 3.6).

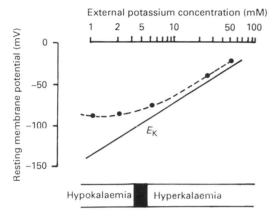

Figure 3.6 Dependence of resting membrane potential of a myocyte on extracellular potassium concentration. The dashed line fits experimental points while the solid line represents E_K, the potassium equilibrium potential (see Nernst equation in the text). At a normal potassium level (black bar) or above, the membrane potential is smaller than E_K due to an inward background current (i_b). In hypokalaemia the deviation increases because potassium conductance declines, making i_b more significant. (After Page, E. (1962) *Circulation Research*, **26**, 582)

Non-equilibrium due to background currents

Figure 3.6 shows that the resting membrane potential always falls a little short of the potassium equilibrium potential. This is due to a small inward current of positively charged ions, mainly sodium ions, which is known as the 'inward background current'

(i_b). Although the permeability of the resting membrane to sodium is only 1/10th to 1/100th of its permeability to potassium, both the electrical gradient and the chemical gradient for Na^+ (Table 3.1) are directed into the cell, and their sum, the 'electrochemical gradient' drives a small inward current of Na^+ into the cell (see Figure 3.5). As a result, the resting membrane potential is 10–20 mV more positive than the potassium equilibrium potential. Moreover, because the resting potential is not quite negative enough to fully counteract the outward diffusion of K^+ ions, there is a continuous trickle of K^+ out of the cell, producing an 'outward background current' (i_K). In work cells, the outward current I_K is equal to the inward current I_b, so the resting membrane potential is stable despite the continuous slow exchange of potassium for sodium.

Ohm's law and the importance of relative ionic permeabilities

The size of the background currents is given by Ohm's law, which states that current (i) is proportional to the potential difference (ΔV) and to the electrical conductance (g – the reciprocal of resistance): $i = g \cdot \Delta V$. In the case of a cell membrane, the conductance is proportional to its ionic permeability (see Appendix). For potassium ions the potential difference driving the current is the difference between the resting membrane potential E_m and the potassium equilibrium potential E_K, so the background potassium current i_K is, by Ohm's law:

$$i_K = g_K(E_m - E_K) \tag{3.2}$$

where g_K is the membrane's potassium conductance. Similarly, the inward background current of sodium ions, i_b, is by Ohm's law:

$$i_b = g_{Na}(E_m - E_{Na}) \tag{3.3}$$

where E_{Na} is the sodium equilibrium potential (+41 mV). If the resting potential is stable, the inward and outward currents must be equal, so Equations 3.2 and 3.3 are equal. Combining them we get a simple expression for the resting potential:

$$E_m = \frac{(E_K + E_{Na} \cdot g_{Na}/g_K)}{(1 + g_{Na} \cdot /g_K)} \quad (3.4)$$

This is a simplified form of the more complex Goldman constant field equation (which takes account of chloride and calcium currents too), but it suffices for our purpose here, which is to highlight the importance of the *ratio of sodium permeability to potassium permeability* in setting the membrane potential. The expression tells us that the resting potential is a potassium equilibrium potential (−94 mV) modified by a fraction of the sodium equilibrium potential (+41 mV). The fraction in question is one-tenth if the ratio of g_{Na} to g_K is 1:10. This simplified constant field equation therefore predicts that the resting potential should be (−94 + 4.1)/1.1, or −82 mV, and this is reasonably close to the potential in many work cells.

Function of sarcolemmal ion pumps

It should be clear from the above account that the cell is in effect a chemical battery, and that the chemical which powers the battery, potassium, is slowly but continuously leaking out of the cell. Unchecked, the concentrations of both potassium and sodium would eventually equilibrate across the cell membrane, leaving the battery flat. This is prevented, however, by active pumps in the sarcolemmal membrane, whose function is to preserve the chemical composition of the interior. A *sodium-potassium exchange pump* simultaneously transports sodium ions out of the cell and potassium ions into the cell (see Figure 3.5), and the pump rate is enhanced by a rise in intracellular Na^+ or extracellular K^+ concentration. The exchange of Na^+ for K^+ is not quite 1 for 1; slightly more sodium ions are pumped out than potassium in, and the pump is therefore electrogenic. It must be stressed however that this effect makes only a minor contribution to the membrane potential; blocking the Na^+–K^+

pump by ouabain causes only a small immediate decrease in membrane potential, because the chief factor generating the resting potential is the potassium concentration gradient. Pumping is an active process and consumes metabolic energy in the form of ATP.

The sarcolemmal membrane also possesses calcium pumps which expel intracellular calcium ions that have entered the cell during the action potential (see below). The predominant pump is a *calcium-sodium exchange pump*, in which the passage of 3 extracellular sodium ions across the membrane causes the expulsion of 1 intracellular Ca^{2+} ion. The entry of the excess Na^+ ion contributes to the inward background current. The Na^+–Ca^{2+} exchange is not powered directly by ATP but is driven by the 'downhill' sodium concentration gradient, rather as a water-wheel is turned by a water gradient. It therefore depends indirectly on the Na^+–K^+ pump, since it is the Na^+–K^+ pump that sets up the sodium gradient. This point is important for understanding the effects of digoxin (Section 3.5). In addition to the sodium-driven calcium pump, the membrane may also possess a few calcium pumps powered directly by ATP. Together the calcium pumps maintain the intracellular Ca^{2+} at an extremely low concentration in resting myocytes, namely 10^{-7} M.

3.4 Action potential of a work cell

The action potential, which triggers contraction, is an abrupt reversal of the membrane potential to a positive value (see Figure 3.4). In a work cell it is normally initiated by the action potential of an adjacent cell, which draws charge passively from the resting membrane (see later; Figure 3.15) and thereby reduces its potential. When the potential reaches a threshold value between −70 mV and −60 mV, the membrane's ionic permeability suddenly changes; the cell very rapidly depolarizes (loses its negative charge) and overshoots to a positive potential of $+20\,mV$ to $+30\,mV$. The membrane then immediately begins to repolarize, but when it reaches zero to $-20\,mV$ it becomes relatively stable for a long period (200–400 ms). This stage is called the 'plateau', and it causes the cardiac action potential to last far longer than a nerve or skeletal muscle action potential (1–4 ms). Finally the membrane repolarizes, though only 1/1000th of the rate of depolarization, to regain its resting potential.

Action potentials differ somewhat in form between the various cardiac cells (see Figure 3.4). Atrial potentials last 150 ms and are triangular in most species; ventricular potentials are longer (400 ms) and more rectangular owing to a more distinct plateau at approximately $0\,mV$; and Purkinje cells have the longest potentials, up to 450 ms, with a distinct initial spike followed by a long plateau at approximately $-20\,mV$.

Rapid depolarization and sodium ions

The action potential is generated by a sequence of changes in sarcolemmal permeability to Na^+, Ca^{2+} and K^+, which allows ionic currents to flow passively down their electrochemical gradients. The depolarization spike is caused by an extremely rapid increase in permeability to sodium ions; at the threshold potential, voltage-sensitive sodium channels (fast channels) open very quickly and increase the sodium conductance of the sarcolemma around 100 times (see Figures 3.7 and 3.8). This allows a rapid flux of sodium ions into the cell (the first inward current, i_{Na}) and drives the potential towards the sodium equilibrium potential (E_{Na}, $+41\,mV$, see Table 3.1). The membrane potential does not quite reach E_{Na} because an outward potassium current is still flowing. The situation at the overshoot is in effect a mirror image of the resting situation, and for a sodium:potassium conductance ratio of 10:1 Equation 3.4 predicts an overshoot potential of $+29\,mV$.

The overshoot is brief because the fast channels are self-inactivating. Their patency is controlled by two different 'gates', which are probably charged intramembranous particles. The 'm' or activating gate opens quickly at threshold potential, producing the sudden rise in sodium permeability. The 'h' or inactivation gate begins to close at the same time as the m gate opens, but it moves less quickly; it inactivates the channel automatically after a few milliseconds. The membrane potential then begins to fall again, owing to an outward current of K^+ ions.

Plateau and calcium ions

Events up to this point have been similar to those in a nerve. Next, however, the myocyte displays its unique feature, the plateau. This is produced by a small but sustained inward current of positive ions, the second inward current (i_{SI}), which prevents the myocyte from repolarizing rapidly like a nerve. The current consists mainly of calcium ions flowing into the cell down their electrochemical gradient, plus a small contribution from sodium ions. The calcium current is caused by a rise in sarcolemmal permeability to calcium (see Figure 3.8) which is due to the opening of 'slow channels' selectively permeable to calcium ions. The slow channels are voltage-controlled channels which begin to activate slowly when the cell depolarizes beyond $-35\,mV$ (i.e. during rapid depolarization) and they stay open for 200–400 ms. The total conductance of the slow channels is much less than that of the fast sodium channels, and the inward Ca^{2+} current is quite small, but it is sufficient, almost, to counterbalance the ever-present outward K^+ current. In this way it almost stabilizes the potential at $0\,mV$ to $-20\,mV$. The outward K^+ current is itself reduced during the plateau owing to a fall in the membrane permeability to potassium (see Figure 3.8): this can be viewed as an economy measure, minimizing the number of potassium and calcium ions exchanged during the long plateau and so reducing the eventual energy cost of the action potential.

The existence of the two distinct inward currents, i_{Na} and i_{SI}, has been proved by the use of tetrodotoxin, which blocks only the fast, sodium channels (see Figure 3.9). Figure 3.9 also illustrates an important

action of adrenaline, namely, it increases the size of the calcium current. This contributes to the increase in contractile force produced by adrenaline.

The inward current during the late part of the plateau may be due partly to sodium ions passing in through the sodium–cal-

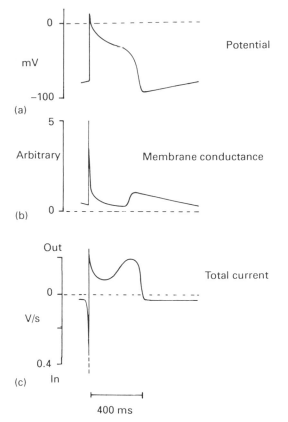

(a)
(b)
(c)

400 ms

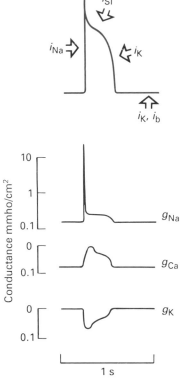

1 s

Figure 3.7 Changes in net electrical conductance of the myocyte cell membrane during an action potential (arbitrary units). The myocyte shown here is a Purkinje cell with an unstable resting membrane potential. (c) Shows net electrical current, with inward current directed downwards and going off the scale during peak depolarization. The current curve is calculated from the slope of the potential record, current being equal to rate of change of voltage multiplied by membrane capacitance. (Weidmann's seminal observation of 1956, redrawn from Noble, D. (1979), see Further reading, by permission)

Figure 3.8 Membrance conductance to individual cations during an action potential in a stable myocyte. The total membrane conductance shown in Figure 3.7b can be dissected into three main components, g_{Na}, g_{Ca} and g_K. The ionic currents which result from these conductances are marked on the action potential at the top. The plateau current, i_{SI}, is carried partly by calcium ions due to the rise in g_{Ca}, and partly by sodium ions due to the sodium-calcium exchange pump (see text). Using the voltage-clamp technique, extracellular ion substitution and blocking agents, it has been found that both g_{Ca} and g_K can be further subdivided. (After a review by Noble, D. (1984), see Further reading list, by permission)

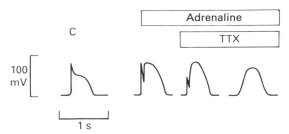

Figure 3.9 Effect of adrenaline (5.5 μM) and tetrodotoxin (TTX, 3 μM) on the action potential of calf Purkinje fibre. The control recording (C) shows a normal Purkinje action potential. Adrenaline enhances the plateau phase by increasing the inward calcium current. Tetrodotoxin, a poison from the Japanese puffer fish, abolishes the initial spike depolarization because it blocks the fast sodium channels. The remaining action potential is quite like that in the SA node (see Figure 3.4). (After Carmeliet, E. and Vereeke, J. (1969) *Pfluger's Archiv* **313**, 303–315)

cium exchange pump (see earlier). One observation favouring this view is that removal of sodium from the extracellular fluid shortens the plateau phase.

The long plateau is important for two reasons:

1. The cell is electrically inexcitable, or 'refractory', during the long period of depolarization (200–400 ms), and since active contraction lasts only 200–250 ms, contraction is weakening by the time the cell becomes repolarized and re-excitable. Consequently, the mechanical response of the myocardium is normally confined to a single twitch (Figure 3.10); a fused series of twitches, such as produce a sustained contraction in skeletal muscle, is not possible in myocardium. A sustained myocardial contraction would of course be fatal.

2. The cardiac action potential does more than just initiate contraction: the plateau phase also directly influences the strength of contraction, because the influx of calcium ions influences the intracellular concentration of calcium. Large plateau currents, such as those

stimulated by adrenaline and noradrenaline, are associated with more forceful contractions.

Repolarization and potassium ions

The potassium conductance gradually increases towards the end of the plateau phase (see Figure 3.8) and, as the slow channels inactivate, the outward K^+ current begins to dominate, producing repolarization to resting potential.

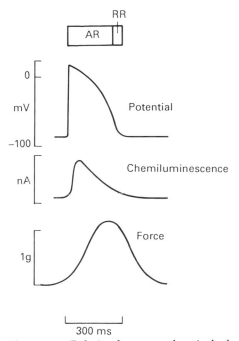

Figure 3.10 Relation between electrical, chemical and mechanical events in the heart. The curve labelled force shows the tension developed by an isolated papillary muscle at constant length. The chemiluminescence curve shows the light emission by cells loaded with aequorin, recorded as current from a photomultiplier; this is in effect a record of free sarcoplasmic calcium ion concentration. Note the time sequence of the peaks – electrical, chemical, mechanical. AR is the absolute refractory period, extending to half-way through repolarization. RR is the relative refractory period (−50 mV to full repolarization) during which the myocyte can only be excited by a larger stimulus than normal

Table 3.2 Ionic currents in a myocardial work cell

Current	Ion	Direction*	Function	Blocker
i_{Na}	Na^+	Inward	Rapid depolarization	Tetrodotoxin
i_{SI}	Ca^{2+} (mostly)	Inward	1. Plateau maintenance 2. Excitation-contraction coupling	Mn^{2+} Verapamil Nifedipine
i_K	K^+	Outward	1. Repolarization 2. Resting membrane potential	Ba^{2+}
i_b	Na^+ (mostly)	Inward	Inward background current keeping resting membrane potential below E_K	–

* 'Inward' means from the extracellular to the intracellular compartment

Quantity of ions exchanged per action potential

It must be stressed that all the ionic currents are small and the change in intracellular ion concentration resulting from an action potential is tiny. Students often assume, wrongly, that the 'rush' of sodium into the cell must raise intracellular Na^+ very substantially – but this is to mistake speed for quantity. In reality only about 40 million sodium ions enter a single myocyte during depolarization, and as the cell contains around 200 000 million sodium ions, the intracellular concentration increases by only 0.02%. For intracellular potassium, the change is only 0.001% (see Appendix, 'Ion exchange'). The Na^+–K^+ and Na^+–Ca^{2+} pumps are therefore able to restore the chemical composition of the sarcoplasm for only a modest expenditure of metabolic energy.

3.5 Excitation-contraction coupling and the calcium cycle

The link between electrical excitation and muscle contraction is provided by calcium ions. The arrival of an action potential causes the sarcoplasmic concentration of calcium ions to rise sharply, from $0.1\,\mu M$ to around $5\,\mu M$, and some of the calcium binds to troponin C to activate the contractile proteins (see Figure 3.3). The correlation between free intracellular calcium ions and contraction has been elegantly demonstrated by the use of aequorin, a protein extracted from luminescent jelly-fish. Aequorin emits a blue light in the presence of free calcium ions, and when aequorin is microinjected into myocytes, a faint blue flash is emitted immediately before each contraction (see Figure 3.10 and 3.12). To find the critical calcium concentration for contraction, Fabiato and Fabiato performed a classic experiment in 1975 which involved stripping the sarcolemma from the myocyte so that the intracellular calcium concentration would equilibrate with that in the bathing fluid. The 'skinned' cell was found to relax when the bathing fluid contained $0.1\,\mu M$ Ca^{2+}, contract slightly at $1\,\mu M$ Ca^{2+} and contract with maximum force above approximately $10\,\mu M$ Ca^{2+} (see Figure 6.4). In the intact cell, the level of free calcium ions during excitation is in the range 1–$10\,\mu M$, and a typical concentration of, say $5\,\mu M$, produces only partial activation of the contractile proteins, i.e. it activates only a fraction of the potential crossbridges. If the free cytosolic calcium ion concentration is increased (e.g. by adrenaline), more cross-bridges are activated and contractile force increases.

How does the action potential produce a 50-fold rise in sarcoplasmic free Ca^{2+} concentration? As the right side of Figure 3.11 shows, the calcium ions arrive from at least two sources: the store of bound calcium in the sarcoplasmic cisternae and the second inward current of the plateau.

Sarcoplasmic cisternae: an internal source of calcium

The cisternae of the sarcoplasmic reticulum contain a store of calcium ions linked to special quick-release sites. As the spike of depolarization travels into the cell along the

T-tubule, it causes the cisternae to release calcium ions from the quick-release sites. Some additional Ca^{2+} may also be released from sites on the inner surface of the sarcolemma. The ions diffuse the micro-metre or so into the sarcomere very rapidly, and the cell begins to develop tension within a few milliseconds (see Figure 3.10).

Stimulated by the raised sarcoplasmic calcium level, the sarcoplasmic reticulum then actively pumps calcium ions back into its interior (see Figure 3.11 left side). In combination with the sarcolemmal Na^+–Ca^{2+} pumps, this reduces the calcium concentration in the sarcoplasm and ter-minates the contraction. Autoradiographic studies indicate that the uptake sites on the sarcoplasmic reticulum are some distance from the release sites, so the stored calcium has then to be transported back to the release sites near the Z lines. This is a

Extracellular fluid

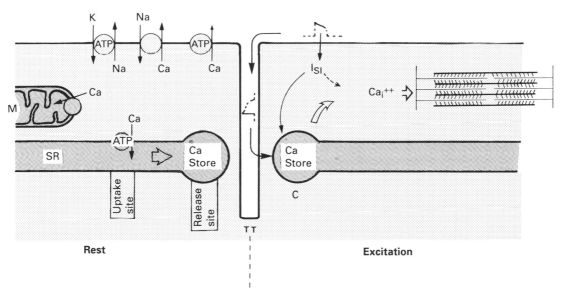

Figure 3.11 Sketch of the calcium cycle. Right side: excitation. Transverse tubule (TT) transmits excitation to the cisterna (C) near the Z line, which releases Ca^{2+} ions from specific quick-release sites. The second inward current (i_{SI}) supplies additional calcium, as well as producing calcium-induced calcium release. Left side: restoration of calcium levels. Calcium concentration is returned to baseline by the sarcolemmal sodium-calcium exchange pump and by the sarcoplasmic reticulum pump (SR), which is some distance from the release sites. The mitochondrial store (M) may be important in the long-term buffering of intracellular calcium

relatively slow process taking 200 ms or more during diastole. Because the restocking process is slow, the quantity of calcium available at the release sites (and therefore the force of contraction) is influenced by the interval between beats, as described in Section 6.8 (the Bowditch effect).

Second inward current: an external source of calcium

During the plateau, an influx of extracellular calcium ions boosts the intracellular calcium content. The importance of extracellular calcium ions was discovered by Sidney Ringer in 1883, and as with many seminal discoveries chance played a part. It was the job of Ringer's assistant to prepare solutions of sodium and potassium chloride, and these maintained the beating of an isolated frog heart for many hours; but when the assistant was unavailable and Ringer had to prepare the 'same' solution himself, the hearts quickly weakened and failed. Ringer discovered that whereas he had used distilled water as solvent, his assistant had been using the local London tap water – which drains from limestone and chalk hills and has a considerable Ca^{2+} content.

Thus, extracellular calcium enhances the contractile force of the heart. This is evidently due to its influx during the second inward current, since the force of myocardial contraction correlates with the size of the current; and if the calcium current is increased by adrenaline, contractile force increases too. The way that the calcium current influences contraction is, however, less direct than might be first supposed, for the number of calcium ions entering the myocyte during a single action potential is actually too small to have much direct effect on intracellular calcium concentration. Rather, the effect of the calcium current on contraction arises in two indirect ways. First, the arrival of the extracellular calcium ions in the sarcoplasm helps to stimulate the release of calcium from the internal store, a phenomenon called 'calcium-induced cal-

cium release'. Second, the influx of extracellular calcium ions increases the amount of intracellular calcium available for re-uptake into the store, which becomes available in subsequent beats. This focuses our attention on the question of what governs the size of the calcium store.

Size of the internal calcium store

The size of the intracellular store of calcium depends on the balance between the influx of extracellular calcium ions during the plateau and their expulsion by the sarcolemmal pumps during diastole. The size of the intracellular store, and with it the contractile force, depends therefore on (1) the extracellular Ca^{2+} concentration, and (2) the relative duration of systole (the calcium influx phase) and diastole (the calcium efflux phase). This point is considered further in Section 6.8 (the Bowditch effect).

Action of digoxin

Digoxin is a cardiac glycoside produced by foxgloves, and it has been used for over two centuries to treat heart failure because it enhances myocardial power. It achieves this by increasing the level of intracellular calcium, as illustrated in Figure 3.12. Its immediate pharmacological action, however, is to slow down the sarcolemmal Na^+–K^+ exchange pump, by inhibiting the membrane ATPase that powers it. This produces a rise in intracellular sodium concentration and a fall in the sodium gradient across the cell membrane. Since the Ca^{2+}–Na^+ exchange pump is itself driven by the sodium gradient (Figure 3.5), calcium expulsion is slowed and calcium accumulates in the cell.

3.6 Pacemaker and conduction system

Up to this point we have been concerned with events within the work cell. It is time now to turn our attention to the events that lead up to electrical excitation of the work cell, beginning with the structure of the cardiac pacemaker and conduction system.

The sino-atrial node or pacemaker

The mammalian heart beat is initiated by the sino-atrial or 'SA' node, a strip of modified muscle roughly 20 mm long × 4 mm wide in man, located on the posterior wall of the right atrium close to the super vena cava (Figure 3.13). The sino-atrial node is so-called because it evolved from the sinus venosus, an antechamber to the right atrium in lower vertebrates. The node is composed of small myocytes with only scanty myofibrils and an electrically unstable cell membrane (see later). As a result of their unstable resting membrane potential, the nodal cells generate an action potential roughly once every second. This excites the adjacent atrial work cells and a wave of depolarization then spreads across the two atria, passing from cell to cell at a rate of approximately 1 m/s and initiating atrial systole.

Atrioventricular node

After passing down the atrial septum the electrical impulse reaches the atrioventricular node (AV node), which is a small mass of cells and connective tissue situated in the lower, posterior region of the atrial septum. The AV node marks the start of the only electrical connection across the annulus fibrosus, which otherwise completely insulates the atria from the ventricles. The impulse is delayed in the node for approximately 0.1 s (at resting heart rates), owing to the complex circuitry of the cells and their small diameter (2–3 μm), which reduces the conduction velocity to only 0.05 m/s. The resulting delay is functionally very important because it allows the atria sufficient time to contract before the ventricles are activated.

The main bundle (bundle of His) and its branches

A bundle of fast-conducting muscle fibres, called the bundle of His, next conveys the electrical impulse from the AV node across the annulus fibrosus into the fibrous upper

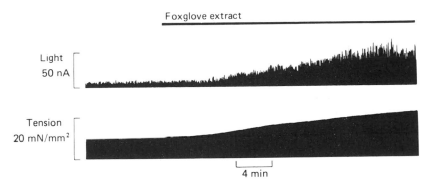

Figure 3.12 In 1785 William Withering reported that a folklore remedy based on an infusion of foxglove leaves (*Digitalis*) was beneficial in the treatment of 'dropsy' (cardiac failure). This experiment, to mark the bicentennial, shows how the extract increases both twitch force and the concentration of free intracellular calcium ions during contraction in ferret papillary muscle. The light emission from aequorin-injected muscle (top trace) is a function of calcium concentration. (From Allen, D. G., Eisner, D. A., Smith, G. L. and Wray, S. (1985) *Journal of Physiology*, **365**, 55P, by permission)

part of the interventricular septum. Here the bundle turns forwards and runs along the crest of the muscular septum (see Figure 3.13), giving off the so-called 'left bundle branch', which really comprises two sets of fibres, one anterior and one posterior. These course down the left side of the septum to supply the left ventricle. The remaining bundle, the right bundle branch, runs down the right side of the septum and supplies the right ventricle. The bundle fibres are wide, fast-conducting myocytes arranged in a regular end-to-end fashion. They terminate in an extensive network of large fibres in the subendocardium which were described by the Hungarian histologist Purkinje in 1845. The Purkinje fibres are the widest cells in the heart and their large diameter (40–80 µm) endows them with a high conduction velocity (3–5 m/s); their role is to distribute the electrical impulse rapidly to the endocardial work cells. From the endocardium the impulse spreads from work cell to work cell at approximately 0.5–1 m/s and excites the entire ventricular wall. The function of the conduction system is thus to excite the whole ventricular mass as near simultaneously as possible.

Dominance

There are other potential pacemaker sites in the heart besides the SA node: the SA node normally sets the heart rate simply because its cells have the fastest intrinsic rate of firing and thus 'get there first'. The cells of the bundle of His are also capable of spontaneous firing, albeit at the slower rate of 40 per min and some Purkinje cells can generate their own rhythm too, though even more slowly (approximately 15 per min). There is thus a gradient of intrinsic

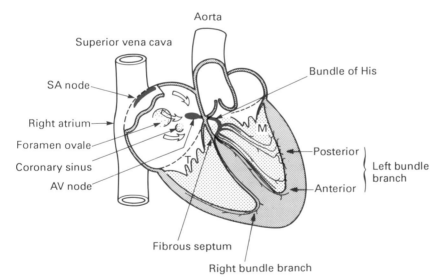

Figure 3.13 Diagram of the cardiac conduction system, with the heart cut obliquely in half. The posterior fibres of the left bundle branch are seen *en face*, running just within the septal wall, then curling round within the left ventricular wall. The broad arrows indicate preferential conduction routes through atrial muscle, called the anterior, middle and posterior internodal paths. Note the proximity of the valves (T, tricuspid; M, mitral) to the bundle of His; valvular lesions can affect the main bundle. The foramen ovale is the remnant of the fetal connection between right and left sides. The coronary sinus is the outlet of the main coronary vein

firing rates along the electrical system. The lower cells are normally excited from the SA node, however, before they have time to fire spontaneously, and this is called 'dominance' by the SA node. The existence of an alternative, albeit slower pacemaker is revealed in a pathological condition called 'heart block', in which there is a blockage of the normal electrical connection across the annulus fibrosus. This prevents the SA node from dominating the bundle of His, and cells in the bundle then take over the pacemaker role, driving the ventricles at their own intrinsic rate of about 40 beats/min (see Section 4.8).

3.7 Nodal electricity

The pacemaker potential

As pointed out earlier, myocytes can be divided into work cells with stable resting membrane potentials and pacemaker cells with unstable membrane potentials. The resting membrane potential of a SA node cell is only about $-60\,mV$, and it decays spontaneously as illustrated in Figures 3.4 and 3.14. This slowly declining potential is called the pacemaker potential (or prepotential) and when it reaches threshold (approximately $-40\,mV$ in nodal cells) it triggers an action potential, which sparks off the next heart beat. The slope of the pacemaker potential determines the time taken to reach the threshold value, so the slope governs heart rate; the steeper the slope the sooner threshold is reached and the shorter the time between beats. Since the pacemaker slope is steeper in SA node cells than elsewhere in the electrical system, the SA node has the fastest intrinsic firing rate and initiates each heart beat.

The decay of the pacemaker potential is caused by a gradual fall in membrane permeability to potassium ions, which is reflected in a fall in total membrane conductance (see Figure 3.7). As a result the outward background current i_K falls progressively, allowing the inward background current (termed i_f in nodal cells) to depolarize the cell slowly (see Figure 3.14). As the potential approaches $-40\,mV$, some low-threshold voltage-gated channels permeable to calcium ions begin to open, so a small inward current of Ca^{2+} ions contributes to the final third of the pacemaker potential.

Nodal action potentials

The nodal action potential is slow-rising and small in amplitude, somewhat like the

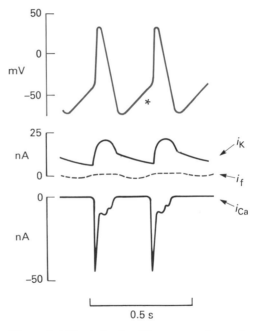

Figure 3.14 Ionic basis of the pacemaker potential and action potential in a sino-atrial cell. The slope of the pacemaker potential (asterisk) determines the interval between heart beats. Inward currents are shown as negative values, outward currents as positive values. The 'funny' inward background current i_f produces depolarization of the pacemaker potential, and is carried by sodium ions (and some calcium ions in the late phase). The main inward current labelled i_{Ca} is due mostly to calcium ions, but can be subdivided into further currents. (After Noble, D. (1984) see Further reading list, by permission)

potential of a tetrodotoxin-blocked work cell (compare Figures 3.4 and 3.9). This is because the nodal cell lacks functional fast sodium channels; its action potential is generated solely by i_{SI}, the slow inward current of predominantly calcium ions (see Table 3.3).

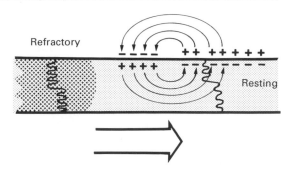

Figure 3.15 The spread of excitation through myocardium by local currents acting ahead of the action potential. The external current flows through the extracellular fluid and the internal current through the sarcoplasm and intercalated junctions. This discharges the cell membrane, which behaves like a lipid capacitor

The transmission of excitation

The spread of excitation from the SA node into the atria, conduction system and ventricles is mediated by local electrical currents acting ahead of the action potential. In the 'active' depolarized region, the exterior of the cell membrane is negatively charged with respect to the interior, while in the resting zone ahead it is positively charged (see Figure 3.15). The two regions are connected by a conducting medium, the extracellular fluid, so positive charge flows from the outside of the resting membrane, depolarizing it. An intracellular current flows in the opposite direction along the cell axis via the gap junctions of the intercalated discs and depolarizes the inside of the membrane. The process is in fact the discharging of a capacitor, the lipid cell membrane. When the resting membrane

has been depolarized to threshold it generates an action potential and the entire process moves on; and since the membrane to the rear is refractory, the excitation progresses unidirectionally. The rate of conduction is greater in wider cells, because they have a lower axial resistance. It is also greater in cells with large, rapid-rising action potentials, because these create bigger propagating currents.

Table 3.3 Ionic currents in nodal pacemaker cells

Current	Ion	Direction	Function	Blocker
(i_{Na})	–	–	Absent	–
i_{SI}	Ca^{2+} (and some Na^+)	Inward	1. Slow action potential 2. Last one-third of pacemaker potential	Mn^{2+}, verapamil, nifedipine
i_K	K^+	Outward	1. Decay during pacemaker potential generates unstable resting membrane potential 2. Repolarization	
i_f†	Na^+ (mainly)	Inward	Supplies inward depolarizing current during pacemaker potential	

† I_f is a 'funny' current

3.8 Nervous control of heart rate

The heart rate of a resting adult human is 50–100 beats/min, while in small mammals it is faster (e.g. rabbit 200–250/min) and in large mammals it is slower (e.g. horse 30–50/min). The pacemaker is innervated by autonomic nerves and its intrinsic rate is continuously modified by activity in these nerves. Increased activity in the sympathetic nerves innervating the SA node speeds up the heart rate (tachycardia), while increased activity of the parasympathetic nerves slows it down (bradycardia). Sympathetic and parasympathetic fibres also innervate the AV node, shortening or lengthening the transmission delay respectively. Both sets of autonomic nerve are continuously active at rest but vagal inhibition predominates: if both systems are blocked the intrinsic heart rate turns out to be around 105 beats/min in a young adult. Physiological alterations of heart rate are usually due to reciprocal changes in autonomic nerve activity; for example, the tachycardia of exercise is induced by both an increase in sympathetic activity and a simultaneous decrease in vagal activity. Pacemaker rate is also sensitive to temperature, and during a fever the heart rate increases by approximately 10 beats/min per °C. The temperature effect is put to practical use during open heart surgery, where cooling is used to slow the heart.

Mechanism of action of parasympathetic fibres

The parasympathetic nerve terminals act by releasing a neurotransmitter substance called acetylcholine, which binds to receptor molecules in the cell membrane, the muscarinic receptors. (Acetylcholine, the first neurotransmitter ever to be identified (by Otto Loewi in 1921), was actually discovered in fluid taken from a frog heart during vagal stimulation: the fluid was found to be capable of slowing an isolated frog heart.)

Receptor activation produces a virtually immediate bradycardia due to two effects: (1) the pacemaker potential becomes more negative (hyperpolarization), and (2) the rate of upward drift of the pacemaker potential is reduced (see Figure 3.16). As a result of these two changes the potential takes longer to reach threshold and the interval between beats increases.

The hyperpolarization is induced by a rise in membrane permeability to potassium due to the opening of additional K^+ channels: this increases the outward background current and shifts the membrane potential closer towards the potassium equilibrium potential. The opening of the additional K^+ channels is mediated by an intramembrane protein called a G-protein, which may directly link the muscarinic receptor to the K^+ channel. The reduced slope of the pacemaker potential is due to a reduction in the depolarizing inward current i_f: the effect on slope lasts longer than the hyperpolarization, and this is not yet fully understood.

Examples of vagal bradycardia in man include a slowing of the heart during each expiration (sinus arrhythmia, see Figure 4.9), slowing of the heart at the onset of fainting (see Chapter 15), and during diving (see Chapter 14). An extreme example of vagal bradycardia has even given rise to an everyday expression, 'playing possum'; to fool its attackers the opossum feigns death by collapsing, with a profound bradycardia.

Quoth Fox to Brer Possum
"You're due in my antrum".
Opossum smiled; he had a hunch
His flaccid apnoe-a
And bradycardee-a
Would leave Fox in no mood for lunch.

Mechanism of action of sympathetic fibres

The sympathetic nerve terminals act by releasing the neurotransmitter noradrenaline ('norepinephrine' in the American literature). Noradrenaline binds to cell membrane receptors called β_1-adrenoreceptors, and over the course of

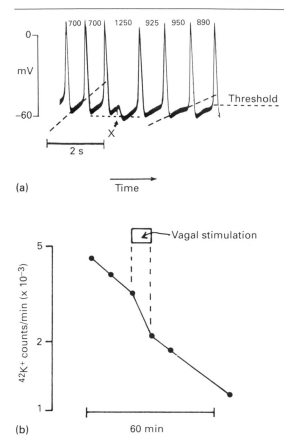

(a)

(b)

Figure 3.16 (a) Illustrates bradycardia induced by vagal stimulation of the sino-atrial node of a kitten, and changes in pacemaker potential. The vagus nerve is stimulated briefly at point X. Note the brisk hyperpolarization of the resting membrane (not well maintained) and the reduced slope of the pacemaker potential (dashed lines). Numbers at top are beat interval in ms. (b) Shows the change in radiolabelled potassium content of frog sinus venosus cells. Vagal stimulation increases the rate of efflux of K^+. ((a) After Jalife, J. and Moe, G. K. (1979) *Circulation Research*, **45**, 595–608; (b) From Hutter, O. F. (1961) *Nervous Inhibition* (ed. Florey), Pergamon Press, Oxford, by permission)

several beats this leads to an increase in firing rate (see Figure 3.17). The hormone adrenaline (epinephrine) acts similarly. Adrenaline and noradrenaline, known collectively as the catecholamines, not only increase the heart rate (their chronotropic action) but also increase the force of myocardial contraction (inotropic action).

The *chronotropic effect* of the catecholamines is mediated by an increase in the rate of rise of the pacemaker potential, which takes less time to reach threshold (see Figure 3.17). The steeper rise is due to an increase in the inward background current i_f, which is carried by Na^+ and Ca^{2+} ions. If the heart is to function effectively at the higher rate, however, all the phases of the cardiac cycle must be shortened (see Section 2.4), and catecholamines have three additional chronotropic effects which help achieve this. (1) They shorten the conduction delay in the AV node. (2) They shorten the plateau of the work cell action potential by increasing the outward potassium current: this shortens systole. (3) They increase the rate of relaxation of myocytes by stimulating the cisternal pumps to take up free cytosolic Ca^{2+} more rapidly. These additional effects help to preserve the diastolic period available for refilling.

The *inotropic (strengthening) effect* of the catecholamines is mediated by an increase in the inward calcium current during the plateau (see Figures 3.9 and 3.17). This

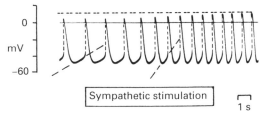

Figure 3.17 Effect of sympathetic stimulation on the pacemaker potential of the frog sinus venosus during the boxed interval. The dashed gradients highlight the increased slope of the pre-potential. The upper double-dashed line draws attention to the increased size of the action potentials, which is due to the enhancement of the inward calcium current by noradrenaline. Note also the relatively slow onset of the tachycardia. (After Hutter, O. F. and Trautwein, W. (1956) *Journal of General Physiology*, **39**, 715–733, by permission)

enhances the intracellular calcium store over the course of several beats.

The biochemical events that link receptor activation to changes in ionic channels are an area of very active current research (see Appendix, 'Second messengers'). The effects of β_1-adrenoreceptor activation seem to be mediated by an intramembrane protein, the G_s-protein, which activates a membrane-bound enzyme, adenylate cyclase. The latter catalyses the formation of an intracellular 'second messenger', cyclic adenosine monophosphate (cAMP), which activates protein kinase. Protein kinase is an intracellular enzyme which, via phosphorylation, influences the number of functional calcium channels in the sarcolemma, and the activity of the cisternal calcium pumps.

3.9 Effect of selected extracellular factors

Severe extracellular electrolyte disturbances will in general be 'A Bad Thing' for cardiac function. As Ringer showed, hypocalcaemia reduces myocardial contractility, while extreme hypercalcaemia arrests the heart in systole. But it is probably hyperkalaemia (a raised concentration of extracellular potassium ions) that is most often a problem in clinical practice.

Hyperkalaemia

The normal potassium concentration in extracellular fluid is 3.5–5.5 mM, and a level of only 7.5 mM K^+ can arrest the heart in diastole. Chronic hyperkalaemia can arise gradually during renal failure, acidosis or potassium overloading and its direct effect is to reduce the resting membrane potential by lowering the potassium equilibrium potential (see Figure 3.6). The action potential is altered too, because the gradual reduction in resting potential allows plenty of time for the inactivation gates (h gates) of

the fast Na^+ channels to close. At a resting potential of −70 mV, about half the 'h' gates are closed, while at −50 mV they are all closed. As a result the action potential has a sluggish rise and a small amplitude (see Figure 3.18). Small action potentials produce only small propagating currents, so the conduction of a small action potential is easily blocked, leading to arrhythmias and heart block. Moreover, the action potential becomes shorter, which reduces the total influx of Ca^{2+} ions and leads to a weakening of the heart beat.

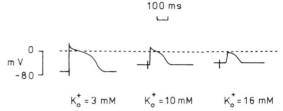

Figure 3.18 Effect of hyperkalaemia on the membrane potential of a Purkinje fibre (see text). The spike to the left of each action potential marks stimulation by an external impulse some distance away. Note the increasing conduction time, reduced resting potential, reduced action potential and slow rate of rise. (After Myerburg, R. J. and Lazzara, R. (1973) In *Complex Electrocardiography* (ed. E. Fisch), Davis Co., Philadelphia

Some drugs affecting cardiac electricity

The actions of many anti-arrhythmic drugs (e.g. quinidine, procainamide, lignocaine) are still ill-understood, but two classes of drug, β-adrenoreceptor blockers and calcium-channel blockers, are better understood. General β-antagonists like propranolol and oxprenolol, and selective β_1-antagonists like atenolol and metoprolol, block the β_1-adrenoreceptors on nodal and work cells. This interrupts the tonic sympathetic drive to the heart, reducing heart rate and contractile force. Since this also reduces cardiac output and work, β-blockers are often used in the treatment of hypertension and angina (ischaemic heart pain due to cardiac work exceeding oxygen supply).

Verapamil and nifedipine (the dihydropyridines) are a relatively new class of drug which act chiefly on high-threshold voltage-gated calcium channels involved in the plateau phase (L-type channels). These channels are partially blocked, impairing the slow inward current of Ca^{2+} ions. This reduces the duration of the action potential and produces a negative inotropic (weakening) effect. These calcium-channel blockers are used to treat certain arrhythmias.

3.10 Summary

Cardiac electricity is a complicated topic and a brief resumé may be helpful. The resting membrane potential approximates to a K^+ equilibrium potential, modified by a slight inward background current of Na^+. Ionic pumps maintain the intracellular composition but do not directly generate the membrane potential. The action potential of work cells arises from a rapid inward current of Na^+, followed by a slow inward current of (chiefly) calcium ions which gives rise to a prolonged plateau. The potential not only excites contraction but also influences contractile force, by affecting the intracellular Ca^{2+} level. In nodal cells, a decaying pacemaker potential is generated by a decaying potassium conductance. This triggers a sluggish action potential which is

generated solely by the slow inward current. Autonomic nerves adjust heart rate by altering the rate of rise of the pacemaker potential and, in the case of parasympathetic fibres, by increasing the resting membrane potential.

Further reading

Bean, B. P. (1989) Classes of calcium channel in vertebrate cells. *Annual Reviews in Physiology*, **51**, 367–384

Davies, M. J., Anderson, R. H. and Becker, A. E. (1983) *The Conduction System of the Heart*, Butterworths, London

Eisner, D. A. and Lederer, W. J. (1985) Na-Ca exchange:stoichiometry and electrogenicity. *American Journal of Physiology*, **248**, C189–202

Fabiato, A. and Fabiato, F. (1977) Calcium release from the sarcoplasmic reticulum. *Circulation Research*, **40**, 119–129

Jewell, B. R. (1982) Activation of contraction in cardiac muscle. *Mayo Clinic Proceedings*, **57**, 6–13

Levy, M. N. and Martin, P. (1984) Parasympathetic control of heart. In *Nervous Control of Cardiovascular Function* (ed. W. C. Randall), New York, Oxford University Press, pp. 68–94

Noble, D. (1979) *The Initiation of the Heart Beat*. Clarendon Press, Oxford

Noble, D. (1984) The surprising heart. *Journal of Physiology*, **353**, 1–50

Sommer, J. R. and Johnson, E. A. (1979) Ultrastructure of cardiac muscle. In *Handbook of Physiology, Cardiovascular System, Vol. 1, The Heart* (ed. R. M. Berne), American Physiological Society, Bethesda, pp. 113–186

Winegrad, S. (1979) Electromechanical coupling in heart muscle. In *Handbook of Physiology, Cardiovascular System, Vol. 1, The Heart* (ed. R. M. Berne), American Physiological Society, Bethesda, pp. 393–428

Chapter 4
Electrocardiography

4.1 Principle of electrocardiography

Electrocardiography is the process of recording the potential changes at the skin surface resulting from depolarization and repolarization of heart muscle; the record itself is called an electrocardiogram (ECG). The method was developed at the turn of the century by Willem Einthoven in Leiden, who invented the string galvanometer, and Augustus Waller in London, whose demonstration to the Royal Society in 1909 provoked protests in Parliament (see Figure 4.1). As explained in Chapter 3, the spread of excitation through the myocardium involves small currents flowing through the extracellular fluid (see Figure 3.15). These extracellular currents create slight potential differences at the body's surface because the extracellular fluid is a continuous conducting medium between the heart and skin. The minute differences in skin potential (approximately 1 mV) are picked up by metal contact electrodes, measured by a millivoltmeter and recorded on a moving paper strip to produce the ECG.

The size of the potential differences at the surface depends on the size of the cardiac extracellular current, which in turn depends on the mass of myocardium being activated. Since the mass of the conduction system is tiny its depolarization does not register on a surface ECG, though nodal and bundle activity can be recorded from an intracardiac electrode introduced by cardiac catheterization. With the usual surface ECG, however, only the activity of atrial and ventricular working muscle is detected.

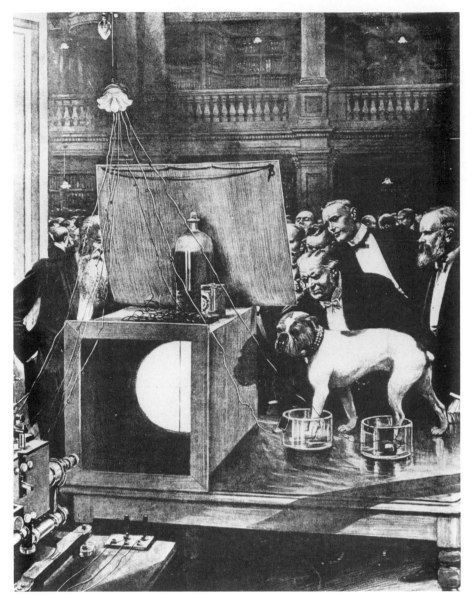

Figure 4.1 The electrocardiogram was demonstrated to the Royal Society by Waller's pet bulldog, Jimmie, in 1909. Jimmie has front and hind paws in pots of normal saline connected to a galvanometer. (From the Illustrated London News, May 22nd 1909.)

The Times newspaper of July 9, 1909 reported that Mr Ellis Griffith (MP for Anglesey) questioned the Secretary of State in Parliament over Waller's 'public experiment' on a dog with 'a leather strap with sharp nails secured around the neck, his feet being immersed in glass jars containing salts . . . connected by wires with galvanometers'. Had the Cruelty to Animals Act (1876) been contravened?

Mr. Gladstone: 'I understand the dog stood for some time in water to which sodium chloride had been added or in other words a little common salt. If my honorable friend has ever paddled in the sea he will understand the sensation. (Laughter) The dog – a finely developed bulldog – was neither tied nor muzzled. He wore a leather collar ornamented with brass studs. Had the experiment been painful the pain would no doubt have been immediately felt by those nearest the dog. (Laughter).'

Mr. MacNeill (Donegal South): 'Will the right honorable gentleman inform the person who furnished him with his jokes that there are members in this House who regard these experiments on dogs with abhorrence?' (Hear)

Mr. Gladstone: 'I certainly shall not. The jokes, poor as they are, are mine own.' (Laughter and cheers) (From Waller, A.D. (1910) *Physiology the Servant of Medicine*, Hodder & Stoughton)

4.2 Relation of ECG waves to cardiac action potentials

Figure 4.2 shows some typical human ECG recordings. There are three main deflections per cardiac cycle: the P wave, corresponding to atrial depolarization; the QRS complex, corresponding to ventricular depolarization; and the T wave, corresponding to ventricular repolarization. The two regions where the trace returns to baseline (the isoelectric state) are called the PR interval and ST segment. These ECG features are compared with the underlying cardiac action potentials in Figure 4.3 and with the mechanical events of the cycle in Figure 2.3.

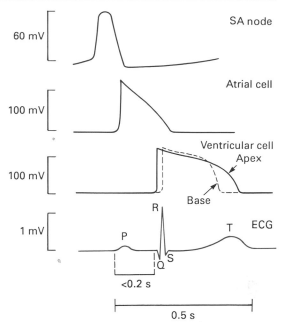

Figure 4.3 Timing of the ECG waves compared with intracellular recordings at different sites, including two sites in the ventricle (apex and base). Note that the ECG voltage scale is much smaller than that for the membrane potentials. Note also that the base (dashed line) repolarizes before the apex, which is why the T wave is upright

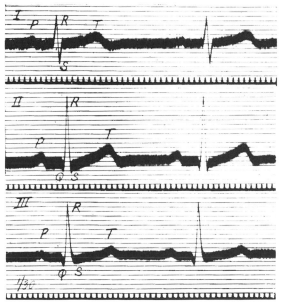

Figure 4.2 ECG of a young healthy adult recorded from Leads I, II and III. Ordinate divisions represent 0.1 mV, horizontal scale marks 1/30th second. (From Sir Thomas Lewis's classic textbook of 1920, *The Mechanism and Graphic Registration of the Heart Beat*, Shaw & Sons London). Lewis's major contributions included the clinical application of the ECG, the recording of the jugular venous pulse and the discovery of the skin triple response (see Chapter 11). The thickness of the baseline is an interference artefact common with the early string galvanometers (cf. Figure 4.9a)

P wave The first event of the cardiac cycle, SA node depolarization, does not register on the ECG because nodal mass is too tiny. The next event, depolarization of the atrial cell mass, produces the P wave, lasting approximately 0.08 s. The P wave coincides with the upstrokes of the atrial action potentials, not with contraction; contraction follows during the PR interval.

PR interval The interval between the beginning of the P wave and the beginning of the QRS complex is always called the PR interval even when it is, strictly-speaking, a P-Q interval. It represents the time taken for excitation to spread over the atria and through the conduction system to reach the ventricular septum. Much of the PR interval is produced by the delay in conduction through the atrioventricular node, which allows time for atrial systole to develop. In a human ECG the total PR interval should not however exceed 0.2 s, and longer intervals indicate a defect in conduction through to the ventricle (see 'heart block', Figure 4.9c). The ECG is isoelectric during the PR interval despite the existence of a potential difference between atria (depolarized) and ventricles (polarized) because the insulating annulus fibrosus breaks the electrical circuit and no current flows.

QRS complex During ventricular excitation a large mass of muscle is activated almost synchronously and this produces a large deflection in the ECG, the QRS complex. The complex lasts 0.1 s or less, a longer period being suggestive of cardiac pathology such as a conduction block in one of the bundle branches. The Q wave is an initial downward spike, the R wave an upward spike, and the S wave a second downward spike: but all three components are not necessarily present in every record – the complex may have just an RS configuration as in Lead I of Figure 4.2, or just a QR configuration, as explained later. Since ventricular depolarization coincides with repolarization of atrial cells (see Figure 4.3) the latter is masked by the QRS complex.

ST segment This coincides with the plateau of the ventricular action potential, and rapid ejection occurs during this period. The first heart sound follows just after the S wave, and the second sound a little after the T wave (see Figure 2.3). Since the ventricle is uniformly depolarized, the ST segment is normally isoelectric. If, however, part of the myocardium is damaged by ischaemia (a poor blood supply, due usually to coronary artery disease) there is an apparent depression of the ST segment, a useful clinical sign illustrated in Figure 4.9g. The electrical effect of ischaemia is to reduce the resting potential of the hypoxic myocytes, and since ventricular polarization is then non-uniform, a current, called the resting injury current,

flows between the healthy myocytes and ischaemic cells. This shifts the baseline of the ECG (i.e. the T-P and P-R regions) and leave the ST segment apparently depressed relative to the new baseline.

T wave, and why it is normally upright Ventricular repolarization produces a broad asymmetrical upright wave, the T wave. At first acquaintance it may seem odd that the T wave is upright when repolarization is a process of opposite sign to depolarization.

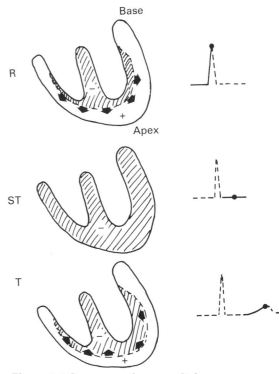

Figure 4.4 Sequence of myocardial depolarization and repolarization. The polarity is shown at three succeeding moments: partial depolarization (R wave), full depolarization (ST segment) and partial repolarization (T wave). Hatched regions are depolarized and the negative sign refers to the external charge. Arrows indicate the direction of advance of the change. Repolarization proceeds in reverse order to depolarization and therefore produces a net electrical difference (dipole) in the same direction as depolarization, creating an upright T wave

The reason is that the myocytes of the base and epicardium have briefer action potentials than the myocytes of the apex and endocardium (see Figure 4.3). Repolarization therefore begins in the base and epicardium, followed by the apex and endocardium. Depolarization on the other hand spreads from apex and endocardium to base and epicardium (see Figure 4.4). Thus repolarization occurs in reverse sequence to depolarization, producing the upright T wave.

Myocardial ischaemia affects not only the ST segment but also the T wave, causing it to invert (see Figure 4.9g).

4.3 Standard limb leads

To understand the shape of the QRS complex we must consider: (1) the position of the recording electrodes relative to the heart, (2) the concept of the heart as an electrical dipole, and (3) the changes in dipole orientation during excitation. Taking the electrode positions first, the ECG is routinely recorded via three electrodes, one on each arm and one on the left leg. In addition, recordings are commonly made from electrodes on the chest wall (the six unipolar precordial leads) but these will not be described here, since our primary concern is with general principles rather than detailed clinical practice). The three limb electrodes can be connected across a voltmeter in three different combinations, called the bipolar limb leads (see Figure 4.5). They are:

Right arm – left arm = LEAD I
Right arm – left leg = LEAD II
Left arm – left leg = LEAD III

Since the limbs act as volume conductors to the trunk, the electrodes form a triangle around the heart, called Einthoven's triangle. The three leads in effect view the heart from three different angles in the frontal plane. Lead I forms the top of the triangle, oriented horizontally across the chest, and this angle is taken as zero. Lead

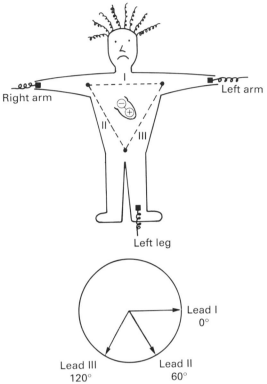

Figure 4.5 Approximate orientation of the standard bipolar limb leads relative to the horizontal (0°). Lead I, right arm-left arm; Lead II, right arm – left leg; Lead III, left arm – left leg

II is oriented more vertically, at roughly 60° to Lead I, and Lead III at roughly 120°. The electronics are arranged so that an upward deflection occurs when the positive pole of a potential difference is directed towards the left arm (Lead I) or left leg (Leads II and III). We must therefore consider next the overall polarity of the heart during excitation.

4.4 Cardiac dipole

At any one instant during the spread of excitation through the ventricle, there exists a resting zone with a diffuse cloud of positive extracellular charges and an excited region with a diffuse cloud of negative extracellular charges (see Figure 4.6a). Just

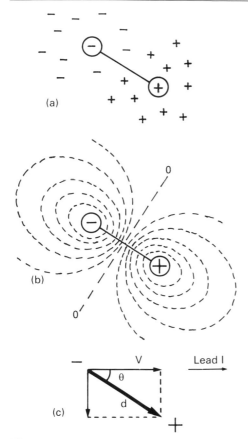

(a)

(b)

(c)

Lead I

V

θ

d

Figure 4.6 Properties of a dipole. (a) Representation of two diffuse groups of opposite charges by a dipole (two points of electrical charge of opposite sign, e.g. the terminals of a battery). (b) Equipotential lines around an electrical dipole. The zero potential run across the middle of the dipole. The potential is inversely proportional to the square of the distance and current flows at right angles to the potential contour. (c) Resolution of a cardiac vector (thick arrow, d) into two components at right angles (thin arrows). The length of the arrow is proportional to vector magnitude, d. The voltage difference detected by Lead I (V) depends on angle θ since V equals d × cosine θ

as a diffuse mass can be represented by a centre of gravity, so a diffuse charge can be represented as a single charge at its electrical centre or pole. During the spread of excitation, the heart has two such poles, one

negative and one positive; it is an electrical dipole.

A dipole is surrounded by positive and negative potential fields, which grow weaker with increasing distance. The potential differences can be measured by a voltmeter aligned across the two poles. If, however, the voltmeter is placed exactly at right angles to the dipole it will register no difference in potential (see Figure 4.6b). This extreme case highlights a crucial point: the potential difference actually recorded depends on the orientation of the recording electrodes relative to the dipole. This is because a dipole is a vector quantity having direction as well as magnitude. The symbol for a vector is an arrow whose length represents vector size and whose direction represents the vector's angle. Like a force vector, the cardiac vector can be resolved into two components at right angles and it is the magnitude of the component directed at the recording lead which is actually registered (see Figure 4.6c). This brings us to the question of the orientation of the cardiac dipole.

4.5 Vector sequence

The size and orientation of the cardiac dipole in the frontal plane changes continuously as excitation spreads through the ventricles. (For simplicity, only orientations in the frontal plane are described here, but as the ventricles are 3-dimensional bodies lying in an oblique, rotated position, the dipole rarely lies purely in the frontal plane.) The first region to depolarize is the left side of the interventricular septum, activated by the left bundle branch (see Figure 4.7). This creates a small dipole directed to the right, about 120° to the horizontal. Next, the remaining septum and most of the endocardium depolarize. The epicardium is still polarized and the bulky left ventricle predominates, creating a large dipole directed to the left, about 60° to the horizontal. The last region to be excited is the base of the ventricles close to the

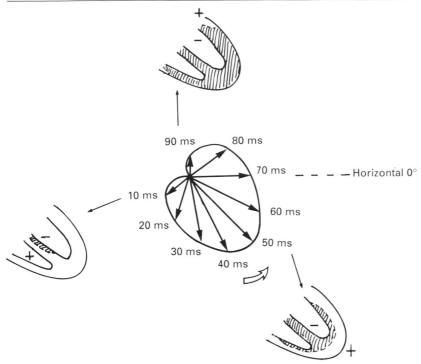

Figure 4.7 Change of the cardiac dipole with time in milliseconds during excitation of the ventricles. Thick arrows represent size and direction of the vector in the frontal plane; the vector swings anti-clockwise. Cross-hatched areas represent depolarized myocardium, which carries a negative external charge. The electrical axis (see text) is about 70° here

annulus fibrosus, so the final dipole is small and directed upwards. As Figure 4.7 shows, this sequence of ventricular activation causes the cardiac vector to swing round in an anti-clockwise direction, and to wax and wane in size over approximately 90 ms.

4.6 Why the QRS complex is complex

From the lead positions and vector sequence it is possible to understand why the QRS wave contains negative waves and why the QRS complex differs from lead to lead. As a simple example let us consider the ventricular dipole at three instants: the beginning, middle and end of excitation, as

'seen' by Leads I and III. In the particular case illustrated in Figure 4.8, the small initial dipole at approximately 120° is directed obliquely away from Lead I. Resolving the vector, we find a small component directed at 180°, which is in the opposite direction to Lead I (0°), so Lead I records a small negative deflection or Q wave. At the same instant Lead III (120°) records a positive deflection – the beginning of an R wave, with no preceding Q wave. After 50 ms the dipole has a large component directed at Lead I, which records a large upward deflection, the R wave. Lead III is at right angles to the dipole at this instant, and so records zero potential difference. By 90 ms the dipole has swung round to −100° in the case illustrated, ausing small negative deflections in both

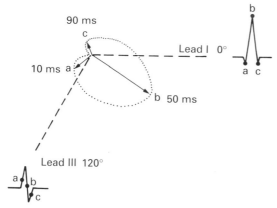

Figure 4.8 Illustration of how the dipole sequence gives rise to different QRS complexes in different leads. The dipole is drawn at three instants (a,b,c) and the component of the vector detected by Leads I and III at these instants is marked on the ECG

Leads I and III, i.e. S waves. Thus the same event produced a QRS complex in Lead I but an RS complex in Lead III. It must be emphasized that there are many variations from the pattern illustrated owing to variations in cardiac orientation.

4.7 Electrical axis of the heart

The direction of the largest dipole is called the electrical axis of the heart, and in Figure 4.7 this is about 60° below the horizontal. The electrical axis can be estimated roughly by comparing the size of the R wave in Leads I, II and III. If for example the largest R wave is in Lead II, the electrical axis is closer to 60° than to 0° or 120°. Conversely the lead with the smallest QRS complex and R and S waves of nearly equal height lies roughly at right angles to the electrical axis. The normal range for the electrical axis is very wide, from −30° to +110°. It depends partly on the anatomical orientation of the heart in the chest, being more vertical in a tall person with a narrow thorax. It also becomes more vertical during each inspiration as the pericardium is pulled down by the descending diaphragm (see Figure 4.9a).

The axis depends, too, on the relative thickness of the walls of the right and left ventricles. Hypertrophy of the left ventricle shifts the electrical axis to the left (left axis deviation) while hypertrophy of the right ventricle produces right axis deviation.

4.8 Palpitations: benign or bad?

The ECG is an invaluable aid to the diagnosis of many cardiac disorders, such as myocardial ischaemia, ventricular hypertrophy and especially irregular rhythms (arrhythmias). One form of arrhythmia which is perfectly normal is *sinus arrhythmia* (see Figure 4.9a), a physiological slowing of the sino-atrial discharge rate during each expiration. Sinus arrhythmia is caused by a phasic rise in vagal activity during expiration and it is especially marked in children and young adults. Other arrhythmias are pathological and may arise from some of the following mechanisms.

An aberrant 'pacemaker' at a site other than the SA node may initiate extra systoles, called *ectopic beats* (see Figure 4.9b). The main bundle or a bundle branch may fail to conduct normally, a condition called *heart block* (Figure 4.9c–e). Complete heart block can cause a sudden collapse, called a Stokes–Adams attack. An abnormal conduction pathway may allow the wave of excitation to travel in a circle (circus). The circle causes a recurring re-entry of excitation just after myocytes emerge from their refractory period, so the process becomes self-perpetuating. The circus mechanism may underlie the relatively harmless condition of *atrial fibrillation* (see Figure 4.9f), and the rapidly fatal condition of *ventricular fibrillation* (see legend to Figure 4.9g for detail). A classic example of the circus process is provided by the Wolff–Parkinson–White syndrome, which is characterized by episodes of *paroxysmal tachycardia* or 'palpitations', where the ventricles beat at over 200 per min. The syndrome is caused

Figure 4.9 A cornucopia of arrhythmias. (a) Sinus arrhythmia in a healthy 19-year-old medical student. Sinus node activity is slowed from 92 per min to 52 per min during expiration due to increased vagal discharge. The change is partly a reflex from pulmonary stretch receptors, but it persists in anaesthetized animals even when ventilation is paralyzed, so the vagal periodicity also arises centrally (see Chapter 13). Note that the size of the R wave is reduced during inspiration as the descending diaphragm pulls the heart into a more vertical orientation. (b) The first example of an extrasystole (ectopic beat, asterisk) recorded by Einthoven. An unstable region within the ventricle has initiated a premature ventricular depolarization. The R wave is broad and ill-synchronized and the T wave inverted because the normal anatomical pathway is not followed. As the ventricle is then refractory, the next impulse from the SA node fails to excite the ventricle and there is a 'compensatory pause' (CP), which the patient may notice: 'My heart keeps missing a beat doctor'. The upper trace is a radial pulse. (c–e). Progressive stages of heart block. First degree heart block (c) is a lengthening of the PR interval, to 0.45 s here, due to slow conduction through the AV node or bundle of His. Second degree block (d) involves an intermittent failure of excitation to pass from atria to ventricles; in this example there are four labelled P waves and only three corresponding QRS complexes. In third degree block (e) excitation fails totally to penetrate from atria to ventricles; the atria and ventricles beat entirely independently of each other, the atria at 72 per min driven by the SA node and the ventricles at 30 per min driven by a pacemaker in the bundle of His or Purkinje system which has emerged from domination. The patient may experience suddent faints (Stokes–Adams attacks) and will need an artificial pacemaker. (f) Atrial fibrillation, not uncommon in the elderly and compatible with a sedentary life. Irregular oscillations (f) replace the normal P waves and the atria undergo a continuous uncoordinated rippling motion, probably due to a circus mechanism. Excitation is transmitted irregularly to the ventricles producing a highly characteristic radial pulse (upper trace) which is 'irregularly irregular' in timing, and in amplitude (due to variation in passive filling time). (g) Ventricular fibrillation, following myocardial ischaemia (indicated by the ST depression and T wave inversion in the first two beats of the trace). After the second sinus

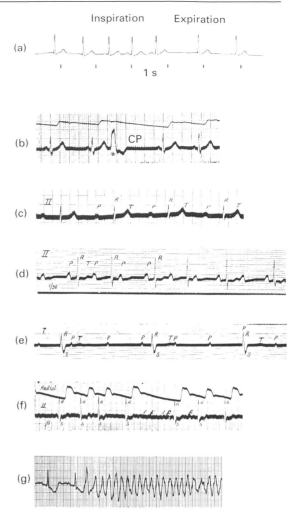

beat a ventricular ectopic occurs during the 'vulnerable period' (latter half of the T wave); the ventricle is vulnerable at this moment because some fibres have repolarized while others are still refractory, so circus pathways can be triggered. This initiates a series of rapid uncoordinated excitations, producing ineffective writhing movements. Causes include ischaemia, anaesthetic overdosage and electrocution. There is no cardiac output or peripheral pulse, and death follows within minutes. 'Cardioversion' by a DC electric shock to the chest sometimes succeeds in restoring sinus rhythm. (Records c–f are from Lewis' classic book, *The Mechanism and Graphic Registration of the Heart Beat*, 1920)

by an anomalous electrical connection across the annulus fibrosus called the bundle of Kent, which allows the ventricular excitation wave to re-enter the atria and re-excite the AV node prematurely. Drugs like quinidine and procainamide prolong the refractory period, which renders re-entry more difficult and terminates some circus arrhythmias. The calcium blocker verapamil shortens the action potential and upsets the timing of re-entry, so it too can terminate a circus movement.

Further reading

Noble, D. (1979) *The Initiation of the Heart Beat*, 2nd edn, Clarendon Press, Oxford

Rowlands, D. J. (1986) The electrocardiogram. In *Oxford Textbook of Medicine* (eds D. J. Weatherall, J. G. G. Leddingham and D. A. Warrell), Oxford University Press, pp. 13.23–13.41

Schamroth, L. (1980) *The Disorders of Cardiac Rhythm*, 2nd edn, Blackwell Scientific Publications, Oxford

Scher, A. M. and Spach, M. S. (1979) Cardiac depolarization repolarization and the electrocardiogram. In *Handbook of Physiology, Cardiovascular System, Vol. 1 The Heart* (ed. R. M. Berne), American Physiological Society, Bethesda, pp. 357–392

Chapter 5
Assessment of cardiac output

Cardiac output is defined as the volume of blood ejected by one ventricle in one minute: it equals the stroke volume multiplied by heart rate. In human subjects the output can be measured by a variety of methods, which either measure the cardiac output as a whole (the Fick principle and the dilution methods) or measure stroke volume and heart rate separately (Doppler and radionuclide methods). The output can also be assessed indirectly by echocardiography and by examining a subject's peripheral pulse. In anaesthetized animals, direct methods requiring surgery can also be employed, such as the placing of an electromagnetic flow meter around the aorta. We will concentrate here, however, on the methods which are currently most important in medicine, beginning with the 'gold-standard' method based on Fick's principle.

5.1 Fick's principle and the measurement of cardiac output

Adolf Fick pointed out in 1870 that the rate at which the circulation takes up oxygen from the lungs must equal the change in oxygen concentration in the pulmonary blood multiplied by the pulmonary blood flow. Since the pulmonary blood flow is of course the output of the right ventricle, this offers a way of determining the cardiac output. The reasoning behind the method is illustrated in Figure 5.1a. The amount of oxygen carried into the lungs in venous blood per minute is the blood flow ($\dot{Q}$) times venous oxygen concentration (C_v);

O_2 in venous blood entering the lungs per minute = $\dot{Q} \cdot C_v$

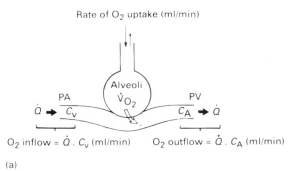

(a)

(b)

Figure 5.1 Fick's principle applied to the measurement of pulmonary blood flow (a). PA, pulmonary artery; PV, pulmonary veins. Fick's principle is quite general and could be applied, for example, to measure crowd flow at a football ground (b). Here the flux of material (money) is in the opposite direction but the same principle applies. If £20 per min is taken and each man emerges poorer by £4, the crowd flow is five men/min. Analogous physiological situations include the passage of carbon dioxide from blood into the alveolar gas, and glucose from blood into muscle

Similarly, the amount of oxygen carried out of the lungs per minute in the blood equals blood flow times arterial oxygen concentration (C_a);

O_2 in arterial blood leaving the lungs per minute = $\dot{Q} . C_a$

The amount of oxygen taken up by the blood during its passage through the lungs therefore equals $\dot{Q} . C_a - \dot{Q} . C_v$ over one minute. In a steady state, this oxygen uptake must equal the loss of oxygen from the alveolar gas in the lungs over one minute ($\dot{V}_{O2}$): oxygen consumption by the alveolar cells can be neglected. We can therefore write:

alveolar oxygen removed per min = blood oxygen gained per min

$$\dot{V}_{O2} = (\dot{Q} . C_a) - (\dot{Q} . C_v)$$

or in terms of the arteriovenous concentration difference $C_a - C_v$:

$$\dot{V}_{O2} = \dot{Q} . (C_a - C_v)$$

This is Fick's expression, and it tells us that the rate of oxygen uptake from alveolar gas (ml/min) equals pulmonary blood flow (litres/min) times the arteriovenous difference in oxygen concentration (ml/litre). Since pulmonary blood flow is actually the output of the right ventricle, we can re-arrange the Fick expression to give the cardiac output:

Cardiac output (l/min) =

$$\frac{\text{Oxygen uptake rate (ml/min)}}{\text{Arterial } O_2 \text{ conc.} - \text{Venous } O_2 \text{ conc. (ml/l)}}$$

A resting man, for example, absorbs around 250 ml/min of oxygen from the alveolar gas ($\dot{V}_{O2}$). Arterial blood contains 195 ml per litre of oxygen, while mixed venous blood in the pulmonary artery contains 145 ml per litre of oxygen. Each litre of blood therefore takes up 50 ml of oxygen, i.e. ($C_a - C_v$) = 50 ml/litre. To take up 250 ml oxygen, 5 litres of blood are required (i.e. 250/50). As this uptake takes 1 min, the pulmonary blood flow must be 5 litres/min.

Practical aspects

Fick's theoretical method only became practicable in man in the 1940s, when progress in cardiac catheterization allowed mixed venous blood to be sampled from the right ventricle. Samples from peripheral veins are unsuitable for this purpose because their oxygen concentration varies; it is 170 ml per litre in renal venous blood, but only 70 ml per litre in coronary venous blood. Venous blood only becomes fully mixed and uniform in the right ventricle outflow tract and pulmonary artery. The problem of obtaining a sample of fully mixed venous blood was finally solved by the German physician Werner Forssman, who in 1929 passed a ureteric catheter through his own arm vein and into the right heart, watching its progress on an X-ray screen. (This brave act founded human cardiac catheterization, won Forssman the disapproval of his head of department and, later, the Nobel prize.)

The Fick method is now as follows. The subject's resting oxygen consumption is measured over 5 to 10 min by spirometry, or by collection of expired air in a Douglas bag. During this period an arterial blood sample is taken from the brachial, radial or femoral artery, and a mixed venous sample is taken from the pulmonary artery or right ventricle outflow tract by a cardiac catheter, introduced through the antecubital vein. The oxygen content of each blood sample is measured and the cardiac output calculated as above.

Limitations of the method

Fick's method is the yardstick by which new methods are usually judged but it has certain limitations. It is slow, and beat-by-beat changes in stroke volume cannot be followed. The method is only valid in the steady state, so the transient early response to exercise cannot be measured. The method is invasive, and cannot be used during severe exercise because the cardiac catheter may provoke arrhythmias in a violently beating heart. An indirect version, in which catheterization is avoided and C_v estimated by analysis of rebreathed gas, is less accurate and rarely used today.

The Fick principle generalized

The Fick principle is quite general and applies to any perfused organ in which material or heat is exchanged at a steady rate. In general terms the flux J of material or heat between the fluid and the perfused organ equals the fluid flow ($\dot{Q}$) multiplied by the concentration change between the inlet (C_{in}) and outlet (C_{out}):

$$J = \dot{Q}\,(C_{out} - C_{in})$$

As Figure 5.1*b* shows, Fick's principle applies even to the influx of money at the turnstile of a football ground. In physiology, Fick's principle is widely used to work out the rate at which an organ consumes nutrients like glucose or fatty acids, from measurements of blood flow and the arteriovenous concentration difference for the nutrient.

5.2 Indicator dilution and thermal dilution methods

Indicator dilution method

In Hamilton's indicator dilution method, a known mass of a foreign substance (the indicator) is injected rapidly into a central vein or into the right heart. The indicator must be one that is confined to the bloodstream and easy to assay, for example the dye indocyanine green, or albumin labelled with radioiodine. The bolus of indicator becomes diluted in the returning venous blood, passes through the heart and lungs and is ejected into the systemic arteries (see Figure 5.2*a*). Samples of arterial blood are taken at frequent intervals from the radial or femoral artery, and the concentration of the indicator in the arterial plasma is plotted against time. For simplicity let us at first suppose that the concentration of indicator in the ejected bolus is uniform, as in Figure 5.2*b*. The concentration against time plot tells us: (1) the time *t* needed for the bolus to pass a given point, and (2) the average concentration of indicator in the bolus over that period. Concentration, *C*, is by definition the injected mass, *m*, divided by the volume of plasma, V, in which the indicator became distributed: $C = m/V$. Therefore:

$$V = m/C$$

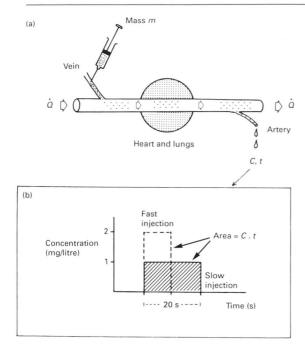

(a)

Mass m

Vein

$\dot{a}$ $\dot{a}$

Artery

Heart and lungs

C, t

(b)

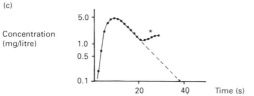

Fast injection

Concentration (mg/litre)

Area = $C . t$

Slow injection

20 s

Time (s)

(c)

Concentration (mg/litre)

5.0
1.0
0.5

0.1

20   40   Time (s)

Figure 5.2 Hamilton's dye-dilution method. (a) arterial concentration C depends on the mass of indicator injected (m) and the volume of blood in which it became diluted. (b) Idealized plot of arterial concentration against time. Although the shape of the plot depends on injection rate, the area C × t (20 mg s per litre) does not; the increase in concentration produced by a fast injection is offset by the shorter duration of the bolus. If the injected mass of dye is 2 mg, the cardiac output of plasma is 2/20 = 0.1 litre/s, or 6 litres/min. For a haematocrit of 0.4 the cardiac output of blood is 10 litres/min. (c) The grim reality. True shape of the concentration curve, with concentration plotted on a logarithmic scale to linearize the decay and allow extrapolation past the recirculation hump (asterisk). Area under the extrapolated curve is used to calculate the cardiac output. (From Asmussen, E. and Nielsen, M. (1953) *Acta Physiologica Scandinavica*, **27**, 217)

For example, if 2 mg of indicator produces a mean plasma concentration of 1 mg/litre, the volume of distribution must be 2 litres. If this volume takes t seconds to pass a fixed point, the left ventricle evidently pumps plasma along at rate V/t (2 litres in 20 s in Figure 5.2b, or 6 litres/min). In other words the cardiac output of plasma can be calculated as:

Cardiac output of plasma =

$$\frac{V}{t} = \frac{\text{mass of indicator } m}{\text{mean concentration } C \times \text{time } t}$$

The output of blood is then easily calculated, being the output of plasma divided by 1 − haemotocrit (the haematocrit is the fraction of blood consisting of cells). Notice that the denominator $C.t$ in the above expression is the area under the plot of concentration against time. Thus:

Cardiac output of plasma =

$$\frac{\text{Mass of indicator}}{\text{Area under } C–t \text{ curve}}$$

In reality the $C–t$ plot is not of course a square wave; it is a curve which rises to a peak and decays exponentially (see Figure 5.2c), but the above expression still applies. The exponential decay is caused by the ventricle ejecting only a fraction of its content with each systole, leaving some indicator behind. Indicator-free venous blood returning to the heart during each diastole dilutes the residual indicator, and this in turn is only partially ejected in the next systole, and so on. After about 15 s, however, the decay curve is disrupted by a 'recirculation hump'. This is caused by blood of high indicator concentration returning to the heart after completing one transit of the myocardial circulation (the shortest route back). To apply the above dilution equation, we must find the area under a $C–t$ curve uncomplicated by recirculation, and this is done by extrapolating the early part of the decay curve, before the recirculation hump. A semi-logarithmic plot facilitates the extrapolation, because it converts the exponential decay into a straight line; the latter is then extrapolated to a negligible concentration (conventionally 1% of the peak value) as in Figure 5.2c. The area under the corrected $C–t$ curve is computed and used to calculate cardiac output.

Pros and cons

The results agree with Fick's direct method to ±5%. The dilution method has an improved time resolution (30 s, cf. >5 min for Fick's method) and can be used in exercise, since ventricular catheterization is not required. The error involved in extrapolating the decay curve is, however, a drawback, and in diseased hearts, where the initial part of the decay curve may be short and distorted, this can be a serious limitation.

Thermal dilution method

This variant of the dilution method is widely used in cardiac departments. Instead of a foreign chemical, temperature is used as the indicator. A known volume of cold saline is injected quickly into the right atrium, right ventricle or pulmonary artery, and the dilution of the cold saline by warm blood is recorded by a thermistor-tipped catheter (Swan–Ganz catheter) in the more distal pulmonary artery. Cardiac output can then be calculated from the area under the temperature-time plot from and the amount of heat (i.e. cold) injected. The major advantage is that the recirculation problem is circumvented because the saline warms up to body temperature long before it returns to the right side. Another advantage is that the ejection fraction can be calculated from the step rise in temperature that follows each refilling of the heart by warm blood. One problem is heat transfer across the walls of the right ventricle and pulmonary artery, which can cause over-estimation of the distribution volume and therefore cardiac output; a computed correction is usually made for this.

5.3 Pulsed Doppler method

This is a relatively new and increasingly popular method in which a pulse of ultrasound is directed down the ascending aorta from a transmitter crystal at the suprasternal notch. Some of the ultrasound is reflected back by the red cells, and this is

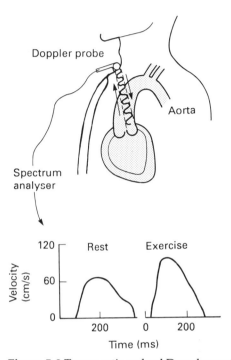

Figure 5.3 Transaortic pulsed Doppler method for measuring the mean velocity of blood across the aorta at each instant of systole. The combined transmitter-receiver unit is positioned over the suprasternal notch and a frequency spectrum analyser produces the velocity-time curve. The area under this curve (velocity × time, i.e. distance) represents the average distance the blood advances along the aorta per stroke (stroke distance). Stroke volume is calculated as the stroke distance × cross-sectional area of the aorta. The maximum rate of rise of velocity (upslope) can be used as in index of contractility. The recordings are from a young subject at rest and during bicycle ergometer exercise at 75 W. (After Innes, J. A., Simon, T. D., Murphy, K. and Guz, A. (1988) *Quarterly Journal of Experimental Physiology*, **73**, 323–341, by permission)

collected and analysed. Since the cells have a high velocity, the frequency of the returning sound waves is different from that of the transmitted signal; this is the Doppler effect, analogous to the change in pitch of a car siren as it speeds past. The average blood velocity across the aorta at each instant is computed from the spectrum of frequencies in the returning signal, and the velocity is plotted against time as in Figure 5.3. To convert the time-averaged velocity (cm/s) to flow (cm^3/s) the diameter of the aorta must be measured by echocardiography (see Section 2.5) and the cross-sectional area, πr^2, then multiplied by mean velocity. The result, aortic flow, represents cardiac output minus coronary blood flow. Although the Doppler method has calibration and 'noise' problems, it has the enormous advantages of non-invasiveness and speed, and can record each individual ejection.

5.4 Examination of the peripheral pulse

The oldest, fastest, cheapest and easiest method of assessing cardiac output, albeit subjectively, is to lay a finger on the radial pulse. Heart rate can be measured and the finger also senses whether the pulse is 'strong' or 'weak', a strong pulse being associated with a large stroke volume (e.g. exercise) and a weak pulse associated with a low stroke volume (e.g. haemorrhage). What the finger actually detects is the expansion of the artery as pressure rises during systole. The rise in pressure, or 'pulse pressure', equals systolic pressure minus diastolic pressure, and this is easily measured with a sphygmomanometer (see Section 7.4). The relation between pulse pressure and stroke volume is illustrated in Figure 5.4. Most of the stroke volume (70–80% at rest) is temporarily accommodated in the elastic arteries, owing to the resistance of the arteriolar system to runoff. The distension of the elastic arteries raises

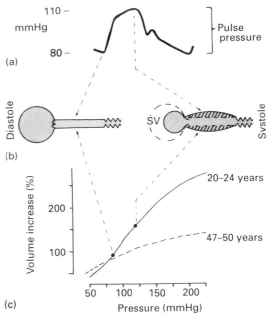

Figure 5.4 Relation of the pulse pressure to stroke volume. (a) Aortic pressure wave. (b) Left ventricle, elastic arterial system and resistance vessels during diastole and systole. SV is stroke volume. (c) Pressure-volume curve of the human thoracic aorta. The volume at diastolic pressure is expressed as 100% here. Note the decline in distensibility with age. Note also the flattening of the pressure-volume curve at high pressures, which indicates that compliance (distensibility) is decreasing. Consequently, a rise in mean pressure would increase the pulse pressure even if stroke volume remained the same ((a) After McDonald, D. A. (1974) *Blood flow in Arteries*, Edward Arnold, London; and (c) after Hallock and Benson (1937) *Journal of Clinical Investigation*, **16**, 597)

the blood pressure and the amount by which pressure rises depends partly on the stroke volume and partly on the distensibility the arterial system. Distensibility or 'compliance' is defined as change in volume per unit change in pressure:

compliance = increase in volume/increase in pressure

Rearranging this, we can see how pulse

pressure relates to stroke volume:

pulse pressure =

$$\frac{\text{stroke volume (minus initial runoff)}}{\text{compliance}}$$

In a young adult, arterial compliance is around 2 ml per mmHg at normal pressures. Unfortunately, however, the compliance is not a constant, and is affected by three factors. (1) High arterial pressures and volumes reduce arterial compliance; this is evident from the curvature of the arterial pressure-volume relation in Figure 5.4. (2) High ejection velocities reduce compliance because the artery wall is a viscoelastic material (see Appendix) and needs time to expand. During exercise the pulse pressure increases proportionately more than stroke volume because there is less time available for viscous relaxation in the wall. (3) Advancing age is associated with arteriosclerosis, a hardening of the artery walls, which reduces compliance and leads to large pulse pressures in the elderly.

Because arterial compliance is so variable, and also because the percentage runoff during early systole varies with peripheral resistance, the pulse pressure offers only an indirect assessment of stroke volume. Nevertheless, within its limitations the peripheral pulse provides an exceedingly convenient indication of changes in cardiac output in an individual patient from day to day.

5.5 Radionuclide angiography and other methods

Radionuclide angiography An intravenous injection of a radionuclide is given and the number of counts emanating from the ventricles is monitored by a precordial gamma camera. The radionuclide is commonly a compound of technetium that binds to red cells. The ejection fraction and stroke volume can be calculated from the difference between the radioactive content of the ventricles in diastole and in systole.

Echocardiography (see Section 2.5 and Figure 6.20) The end-diastolic and end-systolic diameters of the ventricle can be estimated using echocardiography. These measurements can be converted into stroke volume if some assumptions are made about chamber shape.

Electromagnetic flowmeter This technique is used only in animal experiments, because it is necessary to place a curved magnet with its poles directly on either side of the aorta or pulmonary artery. Since blood is an electrical conductor, the flowing blood induces an electrical potential as it cuts the magnetic field; the measured potential is proportional to blood velocity (cm/s). The internal diameter of the vessel must be known to convert mean velocity to flow. Being small, this device can be left inside a conscious animal, transmitting a signal by telemetry (radiowaves), and in this way much has been learned about the regulation of stroke volume of unfettered exercising animals – the subject of the next chapter.

Further reading

Hamilton, W. F. (1962) Measurement of the cardiac output. In *Handbook of Physiology, Cardiovascular System Circulation I* (eds W. F. Hamilton and P. Dow), American Physiological Society, Washington D.C., pp. 551–571

Loeppky, J. A., Greene, E. R., Hoekenga, D. E., Caprihan, A. and Luft, U. C. (1981) Beat-by-beat stroke volume assessment by pulsed Doppler in upright and supine exercise. *Journal of Applied Physiology*, **50**, 1173–1182

Schelbert, H. R., Verba, J. W., Johnson, A. D. *et al.* (1978) Non-traumatic determination of left ventricular ejection fraction by radionuclide angiography. *Circulation*, **51**, 902–909

Chapter 6
Control of stroke volume and cardiac output

6.1 Control of stroke volume

Cardiac output ranges from 4 litres/min to 7 litres/min in a human adult at rest. The output correlates with the body surface area, which is about $1.8\,m^2$ in a 70 kg adult, and the cardiac output per unit surface area (the cardiac index) averages 3 litres/min per m^2. In everyday life, however, the output is continually changing in response to circumstances. Moving from the lying position to standing reduces cardiac output by approximately 20%, while sleep reduces it by approximately 10%. A heavy protein meal or excitement and fear can increase the output by 20–30%. Pregnancy gradually raises the output by 40%. Heavy exercise causes the greatest increase, by as much as four times in untrained students and six times in Olympic athletes (see Chapter 14). Diseased hearts, on the other hand, have a much more restricted range of outputs, as Table 6.1 shows.

Cardiac output is the product of heart rate and stroke volume and is usually altered by

Table 6.1 Output of human heart (l/min ±) standard deviation

	Rest	Exercise
Normal adult	6.0 (± 1.3)	17.5 (± 6.0)
Coronary artery disease	5.7 (± 1.5)†	11.3 (± 4.3)

† The output of the diseased heart was within the normal range at rest but became inadequate during exercise on a bicycle ergometer. Subjects were exercised to 85% of maximum heart rate or to onset of angina; n = 30 (normal) and n = 20 (diseased). After Rerych, S. K., Scholz, P. M., Newman, G. E. et al. (1978) Annals of Surgery, **187**, 449–458)

changes in both factors. In exercise, the rate and stroke volume both increase but in other circumstances they can change in opposite directions. After a haemorrhage, for example, heart rate increases while stroke volume decreases (see Chapter 15). The autonomic nerves controlling heart rate were described in Chapter 5; this chapter concentrates on the control of stroke volume and its coordination with heart rate and vascular factors to determine the cardiac output.

Stroke volume is regulated primarily by two opposing factors: the energy with which the myocytes contract and the arterial pressure against which they have to expel the blood (see Figure 6.1). A highly energetic contraction produces a large stroke volume, other things being equal, while a high arterial pressure opposes ejection and reduces the stroke volume, other things again being equal.

The *energy of contraction* of the myocyte is a variable, regulated quantity, and can be

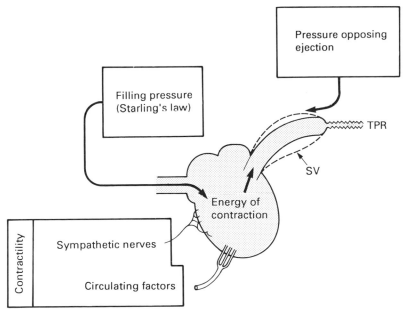

Figure 6.1 Diagram of principle factors regulating stroke volume (SV). The heart and lungs are shown as a single unit. TPR is total peripheral resistance, which influences arterial pressure

increased by two processes. (1) Stretching the cells during diastole enhances their subsequent contractile energy; and since the stretch of the relaxed ventricle depends on the pressure distending it, contractile energy is regulated indirectly by the ventricular end-diastolic pressure. This is known as Starling's law of the heart. (2) The innate strength with which a myocyte contracts from a given initial stretch, or 'contractility' as it is called, can be increased by nervous, hormonal and chemical influences, for example, by noradrenaline or extracellular calcium ions.

The *arterial pressure* opposing ejection has a negative effect on stroke volume. This is because the immediate effect of active tension is not to produce ejection but to raise intraventricular pressure during the isovolumetric phase of the cardiac cycle (see Section 2.2). Ejection cannot begin until ventricular pressure exceeds arterial pressure, and this consumes a substantial part of the energy available per contraction. If arterial pressure is raised, more of the contractile energy is consumed in raising the pressure in the isovolumetric phase and less remains for ejection. Arterial pressure depends partly on total peripheral resistance (TPR), so any rise in TPR tends to diminish the stroke volume.

Stroke volume is thus governed by three factors: (1) *stretch* during diastole, which depends on ventricular end-diastolic pressure, (2) *contractility*, which is modulated by sympathetic nerve activity and other chemical influences, and (3) *arterial pressure*, which opposes ejection and is influenced by TPR (see Figure 6.1). We must now consider each factor more closely, beginning with the relation between passive stretch and contractile energy.

6.2 Contractile properties of isolated cardiac muscle

Before dealing with the intact heart, it is helpful to see how the contraction of an isolated strip of myocardium is affected by stretch and load. Papillary muscle is particularly convenient for this purpose because its fibres run a fairly straight course and it survives well in an oxygenated tissue bath.

Isometric contraction

To study the effect of stretch, the relaxed muscle is stretched to a known length by means of a small weight or preload, and is then stimulated electrically (see Figure 6.2). If the muscle is anchored between two rigid points, excitation cannot cause shortening; it produces tension (force) alone which can be measured by a force transducer. Contraction at constant length is called an isometric contraction, and is very roughly analogous to isovolumetric contraction *in vivo*. The active tension generated during an isometric contraction is found to increase steeply with initial length, as shown by the length-tension relations in Figures 6.2 and 6.3. This demonstrates that stretching the relaxed myocardium enhances its subsequent contractile energy.

Isotonic contraction

To study the ability of muscle to shorten, one end of the muscle is left free to move but is compelled to lift a weight, so that it shortens under a constant tension. This is called an isotonic contraction and the weight is called the 'afterload'. In intact ventricles, the afterload is related to arterial pressure and ventricular radius, as explained late. If the afterload is increased, both the velocity of contraction and the amount of shortening are reduced (see Figure 6.2). If, however, the resting papillary muscle is stretched by raising the preload, and the isotonic contraction repeated, the muscle now contracts with a greater velocity and achieves a greater shortening. These observations reinforce the conclusion that the energy of contraction of isolated myocardium is a function of the resting fibre length.

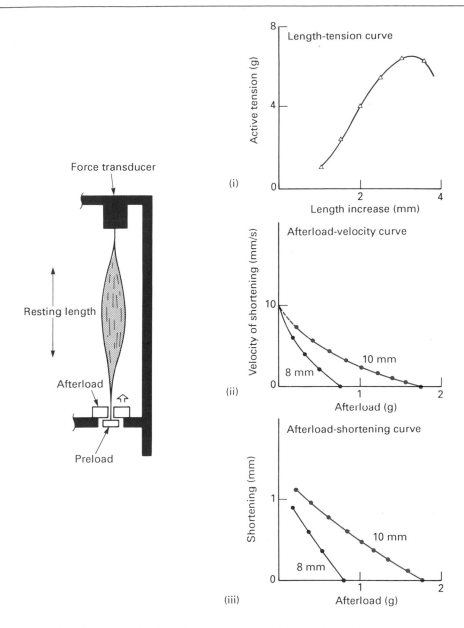

Figure 6.2 Contractile behaviour of isolated myocardial muscle. Left: Simplified diagram of arrangement used to study the contraction of isolated cat papillary muscle. The arrangement shown allows isotonic contraction to occur, the weight labelled 'afterload' being picked up as soon as shortening begins. The weight labelled 'preload' sets the resting length. If the preload is clamped in place, contraction becomes isometric. Right: Three fundamental relations: (i) shows isometric contraction at increasing lengths, (ii) and (iii) show isotonic contractions beginning from two different resting lengths (8 mm and 10 mm). Contractile force, velocity and shortening are all increased by stretching the relaxed muscle. (After Sonnenblick, E. H. (1962) *American Journal of Physiology*, **202**, 931–939)

The sarcomere length-tension relation

The process of stretching a relaxed muscle affects the length of the basic contractile unit, the sarcomere, and this affects its contractile energy (see Figure 6.3). Studies

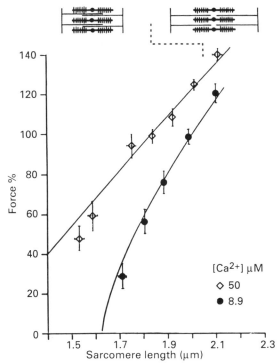

Figure 6.3 Effect of resting sarcomere length on contractile force of 'skinned' myocardium (rat trabecular muscle, *in vitro*). The sarcolemma has been permeabilized by a detergent and the muscle strip exposed to a physiological Ca^{2+} concentration (8.9 μM) or a saturating one that produces maximal contracture at each length (50 μM). Active force is expressed as a percentage of that developed at 2.1 μm length in a reference solution of 4.5 μM Ca^{2+}. The 'physiological' curve closely resembled that of intact, unskinned muscle (not shown). Intact fibres could not be stretched beyond 2.3 μm due to increasing stiffness. Sketches at top show change in actin–actin overlap with stretch: for further explanation, see text. (After Kentish J. C., ter Keurs, H. E. D. J., Ricciardi, L., Bucx, J. J. J. and Noble M. I. M. (1986) *Circulation Research*, **58**, 755–768, by permission)

of sarcomere length by laser diffraction show that maximum contractile energy develops at sarcomere lengths of 2.2–2.3 μm. The sarcomere length in intact hearts at normal end-diastolic pressures (0–9 mmHg) is below this optimal value, so the intact ventricle normally operates on the ascending limb of the length-tension curve. Beyond 2.2–2.3 μm, contractile force decays, but it is very difficult to stretch sarcomeres beyond this point, even *in vitro*, because they become very stiff: it is therefore most unlikely that sarcomere lengths above 2.3 μm are ever produced in intact hearts during life.

Mechanisms underlying the length-tension relation

The question remains, 'How does an increase in sarcomere length enhance the active tension?' Part of the explanation is that at sarcomere lengths below 2.0 μm, the opposing actin filaments overlap each other and this interferes with the formation of actin-myosin crossbridges and hence with force generation (see Figure 6.3 top). When this interference is reduced by stretching the sarcomere to 2.0 μm, contractile force increases.

Exactly the same actin-actin overlap mechanism operates in skeletal muscle, yet the length-tension curve of cardiac muscle is much steeper than that of skeletal muscle, indicating that some additional factor is at work in the myocardium. To investigate this, 'skinned' preparations have been employed, in which the sarcolemma is either removed mechanically or permeabilized chemically so that intracellular calcium ion concentration equilibrates with that in the bathing solution. The upper curve of Figure 6.3 shows the length-tension relation for skinned fibres when all the potential crossbridges at each sarcomere length are activated by a high Ca^{2+} bath (50 μM). A physiological concentration of calcium (8.9 μM) activates only a fraction of the crossbridges, so tension is reduced (lower curve), and this curve closely resembles that for intact, unskinned fibres. The curve for partially activated fibres is much steeper than that for fully activated fibres and gradually approaches the latter curve as sarcomere length is increased. This indicates that the fraction of potential crossbridges activated by physiological concentrations of Ca^{2+} increases with stretch ('length-dependent activation').

Length-dependent activation appears to be due to an increase in the sensitivity of the contractile proteins to calcium with stretch, as shown in Figure 6.4. The curve relating active tension to Ca^{2+} concentration in skinned myocytes is shifted to the left when sarcomere length is increased, and there is a

substantial reduction in the calcium concentration needed to produce 50% of maximal tension. This phenomenon is known as the 'length-dependence of calcium sensitivity'. How the sensitivity to Ca^{2+} is increased is still under investigation, but there is growing evidence that troponin C may be the length-sensor.

We can now turn from isolated muscle to the intact heart and consider the role of the length-tension relation in regulating stroke volume.

6.3 Starling's law of the heart

The length-tension relation in intact hearts was investigated by the German physiologist Otto Frank, who in 1895 studied the effect of diastolic stretch on the contraction of the frog ventricle. The aorta was ligated so that contraction was isovolumetric (i.e. roughly analogous to the isometric state) and fluid was injected into the chamber to stretch the wall. As shown in Figure 6.5, distension during diastole caused the development of a greater pressure during systole, indicating that the energy of contraction depended on diastolic distension.

Ernest Starling and his co-workers in London followed up this observation using the ejecting mammalian heart, and showed that diastolic stretch influences stroke volume. In Starling's classic experiments (see Figure 6.6), the isolated heart and lungs of a dog were perfused with warm oxygenated blood from a venous reservoir, the height of which above the heart controlled central venous pressure (CVP). CVP is the pressure in the great veins at their point of entry into the right atrium. The pressure distending the right ventricle, the right ventricular end-diastolic pressure (RVEDP), is almost equal to the CVP. (Similarly, pulmonary vein pressure governs left ventricle end-diastolic pressure, LVEDP. A general term for all these pressures is *'filling pressure'*.) The aortic pressure opposing outflow from the left ventricle was held

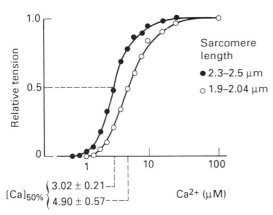

Figure 6.4 Sensitivity of contractile apparatus to calcium ion concentration. The graph shows the effect of bath Ca^{2+} concentration on the isometric force developed by chemically skinned rat ventricular muscle, at two different resting sarcomere lengths. Active tension is expressed as a fraction of the maximum tension at each sarcomere length. An increase in sarcomere length increases the sensitivity of the system to calcium, as indicated by the leftward shift of the curve. The figures below the graph show that the Ca^{2+} concentration required to produce 50% of the maximum response was reduced by stretch. (After Hibberd, M. G. and Jewell, B. R. (1982) *Journal of Physiology*, **329**, 527–540, by permission)

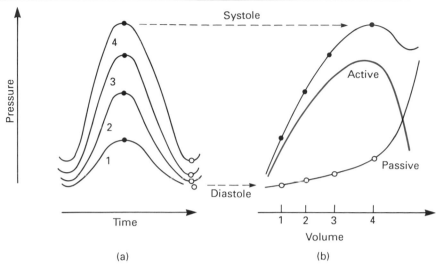

Figure 6.5 Effect of diastolic volume on energy of contraction, as measured by the systolic pressure generated in an isovolumetric frog ventricle (aorta ligated). (a) Active pressure generated between diastole (open circles) and systole (closed circles) increases as ventricular volume is raised from 1 to 4 (arbitrary units). (b) Effect of volume plotted out. Bottom curve shows passive pressure-volume relation: note the increasing stiffness as the ventricle is distended. Top curve shows systolic pressure as function of diastolic volume. Middle curve shows pressure generated actively, i.e. systolic pressure minus diastolic pressure. (After Otto Frank's seminal experiment of 1895)

constant by a variable resistance called a Starling resistor, and the combined stroke volumes of the two ventricles were recorded by a bell cardiometer. Being isolated and denervated, the heart-lung preparation was free of nervous or hormonal influences. The major findings with the heart-lung preparation are as follows.

Active response to central venous pressure

When CVP is increased, ventricular end-diastolic pressure rises, and this increases the end-diastolic volume (see Figure 6.7a). It should be noted that this effect is non-linear and tails off at high end-diastolic pressures (Figure 6.5b). The stretched ventricle develops a greater contractile energy, which results in the ejection of a greater stroke volume, provided mean arterial pressure is fixed. Although a rise in CVP initially increase the filling pressure only on the right side of the heart, the left ventricular stroke volume increases too within a few beats, because the increased right ventricular output raises the pressure in the pulmonary vessels, which in turn raises the filling pressure for the left ventricle.

Active response to arterial pressure

The direct effect of arterial pressure is to oppose ejection (see Section 6.7), but when arterial pressure is raised in a heart-lung preparation, stroke volume declines only transiently and is restored within a few beats (see Figure 6.7b). The reason is that, since output is transiently reduced by the pressure load while input continues un-

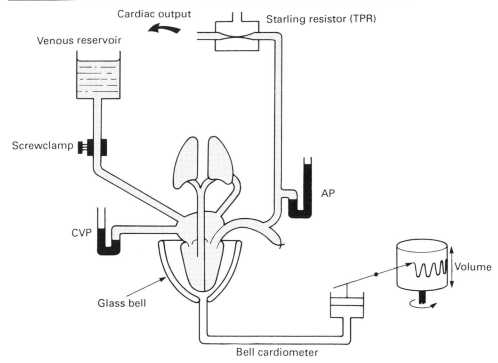

Figure 6.6 Simplified diagram of isolated dog heart-lung preparation of Starling. The height of the venous reservoir and the screwclamp- regulated central venous pressure (CVP). CVP and arterial pressure (AP) were measured by manometers and AP was held constant by a variable resistance equivalent to the total peripheral resistance (TPR). Ventricular volume was measured by Henderson's bell cardiometer (an inverted glass bell) which is attached to the atrioventricular groove by a rubber diaphragm. Beat-by-beat volume changes were recorded on a rotating smoked drum. Mean cardiac output was measured directly by collecting blood in a measuring cylinder beyond the resistor. Warming coil and air capacitance are omitted from diagram. (After Knowlton, F. P. and Starling, E. H. (1912) *Journal of Physiology,* **44**, 206–219)

changed, the ventricle distends; this increases its contractile energy and restores stroke volume.

The ventricular function curve

A graph whose ordinate is stroke volume or any other measure of contractile energy, and whose abscissa is filling pressure or any other index of resting fibre length is called a ventricular function curve, or *Starling curve* (see Figure 6.8). The curve appears in many guises. For the abscissa, CVP is often chosen because human CVP is easily measured by catheterization and is an important regulator of average fibre length; however, its relation to fibre length is indirect and non-linear. Other indices of stretch include RVEDP, LVEDP, ventricular end-diastolic volume measured by 2-plane cineangiography or radionuclide angiography, and ventricular diameter measured by echocardiography: all are indirect indices of resting fibre length. For the ordinate, stroke volume can serve as an index of contractile energy if mean arterial pressure is held

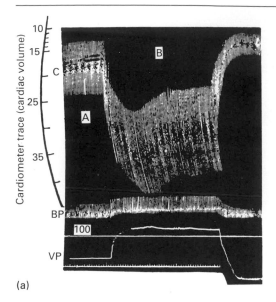

(a)

(b)

Figure 6.7 Volume of the ventricles recorded beat-by-beat by a cardiometer in the isolated heart-lung preparation; note the inverted volume scale (ml): diastolic volume is at the bottom of the volume excursion, systolic volume is at the top. Stroke volume is represented by the distance from top to bottom of the trace. (a) Shows a 64% rise in stroke volume upon raising central venous pressure (VP) from 9 cmH$_2$O (period A) to 14 cmH$_2$O (period B), while arterial pressure is kept almost constant. (b) Shows effect of raising arterial pressure (BP) by increasing the artificial peripheral resistance. A high arterial pressure depresses stroke volume for a few beats (white arrowhead), but the continuing venous inflow leads to cardiac dilatation, which increases contractile energy and restores the stroke volume (period C). Both traces (a) and (b) also show a modest reduction in ventricular distension without any fall in stroke volume following the increase in work load, indicating a small rise in contractility (the Anrep effect). (From the original smoked-drum recordings of Patterson, S. W., Piper, H., Starling, E. H. (1914) *Journal of Physiology*, **48**, 465–511, by permission)

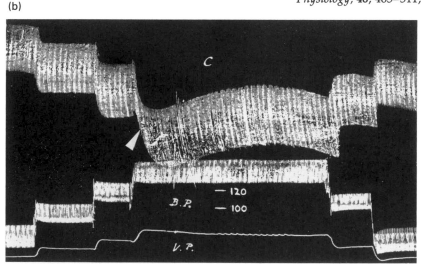

constant, as in the heart-lung preparation. However, it obviously takes more energy to eject blood at a high pressure than at a low pressure, and the product of stroke volume and mean arterial pressure (the stroke work) is a better energy index (see Section 6.4).

Stroke volume and stroke work increase as a curvilinear function of CVP between zero and 10 mmHg, forming the 'ascending limb' of the Starling curve. In the human left ventricle *in situ*, the curve almost reaches a plateau above 10 mmHg LVEDP (see Figure 6.8b). During standing and sitting, human

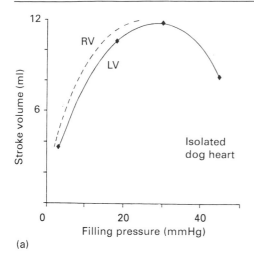

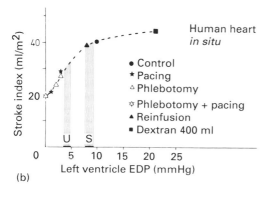

Figure 6.8 The ventricular function curve.
(a) Effect of filling pressure on the stroke volume of an isolated dog heart pumping against a constant arterial pressure. The solid line shows Starling's data for left ventricular stroke volume (LV) and left atrial pressure. The dashed line shows how the right ventrticle has a similar shaped curve but a slightly lower filling pressure. (b) Human ventricular function curve. Left ventricular end-diastolic pressure was varied *in vivo* by phlebotomy (venous blood aspiration) and atrial pacing, and the effect on stroke volume per unit body surface area observed (stroke index, SI). Normal range of human LVEDP in supine position (S) and upright position (U) are shown. Note that the human ventricle *in situ* reaches a virtual plateau above 10 mmHg. A similar curve is obtained if stroke work is plotted. (After Parker, J. D. and Case, R. B. (1979), see Further reading list, by permission)

LVEDP is 4–5 mmHg, and the heart is on the ascending limb of the curve, while in a supine subject (LVEDP 8–9 mmHg) the heart operates close to the plateau. In isolated dog hearts, the curve peaks at about 20 mmHg and then falls off (see Figure 6.8a), but such preparations are probably never completely normal and it is doubted whether human hearts ever reach this 'descending limb'. Stroke volume declines in the over-distended preparation partly because the distended atrioventricular valves begin to leak and partly because the reduced curvature of the cardiac wall impairs the conversion of active tension into pressure (Laplace's law, see later).

The shape of the Starling curve is similar for the right and left ventricles except that the left ventricle has slightly higher filling pressures (see Figure 6.8 and Table 2.1). This is because the left ventricle has thicker, less distensible walls, and LVEDP has to be 4–5 mmHg higher than RVEDP to produce an equivalent stretch and output.

The results shown in Figure 6.8 establish that *the greater the stretch of the ventricle in diastole, the greater the stroke work achieved in systole*. As Patterson, Piper and Starling concluded in 1914, 'The energy of contraction of a cardiac muscle fibre, like that of a skeletal muscle fibre, is proportional to the initial fibre length at rest'. This deduction, now honoured as Starling's 'law of the heart', has been amply confirmed by the direct studies illustrated in Figures 6.2 and 6.3.

Laplace's law and wall tension

When one attempts to relate the pumping behaviour of the intact ventricle to the contractile properties of isolated muscle, the radius of the cavity is important as well as CVP and arterial pressure. In any hollow chamber it is the radius that relates wall tension to internal pressure, as pointed out in 1806 by a French mathematician, the Marquis de Laplace – in a treatise on celestial mechanics! Laplace's law states that the pressure P within a sphere is proportional to the wall tension (which equals stress, S, times wall thickness, w), and is inversely proportional to the radius, r:

$$P = 2S \cdot w/r \qquad (6.1)$$

The involvement of radius is readily understood by considering the wall's curvature (see Figure 6.9): as radius increases, curvature is reduced, so a smaller component of

the wall tension is angled towards the cavity, generating less pressure. Thus, the curvature of the ventricle wall determines how effectively the active wall tension is converted into intraventricular pressure. Since the Frank–Starling mechanism requires some increase in chamber size, it also involves a small fall in mechanical efficiency, and this becomes a dominating effect in grossly dilated, failing hearts (see Chapter 15).

Laplace's law can be re-arranged to give $S = Pr/2w$, and this form helps us to understand what governs the afterload on myocytes in an intact heart. The afterload is the stress, S, during systole, and from the statement $S = Pr/2w$ we see that it depends not only on arterial pressure but also on chamber radius and wall thickness. Since both pressure and radius decline in the later stages of ejection, there is a gradual reduction in afterload, which facilitates late ejection. Any contraction in which afterload changes is not a truly isotonic one, and is called auxotonic.

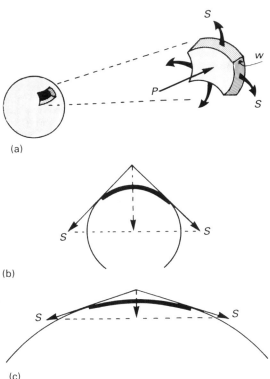

(a)

(b)

(c)

Figure 6.9 Relation between wall stress (S), pressure (P) and curvature of a hollow sphere; stress is force (tension) per unit cross-sectional area of wall (w). (a) Depicts a hollow sphere, such as a tennis ball, with an 'exploded' segment showing the two circumferential wall stresses. (b) Shows, in two dimensions, how the wall stresses (tangential arrows) give rise to an inward stress equal and opposite to pressure. Arrow length is proportional to stress magnitude. The thick line represents a muscle segment exerting tension. (c) Shows how an increase in radius reduces curvature, which in turn reduces the inward component of the wall stress

6.4 Stroke work and the pressure-volume loop

The energy expended in systole results partly in heat formation and partly in external mechanical work in the form of an increase in the pressure and volume of blood in the arterial system. Mechanical work is, by definition, the applied force (F) times the distance it moves (L); 1 joule of work equals 1 newton force displaced over 1 metre. This definition has to be adapted for a fluid, where force is applied not by a point but by a surface such as the ventricle wall. The active force exerted on the blood by the ventricle in systole equals the rise in pressure ΔP times the wall area A (since pressure is force per unit area). If the wall moves an average distance L, a volume (ΔV) equal to $L \times A$ is displaced into the aorta

(Figure 6.10). Thus the work performed per beat, or stroke work (W) is:

$$W = F \times L = (\Delta P \times A) \times L = \Delta P \times (A \times L) = \Delta P \times \Delta V$$

In other words, stroke work equals the rise in ventricular blood pressure × volume ejected (stroke volume).

Ventricular blood pressure varies throughout the ejection phase, so to evaluate stroke work properly it is necessary to construct a graph of ventricular pressure against volume, as in Figure 6.10. Line A-B in (a) shows ventricular filling. In the initial rapid-filling phase pressure is falling because elastic recoil of the relaxing ventricle exerts a suction effect. In the later slow-filling phase, a rise in pressure drives the increase in volume, so the line coincides with the passive pressure-volume curve of the relaxed ventricle (see Figure 6.10b). During isovolumetric contraction (line B-C) the heart is 'working' very hard in the

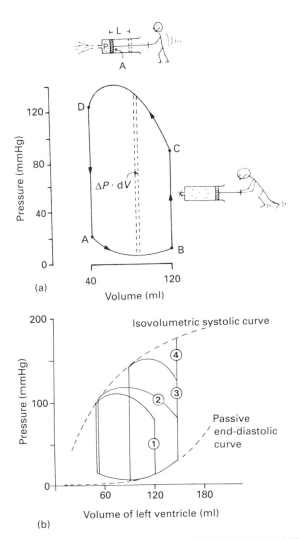

(a)

(b)

Figure 6.10 (a) Pressure-volume cycle of the human left ventricle. A, opening of mitral valve; AB, filling phase; B, closure of mitral valve at onset of systole; BC, isovolumetric contraction; C, opening of aortic valve; CD, ejection phase; D, closure of aortic valve; DA, isovolumetric relaxation. The mechanical work performed equals the sum of all the $\Delta P.dV$ strips within the loop, i.e. total loop area. The sketches indicate how the isovolumetric phase produces no external work despite a large energy expenditure by the myocardial manikin. (b) Factors influencing the pressure-volume cycle. The lower border is set by the passive pressure-volume curve of the relaxed ventricle. The upper boundary is set by the systolic pressure that would be produced in a purely isovolumetric contraction from a given end-diastolic volume; this line represents the Frank–Starling mechanism. Loop 1 represents a control state. Loop 2 shows the effect of increasing the end-diastolic volume; stroke volume increases, provided arterial pressure is held steady. If, however, arterial pressure is raised (Loop 3), stroke volume decreases, provided end-diastolic volume is held steady. Line 4 depicts a purely isovolumetric contraction

every-day sense of the word (consuming metabolic energy and oxygen to generate force), but since no blood is transported out of the system, the ventricle accomplishes no external work. This phase can be likened to a man trying to push over a house; he accomplishes no external work but consumes a lot of oxygen in the process. At point C the aortic valve opens; the height of line BC is set by the diastolic arterial pressure. In the ejection phase (C-D) external work is accomplished. At point D (end-systolic pressure) the aortic valve closes, and line D-A represents isovolumetric relaxation. Since the total stroke work is the sum of the pressure gain × displaced volume at each instant ($\int \Delta P . dV$), it equals the total area within the pressure-volume loop.

Figure 6.10b shows how the Frank–Starling mechanism affects the pressure-volume loop. The loop is confined within two lines. The lower confine is the end-diastolic pressure-volume curve as explained above; each contraction begins from this line. The upper confine is the curve relating systolic pressure to end-diastolic volume in a purely isovolumetric contraction (see Figure 6.5b). This line depicts the contractile energy available due to the Frank–Starling mechanism. If ejection were prevented, ventricular pressure would just reach this upper confine. Loop 1 represents a normal cycle, in which ejection occurs. The aortic valve closes when the end-systolic pressure and volume reach the upper confining line. In loop 2, the end-diastolic volume has been raised, increasing the contractile energy by the Frank–Starling mechanism. Stroke volume increases, providing that the arterial pressure opposing ejection is prevented from rising. If, however, the arterial pressure is raised, as in loop 3, more of the available energy goes into raising ventricular pressure and stroke volume decreases, though the stroke work (loop area) is unchanged. If ejection is prevented totally, as in line 4, stroke volume and stroke work are zero but maximum systolic pressure is generated.

6.5 Control of ventricular filling and central venous pressure

Because of the Frank–Starling mechanism, ventricular end-diastolic volume has an important effect on stroke work, and the factors that regulate end-diastolic volume are therefore very important physiologically. End-diastolic volume (EDV) depends primarily on the distensibility of the ventricle and on the transmural pressure distending the relaxed chamber: transmural pressure is the internal pressure minus the external pressure (intrathoracic pressure). As noted earlier, ventricular distensibility decreases with stretch (rather like the distensibility of a bicycle tyre), so EDV becomes less and less sensitive to diastolic pressure above approximately 10 mmHg (see Figure 6.10b).

Pressure outside the heart

Intrathoracic pressure falls from about $-5 \, cmH_2O$ at the end of expiration to $-10 \, cmH_2O$ at the end of inspiration. Inspiration thus produces a suction effect around the heart and central veins, enhancing right ventricular filling. Conversely, when intrathoracic pressure becomes positive, as for example during a forced expiration (Valsalva manoeuvre; see Chapter 14), ventricular filling is reduced. Pressure outside the ventricle also becomes positive in patients with constrictive pericarditis and pericardial effusions, impairing filling and output.

Pressure inside the heart: control of central venous pressure (CVP)

End-diastolic pressure in the right ventricle is nearly equal to central venous pressure, so the latter plays a key role in regulating stroke volume. Central venous pressure is set by the following factors:

1. Blood volume. About two-thirds of the entire blood volume is located in the

venous system, so the greater the blood volume, the greater the average venous pressure. Conversely, haemorrhage or dehydration reduces the blood volume and lowers CVP, unless compensated for by venoconstriction (see later).

2. Gravity. Gravity, venous tone and the muscle pump together govern the *distribution* of venous blood between peripheral veins and thoracic veins. In a standing man, gravity redistributes around 500 ml of blood from the intrathoracic vessels into the veins of the lower limbs (venous 'pooling'; see Section 7.7). This reduces the CVP, and stroke volume declines. Conversely, lying down redistributes venous blood from the lower limbs into the thoracic vessels, and stroke volume increases.

3. Peripheral venous tone. The veins of the skin, kidneys and splanchnic system are innervated by sympathetic nerves capable of exciting venoconstriction. The nervous system can thus exert some control over the proportion of blood in the peripheral veins and thereby influence CVP. Venoconstriction occurs during exercise, anxiety states, deep respiration, haemorrhage, shock and cardiac failure. Conversely, venodilatation occurs in skin vessels under hot conditions for reasons of temperature regulation, and incidentally lowers the CVP.

4. The muscle pump. Rhythmic exercise repeatedly compresses the deep veins of the limbs and displaces venous blood centrally (see Section 7.7). This helps to enhance CVP and stroke volume during dynamic exercise. At the opposite extreme, guardsmen standing at attention for long periods in hot weather have an embarrassing propensity to faint, partly because their muscle pump is inactivated. Combined with gravitational venous pooling and heat-induced venodilatation, this reduces the CVP and stroke volume, leading to cerebral hypoperfusion (see Posture, Chapter 14).

5. Respiration. During inspiration, intrathoracic pressure becomes more negative and intra-abdominal pressure more positive. This increases the venous pressure gradient from abdomen to thorax and promotes filling of the central veins.

6. Cardiac output. The pumping action of the heart transfers blood from the venous system into the arterial system, and this not only raises arterial pressure but also simultaneously lowers the central venous pressure (see Figure 6.11). If uncorrected, the fall in CVP and rise in arterial pressure then act as a brake on output. This important negative effect is considered further in Section 6.9.

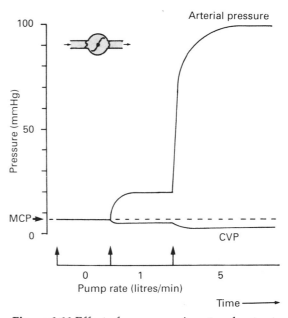

Figure 6.11 Effect of a pump on input and output pressure: volume-transfer lowers the input pressure, as well as raising output pressure. At zero pumping rate, the central venous pressure (CVP) and arterial pressure would be equal (mean circulatory pressure, MCP). When volume transfer rate is increased, CVP changes, though not as much as arterial pressure because veins are more compliant than arteries. The curves are based on the compliance of the venous system being 20 times that of the arterial system, and on a TPR of 20 mmHg per min per litre. (After Bern, R. M. and Levy, M. N. (1981) *Cardiovascular Physiology* C. V. Mosby Co., by permission)

6.6 Operation of Starling's law in man

The Frank–Starling mechanism has many important effects, the most vital being to balance the outputs of the right and left ventricle (see later). It also contributes to an increase in stroke volume during upright exercise if CVP rises (see Chapter 14). It mediates postural hypotension (a fall in cardiac output and blood pressure leading to dizziness following a fall in CVP in the upright posture, see Chapter 14), and the arterial hypotension which follows haemorrhage and other pathological events that lower CVP (see Chapter 15), and it causes a fall in stroke volume during the Valsalva manoeuvre (forced expiration, see Chapter 14).

To return to the mechanism's single most important role, it is vital that right ventricular output equals left ventricular output, except transiently for a few beats. A mere 1% imbalance between the outputs, such as a right ventricular output of 5.05 litres/min and a left output of 5.00 litres/min, would if sustained, raise the pulmonary blood volume from its normal level of 0.6 litres to 2.1 litres in half-an-hour, causing pulmonary congestion and oedema. In heavy exercise, with outputs around 25 litres/min, a sustained imbalance would be catastrophic within minutes. The Frank–Starling mechanism, however, preserves a balance between the two outputs. If right ventricular output transiently exceeds that from the left ventricle, pulmonary blood volume increases slightly and raises pressure in the pulmonary veins, which constitutes the left ventricle filling pressure. This distends the left ventricle and, by the length-tension mechanism, raises left ventricular output. The opposite happens if left ventricular output transiently exceeds right output, and the two outputs are thus kept equal in the long term. Common situations that provoke a transient imbalance include standing up, where the right output drops below the left output for a few beats (see 'Gravity' above) and inspiration, where right output tran-

siently exceeds left output owing to the operation of the respiratory pump and the fall in pulmonary vascular resistance as the lungs expand.

Caution: 'venous return' in the intact circulation

'Venous return' is the flow of blood into the right side of the heart, and it is driven by the pressure drop between the capillaries and central veins. In the intact circulation venous return must equal the cardiac output in the steady state because the circulation is a closed system of tubes: any inequality can only be transient. If the equality constraint is remembered, the term 'venous return' can be useful. In general, however, it is better to avoid the notion that venous return 'controls' cardiac output, because this is a circular (literally) and unhelpful viewpoint. In the steady state, venous return *is* the cardiac output, simply observed in veins rather than arteries. Venous return is thus directly dependent on cardiac output. Central venous pressure by contrast is essentially an independent variable and can regulate the stroke volume. Venous return 'controls' cardiac output only in the sense that transient inequalities between the two alter the CVP.

Guyton's graphical analysis of output control

The pivotal role of CVP is well illustrated by an ingenious analysis of the circulation devised by Guyton. His 'cardiac output curve' (see Figure 6.12) shows how CVP raises the output by the Frank–Starling mechanism, provided heart rate and arterial pressure are unchanged. The direct hydraulic effect of CVP is, by contrast, to oppose venous return, because a high CVP reduces the pressure gradient driving blood from the capillaries into the great veins. This effect is shown as a 'venous return curve' in Guyton's plot. If flow ceased altogether, pressures would in principle equilibrate throughout the circulation to produce a 'mean circulatory pressure' (approximately 7 mmHg), so the venous return curve is shown falling to zero flow at approximately 7 mmHg. Since the axes for the output curve and return curve are the same, the two curves can be plotted on the same graph. Cardiac output and venous return must be equal when the circulation is in a steady state and this happens at only one point – the point of intersection of the two curves, where CVP creates the same output and return. Thus Guyton's analysis illustrates how CVP acts as the independent variable governing both output and return. Guyton's graphical method can also be used to analyse altered states, as illustrated in Figure 6.12.

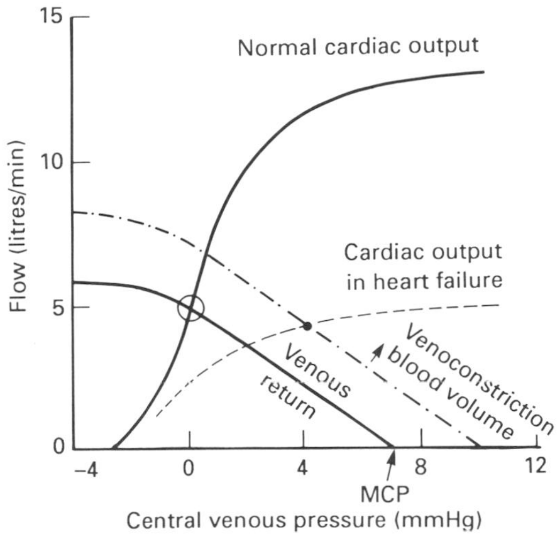

Summary of Starling's law of the heart

Few summaries of the operation of Starling's law in man could be more memorable than Professor Alan Burton's rhyme, 'What goes in, must come out'.

> The great Dr. Starling, in his Law of the Heart
> Said the output was greater if, right at the start,
> The cardiac fibers were stretched a bit more,
> So their force of contraction would be more than before.
> Thus the larger the volume in diastole
> The greater the output was likely to be.
>
> If the right heart keeps pumping more blood than the left,
> The lung circuit's congested; the systemic – bereft.
> Since no-one is healthy with pulmo-congestion,
> The Law of Doc. Starling's a splendid suggestion.
> The balance of outputs is made automatic
> And blood-volume partition becomes steady-static.
>
> When Guardsmen stand still and blood pools in their feet
> Frank–Starling mechanics no longer seem neat.
> The shift in blood volume impairs C-V-P,
> Which shortens the fibers in diastole.
> Contractions grow weaker and stroke volume drops
> Depressing blood pressure, and down the Guard flops.
>
> But when the heart reaches a much larger size,

Figure 6.12 Guyton's analysis of the circulation. The cardiac output-CVP relation follows Starling's law of the heart, at constant heart rate. The venous return curve reflects blood flow from the peripheral vasculature into the central veins, where a high pressure opposes return (see text). MCP is mean circulatory pressure at zero flow. The normal operating point is where the two thick lines cross (output equals return; open circle). The output curve can be shifted upwards by increased contractility or reduced peripheral resistance (not shown) and downwards by impaired contractility, as in heart failure (dashed curve). The venous return curve is shifted upwards if MCP is increased by venoconstriction or increased plasma volume; this happens in cardiac failure, due to sympathetic activity and renal retention of fluid (see Chapter 15). The new intersection (filled circle) represent the new steady state in cardiac failure; an almost normal output achieved at a raised CVP. (After Guyton' A. C., Jones, C. E. and Coleman, T. G. (1973) *Circulatory Physiology: Cardiac Output and its Regulation*, Saunders, Philadelphia)

This leads to heart failure, and often,
   demise.
The relevant law is not Starling's, alas,
But the classical law of Lecompte de
   Laplace.
Your patient is dying in decompensation,
So reduce his blood volume or call his
   relation.

(From *Physiology and Biophysics of the Circulation* (1972). Year Book Medical Publishers, Chicago, by courtesy of the publishers. With apologies to Alan Burton's spirit for the addition of verse 3.)

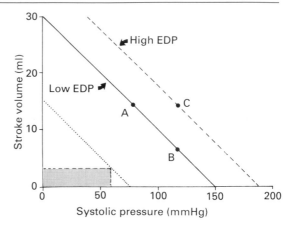

Figure 6.13 Pump function curves. Dotted line: output of a laboratory roller pump *versus* outflow pressure. The product of volume and pressure at any point (area of shaded rectangle) represents the pump's stroke work. Solid line: left ventricular stroke volume as a function of pressure opposing ejection in the isolated dog heart, with end-diastolic pressure (EDP) held constant. Raising the ejection pressure reduces the stroke volume from A to B. Dashed line: if EDP is raised, the Frank–Starling mechanism shifts the pump function line to a higher energy level and stroke volume is restored (C) as in Starling's experiment (see Figure 6.7b). The increased pressure intercept at zero output corresponds to the increase isovolumetric pressure with distension in Frank's experiment (see Figure 6.5). (After Elzinga, G. and Westerhof, N. (1979) *Circulation Research*, **32**, 178–186, and Weber, K. T., Janicki, J. S. and Hefner, L. L. (1976) *American Journal of Physiology*, **231**, 337–343)

6.7 Effects of arterial pressure on stroke volume

Arterial pressure affects the output of the heart both directly and indirectly.

Direct effect: the pump function curve

As stated in Section 6.1, a high arterial pressure directly opposes ejection. One of the basic properties of any pump, such as a laboratory roller pump, is that the outflow goes down when the pressure at the outlet is raised (see Figure 6.13, dotted line). This relation is called a 'pump function curve' and its intercepts characterize the power of the pump. The cardiac pump obeys the same rule; the direct effect of a high arterial pressure is to depress the stroke volume by increasing the proportion of the available energy that is consumed in the isovolumetric contraction phase (compare pressure-volume loops 2 and 3, in Figure 6.10). Because arterial pressure depends partly on the resistance of the peripheral circulation, stroke volume is influenced by vascular resistance. For example, the stroke volume of failing hearts can be improved by treatment with peripheral vasodilator drugs, which lower vascular resistance and therefore the arterial pressure opposing ejection.

Secondary effects of arterial pressure on stroke volume

If ventricular end-diastolic volume is not held constant there is a further change after arterial pressure is raised: the decrease in ejection allows ventricular diastolic volume to increase. This enhances contractile energy by the Frank–Starling mechanism and helps to maintain stroke volume, as illustrated in Figure 6.7b. The Frank–Starling mechanism thus displaces the pump function curve upwards (see Figure 6.13, dashed line).

In hearts *in situ* a third factor, the baroreceptor reflex, comes into play and leads to a fall in stroke volume. The baroreceptor reflex is described fully in Chapter 13, but in brief it is a nervous reflex triggered by a rise in arterial pressure. The reflex reduces the activity of the cardiac sympathetic nerves diminishing the heart rate and contractility (see Figure 6.14). In the intact heart *in situ* the effect of arterial pressure on stroke volume thus depends on the interplay of three effects (direct opposition, ventricular distension and altered sympathetic drive) and the outcome depends on their balance in the specific situation.

6.8 Regulation of contractile force by extrinsic factors

Meaning of 'contractility' and inotropic state

Contractile energy is affected by chemical influences outside the myocyte (extrinsic regulation) as well as by resting fibre length (intrinsic regulation). A change in contractile energy that is *not* due to changes in fibre length is called a change in contractility. The definition of contractility thus specifically excludes the Frank–Starling mechanism.

The term 'inotropic state' is synonymous with contractility, 'inos' meaning strength. The most important natural inotropic (strengthening) agent is noradrenaline, the

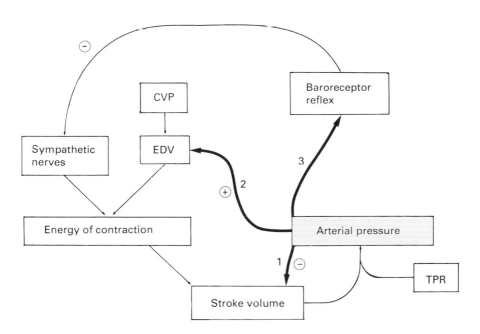

Figure 6.14 Summary of three effects of a rise in arterial pressure on stroke volume. Effect 1: a fall in stroke volume is mediated by increased afterload. Effect 2; a rise in stroke volume is mediated by increased ventricular distension secondary to a transient reduction in stroke volume. Effect 3: a fall in stroke volume is mediated by a reflex reduction in contractility. CVP, central venous pressure; EDV, end-diastolic volume; TPR, total peripheral resistance

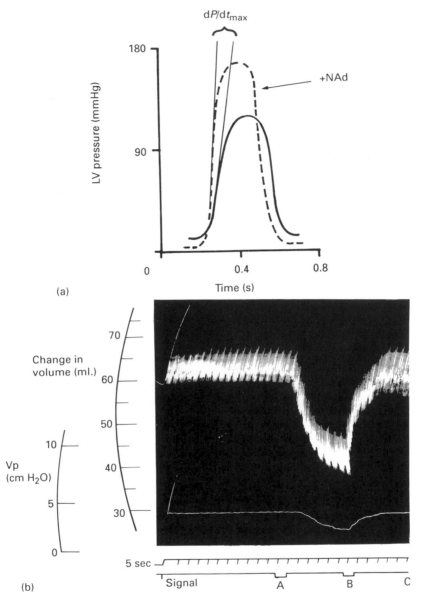

(a)

(b)

Figure 6.15 Effect of local sympathetic release of noradrenaline on cardiac performance. (a) Left ventricular pressure wave shows increased rate of climb (dP/dt_{max}), increased peak pressure, reduced EDP and reduced duration. (b) Combined stroke volume recorded by a cardiometer with heart rate held constant. The left cardiac sympathetic nerves were stimulated continuously from point A onwards. An increase in ejection fraction caused right atrial pressure (Vp) and cardiac volume to fall, and the Frank–Starling mechanism then largely prevented any rise in stroke volume. Enhanced contractility is evident from the maintenance of a normal stroke volume despite a smaller end-diastolic volume. At B the filling pressure was artificially restored to its previous level, allowing the effect of contractility on stroke volume to be fully expressed. (From Linden, R. J. (1968) *Anaesthesia*, **23**, 566–584, by permission)

neurotransmitter released from sympathetic nerve terminals within the myocardium; the other main physiological inotropic factors are circulating adrenaline and extracellular Ca^{2+} ions.

Effect of a positive inotropic agent, noradrenaline

Noradrenaline is released from sympathetic nerve terminals in the ventricle wall and binds to β_1-receptors on the myocytes. This leads, via activation of the G_s protein – cAMP sequence, (see Section 3.8), to an increase in the inward calcium current during the plateau of the action potential, which in turn builds up the intracellular calcium store. Aequorin emission studies show that more free Ca^{2+} is then liberated from the store during depolarization, increasing the proportion of crossbridges activated and generating a greater contractile force. In addition, the Ca^{2+} uptake pumps of the sarcoplasmic reticulum are accelerated resulting in a shorter systole. The effect of noradrenaline is therefore to stimulate a more forceful and shorter systole. This has the following effects on ventricular pressure and volume.

Ventricular pressure rises more rapidly in the isovolumetric phase (see Figure 6.15a) and a higher arterial pressure is produced. The maximum rate of rise of pressure, dP/dt_{max}, can be measured with a transducer-tipped cardiac catheter and may, with caution, be used as an index of myocardial contractility. The need for caution arises from the fact that dP/dt_{max} is affected also by initial fibre length (see Figure 6.5).

Ejection fraction increases, because both the velocity of contraction and the shortening are enhanced by noradrenaline. Ejection fraction is often used clinically as an indirect index of contractility. The enhanced transfer of blood out of the central veins into the arterial system lowers the filling pressure, so heart size is reduced in diastole as well as systole (see Figure 6.15b).

Stroke volume increases transiently as ejection fraction rises but is then limited by the concomitant fall in end-diastolic pressure and rise in arterial pressure. If these counter-productive pressure changes are prevented by experimental intervention, or by peripheral vascular adjustments (as in exercise), a substantial increase in stroke volume occurs (see Figure 6.15b).

The duration of systole grows briefer, which helps to preserve diastolic filling time. The shorter ejection time does not significantly curtail stroke volume because the velocity of shortening is increased.

The net effect of noradrenaline then is to increase arterial pressure and ejection fraction and to reduce ejection time, EDP and heart size. Stroke volume increases substantially only if the effects on arterial pressure and EDP are offset by peripheral circulatory adjustments (see Section 6.9).

Sympathetic innervation and the family of ventricular function curves

Postganglionic noradrenergic fibres innervating the heart arise in the superior, middle and inferior (stellate) cervical ganglia of the two sympathetic chains (see Figure 6.16). The fibres pass as fine cardiac nerves along the outer surface of the great vessels to reach the heart. The left nerves supply mainly atrial and ventricular muscle, while the right nerves supply the pacemaker and conduction system. The sympathetic firing rate increases in response to exercise, orthostasis (standing up), stress and haemorrhage, so that ejection fraction rises in these conditions (see Chapters 14 and 15).

Sympathetic stimulation augments the contractile energy produced at a given end-diastolic length or pressure. The effect of this is to shift the entire ventricular function curve upwards and make it steeper (see Figure 6.17). Sarnoff showed in the 1960s that this shift occurs in graded fashion

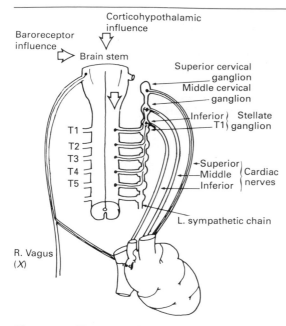

Figure 6.16 Diagram of innervation of the heart by sympathetic and parasympathetic nerves. The right sympathetic chain and nerves innervate mainly the SA node but are omitted for clarity, as is the left vagus

so that the heart has not just one Starling curve but an entire 'family' of curves, depending on the intensity of stimulation. A change in stroke work can be produced either by movement along one Starling curve (involving a change in filling pressure and contractile energy but not contractility), or by movement from one curve to another curve at constant filling pressure (involving a change in contractility alone). *In vivo*, both processes usually operate simultaneously because both the inotropic drive and filling pressure are altered by challenges like exercise and postural change.

Figure 6.18a illustrates the effect of sympathetic activity on the pressure-volume loop. The increase in contractility causes the upper confine, the isovolumetric systolic pressure curve, to steepen and shift upwards. This allows ejection pressure and ejection fraction to increase and the latter causes a decrease in end-diastolic volume

which limits the growth in stroke volume (as in Figure 6.15). Stroke work (i.e. loop area) increases. During hard dynamic exercise (see Figure 6.18b), not only contractility but also the end-diastolic volume is increased by venoconstriction and the muscle pump, resulting in a much larger rise in stroke volume and stroke work.

Note on the parasympathetic innervation of myocardium. The vagal nerves innervate the conduction system and atrial muscle, but the ventricular muscle is not well innervated, except in diving mammals. The vagal neurotransmitter, acetylcholine, binds to muscarinic receptors and exerts a negative inotropic effect (weakening) on atrial contractility. Since few parasympathetic fibres seem to innervate the human ventricle, it is thought that their effect on human stroke

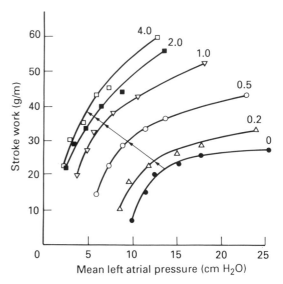

Figure 6.17 Effect of graded cardiac sympathetic nerve activity on the ventricular function curve in an isolated canine heart-lung preparation. As the sympathetic supply is stimulated at increasing frequencies (0 to 4 per s, number beside curve), the released noradrenaline produces a 'family' of curves of increased contractility. (From Sarnoff, S. J. and Mitchell, J. H. (1962) *Handbook of Physiology, Cardiovascular System, Vol. 1, Circulation*, American Physiological Society, Baltimore, pp. 489–532, by permission)

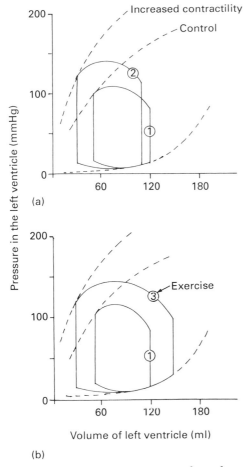

(a)

(b)

Volume of left ventricle (ml)

Figure 6.18 Schematic pressure-volume loops for human left ventricle when myocardial contractility is increased. The upper dashed confine is the relation between systolic pressure and end-diastolic volume for a purely isovolumetric contraction (Frank–Starling mechanism). (a) Loop 1 represents a basal state. Loop 2 represents a state of increased contractility. The Frank–Starling relation is shifted upwards by increased contractility, so a higher ejection pressure is reached and a smaller end-systolic volume is attained. Ejection fraction is increased, so end-diastolic volume falls unless actively regulated. Loop area (stroke work) is increased. (b) During exercise (loop 3), contractility is raised by sympathetic activity and end-diastolic volume is raised by peripheral circulatory adjustments (venoconstriction, muscle pump). The increase in stroke volume is now much greater

volume is slight. In the dog, maximal vagal stimulation reduces ventricular contractility by 15–25% and this is associated with a reduction in noradrenaline level in the coronary venous blood. It seems that, in the dog, the vagal fibres terminate close to sympathetic endings and can inhibit the release of noradrenaline from these endings.

Circulating inotropic factors

The human adrenal medulla secretes the hormones adrenaline and noradrenaline (the catecholamines) in the ratio of approximately 4:1. Adrenaline has the same effect on the heart as noradrenaline, but at physiological levels its cardiac effects are small compared with those of the cardiac sympathetic nerves. β-agonists like isoprenaline and dopamine have similar inotropic and chronotropic effects to the catecholamines.

Calcium ions and certain drugs are again positive inotropic agents but unlike the catecholamines, calcium ions do not shorten systole. Studies with intracellular aequorin show that extracellular Ca^{2+} and the drugs digoxin, caffeine and theophylline all act by raising the concentration of free sarcoplasmic Ca^{2+} ions during excitation (see Figure 3.12); this rise is due to an increased store of calcium in the sarcoplasmic reticulum. In the longer term, thyroxine too exerts a positive inotropic effect.

There are also negative inotropic agents that reduce contractility. They include acetylcholine and cholinergic agonists, β-receptor antagonists such as propranolol and practolol, calcium-channel blockers like verapamil and nifedipine, and barbiturates and many anaesthetics. Acidosis impairs contractility markedly, and in patients with coronary artery disease the combination of myocardial hypoxia and acidosis can seriously reduce the contractility: this results in a feeble cardiac response to exercise (see Table 6.1). Contractility is also severely impaired in chronic cardiac failure of non-coronary origin, as described in Chapter 15.

Inotropic effect of beat frequency (the interval-tension relation)

The American physiologist Bowditch noted in 1871 that the force of beating of a frog heart increases markedly when the interval between beats is reduced. The same phenomenon occurs in isolated papillary muscle stimulated electrically (see Figure 6.19a), and is not due to noradrenaline release. When the frequency of electrical stimulation is raised, the first beat is of low strength but subsequent beats grow more forceful producing a staircase-like recording, until a new steady state is reached. This 'Bowditch

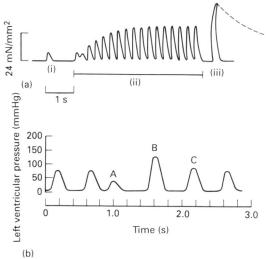

Figure 6.19 Effect of interval between beats on contractility. (a) Isometric twitches of isolated papillary muscle stimulated electrically at (i) 1/s (ii) 4/s and (iii) 1.7/s. During period (ii) the *Bowditch staircase effect* develops. Dashed line indicates how contractile force decays towards control level when low-frequency stimulation (iii) is continued (not on original figure). (b) In an isovolumetrically beating canine left ventricle a premature systole (beat A) is feeble but the beat after the compensatory pause (beat B) is stronger than usual (*post-extrasystolic potentiation*). ((a) from Edman, K. A. P. and Johannson, M. (1976) *Journal of Physiology*, **254**, 565–581; (b) from Berne, R. M. and Levy, M. N. (1981) *Cardiovascular Physiology*, Mosby Co., by permission)

staircase effect' is caused by the reduction in the time available for the expulsion of intracellular calcium by the surface pumps between beats, which leads to a gradual accumulation of intracellular calcium.

As noted above, the first beat after an interval reduction is weaker rather than stronger; conversely, the first beat after a lengthened interval is stronger rather than weaker (see Figure 6.19a). Similarly, when a premature depolarization occurs in the human heart due to an ectopic beat, the resulting systole is weaker than normal and the beat after the compensatory pause is stronger than usual (see Figure 6.19b). The latter effect is called *post-extrasystolic potentiation* and the patient often notices it: 'My heart gives a jump, Doctor'. In intact hearts, an increased filling time contributes to post-extrasystolic potentiation, but this cannot explain the same phenomenon in muscle strips. The effect arises from the 0.2 s to 0.8 s delay between the uptake of sarcoplasmic Ca^{2+} ions by the sarcoplasmic reticulum and their transport back to the quick-release sites in the cisternae (see Figure 3.11). A longer interval between beats allows a more complete transfer of calcium from the uptake sites to the release sites, and therefore a more energetic subsequent contraction.

The effects of interval on the mammalian heart *in vivo* are rather modest and contribute only a little to the enhanced contractility during exercise. It is worth emphasizing that cardiac sympathetic nerve activity is the dominant factor regulating contractility *in vivo*.

6.9 Coordinated control of cardiac output

Minimal effect of an uncoordinated stimulus

Up to this point each factor affecting stroke volume and heart rate has been presented essentially in isolation, but in the intact animal it is usual for several factors to change simultaneously in a coordinated fashion so as to produce an effective regulation of the output. It is illuminating to see how truly ineffective a single uncoordinated drive to output can be. If, for example, the heart rate alone is 'turned up' in a patient with an artificially paced heart, the cardiac output increases remarkably little because stroke volume falls. The reasons for the decrease in stroke volume are that: (1) any increase in pumping transfers blood from the input (venous) side to the output (arterial) side at a faster rate, lowering end-diastolic pressure and raising arterial pressure (see Figure 6.11), both of which impair stroke volume, and (2) an increase in artificial pacing rate shortens diastole but not systole, and this curtails the filling time. These effects are so marked that the cardiac output of a resting subject actually declines at high pacing rates. Thus a change in a single drive to output is remarkably ineffective; a coordinated cooperative change in all the controlling factors is needed to alter output significantly. This is well illustrated by the cardiac response during physical exercise.

Coordinated response to exercise

Cardiac output increases by approximately 6 litres/min for every extra litre of oxygen consumed per min in the human adult. The changes in heart rate and stroke volume have been studied by echocardiography and aortic Doppler flowmetry in man, and by implanted electromagnetic flowmeters around the aorta in dogs. These techniques reveal that cardiac output can be increased by various combinations of tachycardia and increased stroke volume, the precise combination depending on exercise intensity, posture and perhaps species. In the dog, changes in heart rate predominate during light exericse and stroke volume hardly alters, but during maximal exercise the stroke volume increases too. In man, tachycardia is again the major factor increasing

Table 6.2 Typical cardiac response to upright exercise in a non-athlete

	Rest	Hard exercise
Oxygen consumption (litres/min)	0.25	3.0
Cardiac output (litres/min)	4.8	21.6
Heart rate (beats/min)	60.0	180.0
Stroke volume (ml)	80.0	120.0
End-diastolic volume (ml)	120.0	140.0
Residual volume (end-systolic)	40.0	20.0
Ejection fraction	0.67	0.86
Cycle time (s)	1.0	0.33
Duration of systole (s)	0.35	0.2
Duration of diastole (s)	0.65	0.13

(After Braunwald, E. and Ross, J. (1979), see Further reading list and Rerych, S. K., Scholz, P. M., Newman, G. E. *et al.* (1978) *Annals of Surgery*, **187**, 449–458)

the output, the rate increasing in proportion to oxygen consumption and reaching a maximum of about 180–200 beats/min. In upright exercise, the stroke volume can increase substantially too, by 50–100% (see Table 6.2 and Figures 6.20 and 14.4). In the supine position however, where stroke volume is already high at rest, changes in stroke volume are slight.

The changes in rate and stroke volume are coordinated as follows.

1. At the onset of exercise cardiac sympathetic nerve activity increases and vagal activity decreases. This increases the heart rate and shortens both systole and diastole. Augmented atrial contractility helps to offset the effect of the reduced filling time by increasing the atrial contribution to ventricular filling.
2. Ventricular contractility is increased by cardiac sympathetic activity and to a lesser degree by the secretions of the adrenal medulla. This increases the ejection fraction and stroke volume, as can be seen in the echocardiogram of Figure 6.20.

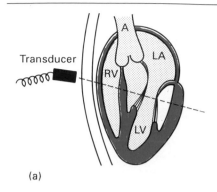

(a)

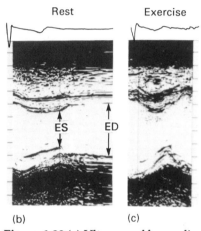

(b)   (c)

Figure 6.20 (a) Ultrasound beam directed across ventricle, in sagittal section. Echocardiograms of human left ventricle during rest (b) and upright exercise (c). Resting end-diastolic dimension (ED) increased on average by 2 mm and end-systolic dimension (ES) fell by 5 mm. The change in dimension during systole, which serves as an index of stroke volume, increased by 24% during exercise. (From Amon, K. W. and Crawford, M. H. (1979) *Journal of Clinical Ultrasound*, **7**, 373–376, by permission)

3. Sympathetic vasomotor nerves induce venoconstriction in the splanchnic circulation, and the skeletal muscle pump compresses veins in the limbs. The resulting shift of blood into the central veins prevents CVP from falling as cardiac pumping increases. CVP may even increase by a mmHg or so in man during upright exercise, raising the end-diastolic volume and enabling the Frank–Starling mechanism to contribute to the increase in stroke volume (see Figures 6.18 and 6.20).

4. Vasodilatation in the exercising skeletal muscle reduces the peripheral vascular resistance, which minimizes any rise in arterial pressure – indeed arterial pressure can fall a little at the start of light exercise. This prevents impairment of the stroke volume by a rise in arterial pressure.

The increased cardiac output of exercise thus involves a coordinated interaction between changes within the heart (rate and contractility) and outside it (CVP and systemic vascular resistance).

Transplanted and artifically paced hearts

Cardiac transplantation inevitably involves denervation but the effect of denervation is less than one might imagine. Indeed, racing greyhounds with denervated hearts are almost as fast as normal animals, with only a 5% reduction in track speed. In the denervated heart there is of course no instant tachycardia on commencing exercise but cardiac output does nevertheless increase. There is a brisk increase in stroke volume, which is caused by a rise in end-diastolic pressure as the leg muscle pump becomes active and by a fall in arterial pressure as the skeletal muscle arterioles dilate. This is followed by a rise in heart rate and contractility over 1–2 minutes as the adrenal glands secrete catecholamines into the plasma. It is only when the adrenergic back-up system is blocked by a β-antagonist that the track performance of the denervated greyhound deteriorates substantially.

Patients with an artificial pacemaker, as well as greyhounds, benefit from the reduplication of cardiac control systems. Even at fixed pacing rates, moderate increases in stroke volume can be produced by increased muscle pumping, peripheral vasodilatation and adrenaline enhanced contractility during exercise. These changes may not make for Olympic records, but more importantly they allow thousands of people to walk in the park, do the shopping and generally lead a full life.

6.10 Cardiac energetics and metabolism

Pressure work and kinetic work

Some of the energy expended in systole is 'useful' in the sense that it performs work on an external system, the arteries, while the remainder of the energy appears as heat. The external, mechanical work takes the form of an increase in the volume, pressure and velocity of blood in the arterial system. The mechanical work involved in displacing a pressurized volume of blood is the product of stroke volume and mean pressure rise (see Section 6.4), and equals the area within the ventricular pressure–volume loop. This 'pressure work' can be estimated roughly as follows. If the left ventricle raises the pressure by 100 mmHg ($1.33 \times 10^4\,\mathrm{N/m^2}$) and ejects 75 cm^3 blood ($0.75 \times 10^{-4}\,\mathrm{m^3}$), it performs very nearly 1 Nm or 1 joule of mechanical work, and the external system gains 1 joule of potential energy (pressure-energy).

In estimating cardiac work, we ought also to take account of the fact that the heart imparts a significant velocity to the blood during ejection, as well as pressure. It therefore imparts *kinetic energy* (KE, energy of motion) as well as potential energy. Kinetic energy depends on velocity v and mass m, and extra work is required to provide it. The total mechanical work of the heart, W_T is therefore:

$$W_T = \Delta P . \Delta V + KE = \Delta P . \Delta V + mv^2/2$$

Under resting conditions, where 0.08 kg of blood is ejected at a mean velocity of only 0.5 m/s, the kinetic energy works out at $0.02\,\mathrm{kg\,m^2\,s^{-2}}$ or 0.01 J. For the left ventricle, this is merely 1% of the external work at basal outputs, and can be neglected. For the right ventricle on the other hand, kinetic energy constitutes approximately 5% of the mechanical work because the blood velocity in the pulmonary artery is almost the same as in the aorta, whereas the pressure rise (and hence pressure work) is only about 20% as great. During heavy exercise, the ejection velocity increases greatly in both the pulmonary artery and aorta (to approximately 2.5 m/s), while pressure increases only slightly. Even in the left ventricle, kinetic energy then accounts for up to 14% of the total external work, and in the right ventricle the figure can reach 50%.

The *power* of the heart is its rate of working, i.e. stroke work × heart rate. Cardiac power ranges from approximately 1.2 W (J/s) at rest to around 8 W in heavy exercise, which is about a fiftieth the power of a small electric lawn mower.

Cardiac efficiency during exercise, emotional stress and cardiac dilatation

The gross mechanical efficiency of any machine equals the external work achieved per unit of energy expended. Cardiac efficiency is rather low at resting outputs, being only 5–10%. Even if allowance is made for the 'baseline' energy costs of cell maintenance and ionic pumping, which account for 25% of the oxygen consumption at basal outputs, the net efficiency is only a little better. The reason for the low efficiency is that the generation of tension during isovolumetric contraction has a high energy cost, yet accomplishes no external work. Myocardial oxygen consumption is in fact dominated by internal work rather than external work, and correlates better with active tension multiplied by the time for which it is maintained (the *tension-time index*) than with stroke work. Similarly, the oxygen consumption of a tug-of-war team depends mainly on the huge tension they exert rather than on the external work involved in shifting the opposition a metre or so.

During dynamic exercise the gross efficiency can improve to around 15% because stroke volume increases relatively more than arterial pressure, producing more external work without much increase in the energy-expensive isovolumetric phase. This is an important point in relation to patients with cardiac disease, allowing them to indulge in gentle exercise such as walking. The opposite side of the coin, however, is that a high blood pressure, whether acute or chronic, raises myocardial oxygen demand and reduces efficiency. An angry emotional scene, which causes a large rise in blood pressure, will raise the tension-time index and increase myocardial oxygen demand sharply – the very thing to be avoided by patients with coronary artery insufficiency. As the celebrated eighteenth-century anatomist John Hunter observed in relation to his own ischaemic heart, 'My life is at the mercy of any rascal who chooses to annoy me'; and the couch on which Hunter expired after a stormy committee meeting can still be seen in the library at St. George's Medical School in London.

Another clinically important aspect of mechanical efficiency involves cardiac dilatation. Dilatation can be gross during chronic heart failure (see Figure 15.4) owing to a chronically raised filling pressure. Laplace's law (Section 6.3; $P = 2Sw/r$) shows that in order to maintain a normal systolic pressure, P, a heart of enlarged radius, r, has to exert a greater contractile force, Sw. This entails a rise in oxygen consumption and a fall in efficiency – the very things a failing heart can least afford. Excessive cardiac dilatation is thus dangerous in heart failure, and it is important to reduce the dilatation by diuretic therapy.

Myocardial oxygen consumption and metabolic substrates

Myocardial metabolism is normally aerobic. The immediate energy source for the actin-myosin machinery is ATP, which is synthesized by oxidative phosphorylation in the abundant mitochondria. The small store of ATP is backed up by a bigger reserve of high-energy phosphate bonds in the form of creatine phosphate. There is also a small store of oxygen as oxymyoglobin. Even so, the supply of oxygen by coronary vessels must keep pace with demand because the heart, unlike skeletal muscle, cannot stop for a rest. At basal outputs, myocardial oxygen consumption is approximately 10 ml/min per 100 g. This represents a high proportion of the oxygen delivered by the coronary blood, and about 65–75% of the coronary blood oxygen is normally extracted. The additional oxygen needed at increased work loads is obtained chiefly by increasing the coronary blood flow (see Chapter 12).

The metabolic substrates of myocardium have been determined by chemical analysis of coronary sinus blood, collected by cardiac catheterization. It appears that myocardium is something of an opportunist, increasing its utilization of whatever substrate is currently most abundant in the bloodstream, for example glucose after a carbohydrate meal or ketone bodies in uncontrolled diabetes. Generally, however, free fatty acids supply 65–70% of the energy requirement (or more during endurance exercise) and the remaining 30–35% is supplied roughly equally by glucose and lactate. Myocardium, unlike skeletal muscle, can oxidize lactate, and this is a useful asset during hard exercise when blood lactate rises at the same time as myocardial energy demand. If myocardium becomes hypoxic, however, there is a switch to the anaerobic production of lactate, leading to local acidosis and impaired contractility.

The oxidation of lactate involves lactic dehydrogenase of a type specific to heart (LDH-H4). When heart muscle dies after a coronary thrombosis, LDH-H4 and other intracellular enzymes, such as creatine phosphokinase and aspartate aminotransferase, escape into the circulation. Their detection in plasma is a valuable diagnostic test for myocardial infarction.

6.11 Summary

Stroke volume is controlled primarily by filling pressure via the Frank–Starling mechanism and by myocardial contractility, which depends on the intracellular calcium level; and it is opposed by arterial pressure. Both the heart rate and contractility are regulated primarily by autonomic nerves, but an increase in output normally requires a coordinated change not only in cardiac nerve activity but also in peripheral resistance and the factors governing CVP.

Further reading

Allen, D. G. and Kentish, J. C. (1985) The cellular basis of the length-tension relation in cardiac muscle. *Journal of Molecular and Cellular Cardiology*, **17**, 821–840

Braunwald, E. and Ross, J. (1979) Control of cardiac performance. In *Handbook of Physiology, Cardiovascular System, Vol. 1, The Heart* (ed. R. M. Berne), American Physiological Society, Bethesda, pp. 533–579

Chapman, C. B. and Mitchell, J. H. (1965) *Starling on the Heart*, Dawsons, London

Drake-Holland, A. J. and Noble, M. I. M. (1983) *Cardiac Metabolism*, Wiley, Chichester

Jewell, B. R. (1982) Activation of contraction in cardiac muscle. *Mayo Clinic Proceedings*, **57**, 6–13

Noble, M. I. M. (1979) *The Cardiac Cycle*, Blackwell, Oxford

Parker, J. O. and Case, R. B. (1979) Normal left ventricular function, *Circulation*, **60**, 4–12

Sagawa, K., Maughan, L., Suga, H. and Sunagawa, K. (1988) *Cardiac Contraction and the Pressure-Volume Relationship*, Oxford University Press, New York

te Keurs, H. E. D. J. and Noble, M. I. M. (1988) *Starling's Law of the Heart Revisited*, Kluwer Academic, Dordrecht

Chapter 7
Haemodynamics: pressure, flow and resistance

7.1 Some hydraulic principles

The science of haemodynamics concerns the relation between blood flow, pressure and hydraulic resistance. The simplest guide to such hydraulic issues is Darcy's law of flow, which is the hydraulic equivalent of Ohm's law of electricity. By studying the flow of water through the gravel beds of the fountains of Dijon, Darcy showed in 1856, that flow in the steady state ($\dot{Q}$) is linearly proportional to the pressure difference between two points ($P_1 - P_2$):

$$\dot{Q} = K \cdot (P_1 - P_2) = \frac{(P_1 - P_2)}{R} \qquad (7.1a)$$

where K is a proportionality coefficient called hydraulic conductance. The reciprocal of hydraulic conductance is hydraulic resistance R, and the resistance arises from internal friction within the moving liquid.

Darcy's law can be applied to channels of any geometry including the branching network of blood vessels. For the systemic circulation, flow equals cardiac output (CO), the driving pressure is mean arterial pressure minus central venous pressure ($P_a -$ CVP), and the resistance is the total peripheral resistance (TPR). Therefore we can write Darcy's law in the form:

$$CO = \frac{(P_a - \text{CVP})}{\text{TPR}} \qquad (7.1b)$$

Since CVP is nearly zero (atmospheric pressure) the expression can be simplified to $CO = P_a/\text{TPR}$, or alternatively $P_a = CO \times$ TPR. The latter expression has been tested experimentally by driving blood through the aorta and peripheral circulation at various rates from an artificial pump. Providing that active changes in peripheral resistance are blocked pharmacologically, the arterial pressure is virtually a linear function of the flow, and the extrapolated line passes close to the origin. (The slight positive pressure obtained by extrapolating to zero flow is considered in Section 7.6.) The slope of the pressure-flow plot represents the systemic peripheral resistance, which is typically approximately 1 mmHg per ml/s in human adults (1 peripheral resistance unit or PRU).

Darcy's law concerns volume flow, the units of which are volume/time, and this must be clearly distinguished from fluid velocity (distance/time). The mean velocity of the fluid is its flow divided by the total cross-sectional area of the channels. Since the latter increases as blood enters the branching microvascular network, velocity decreases progressively (see Figure 1.6), whereas the total flow is unaltered and equals the cardiac output.

The mechanical energy of fluid

Pressure, with which Darcy's law is concerned, is only one of three sources of mechanical energy affecting blood flow; the other two are potential energy and kinetic energy. A more general law of flow, called *Bernoulli's theory*, states that flow between point A and point B in the steady state is proportional to the difference in the fluid's mechanical energy between A and B, mechanical energy being the sum of pressure energy, potential energy and kinetic energy. *Pressure energy* equals pressure × volume (PV), as explained in Section 6.4. *Potential energy* is the capacity of a mass to do work in a gravitational field by virtue of its vertical height above a reference level, such as the heart. The potential energy equals fluid mass (density ρ × volume V) × height (h) × gravitational force (g). *Kinetic energy* is the energy that a moving mass possesses due to its momentum. Kinetic energy increases in proportion to velocity

squared (v^2) and equals $\rho V . v^2/2$. Adding the 3 energies together we have:

Mechanical energy per unit volume =

$$P + \rho gh + \rho v^2/2 \qquad (7.2)$$

The experiment shown in Figure 7.1 demonstrates the interconvertibility of pressure energy and kinetic energy, and proves that flow occurs down a gradient of total energy rather than pressure alone.

While Darcy's law is often sufficient for our needs in vascular physiology, the more general Bernoulli theory has been introduced here to clarify some aspects of haemodynamics which might otherwise seem puzzling. For example, mean arterial

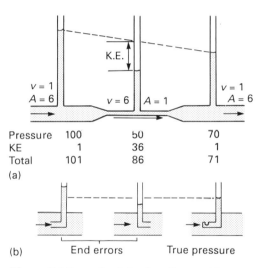

Pressure 100   50   70
KE   1   36   1
Total   101   86   71
(a)

Figure 7.1 Some basic hydraulic considerations. (a) Illustrates how flow is driven by the gradient of total mechanical energy. When the cross-sectional area A narrows, velocity (v) must increase, converting some pressure energy into kinetic energy (KE). When the tube widens, kinetic energy is converted back into pressure energy: flow from the narrow to broad segment is contrary to the pressure gradient but is down the total energy gradient. (b) Kinetic energy error in pressure measurement with an open-ended catheter pointing upstream or downstream. (After Burton, A. C. (1972) *Physiology and Biophysics of the Circulation*, Year Book Medical Publishers, Chicago, by permission)

pressure is typically 95 mmHg above atmospheric pressure in the aorta and 180 mmHg above atmospheric pressure in the foot during standing (Section 7.4), yet blood flows from aorta to foot against a pressure gradient and apparently in defiance of Darcy's law. The explanation is of course that in the upright posture the aortic blood possesses more gravitational potential energy than blood in the foot, in fact about 90 mmHg more, so that the total energy of aortic blood is 185 mmHg relative to the foot, and there is in fact a net energy gradient of 5 mmHg driving flow from the aorta into the foot.

Kinetic energy forms only approximately 1% of the fluid energy in the aorta at rest, and 5% in the pulmonary artery, rising to 14% and 50% respectively at maximal cardiac output (Section 6.10). In the great veins, the kinetic energy forms a greater proportion of the fluid energy because the blood velocity is similar to that in the aorta while blood pressure is much lower: kinetic energy accounts for 12% of the fluid energy in the vena cava at rest and for most of it at maximal flows. On reaching the relaxed ventricle, the returning blood's kinetic energy falls virtually to zero, so there is a kinetic energy gradient from vein to ventricle which aids ventricular filling. Put another way, the momentum of the returning blood contributes to ventricular expansion. Kinetic energy can also be an important consideration when measuring pressure: the measuring catheter should face neither upstream (collecting kinetic energy and therefore overestimating pressure) nor downstream (which has the reverse effect), but should be directed laterally (see Figure 7.1).

It should be noted that the above laws

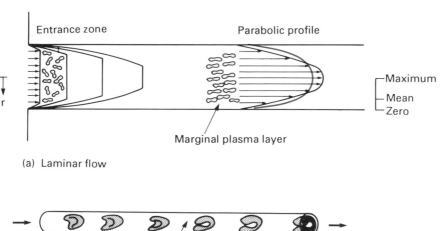

(a) Laminar flow

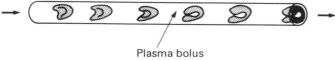

(b) Single-file flow

Figure 7.2 Blood flow patterns in a large vessel (a) and capillary (b). In the laminar flow diagram, arrow length indicates the velocity (v) of each lamina. For a Newtonian fluid in fully-developed laminar flow, velocity is a parabolic function of radial position (r): $v = v_{max}(1 - r^2/R^2)$ where R is tube radius. Mean velocity is $v_{max}/2$. The profile is blunter for a non-Newtonian fluid like blood (dotted line). The gradient of the velocity curve is called the 'shear rate'. In capillaries the red cells deform into parachute/slipper configurations (left) and folded shapes (right). (After Chien, S., Usami, S. and Skalak, R. (1984) see Further reading list)

describe flow that does not vary with time. If flow is pulsatile, as in arteries, the laws can still be applied to the mean flow, if this is not varying with time. To describe an oscillating flow instant by instant, however, is a more complex matter (Section 7.4).

7.2 Nature of flow in blood vessels

Three quite different patterns of flow occur in the circulation: laminar flow, turbulent flow and single-file flow (see Figures 7.2 and 7.3). Laminar flow occurs in normal arteries, arterioles, venules and veins, turbulent flow in the ventricles, and single-file flow in capillaries.

Laminar flow along a cylindrical tube

In laminar flow, the liquid follows smooth, regular streamlines. If the flow is along a cylindrical tube the liquid behaves like a set of thin concentric shells (the laminae), which slide past each other during flow. The lamina in direct contact with the vessel wall is fixed there by molecular cohesive forces (the 'zero-slip' condition) and has zero velocity. The adjacent lamina slides slowly past the non-slip lamina. The next (third) lamina slides past the second lamina, and since the second lamina is itself moving, the third lamina has a higher velocity relative to the tube wall, and so on until maximum velocity is reached at the centre of the tube. The high velocity of red cells in the central stream and the slow velocity of marginal red cells are plainly visible when a small blood vessel *in vivo* is viewed through a microscope.

For a simple fluid like water, the transverse velocity profile is a parabola during fully developed laminar flow (see Figure 7.2). It takes some distance from the tube entrance, however, to establish the parabola, several tube-diameters in fact. In the entrance region itself, the velocity profile is almost flat, i.e. a broad core of fluid flows at almost uniform velocity. This situation exists in the ascending aorta and the near-uniform velocity here facilitates the estimation of aortic flow by the Doppler method (Section 5.3).

With a particulate suspension like blood, the velocity profile is more blunted than a parabola (see Figure 7.2, red line). Also the shearing of lamina against lamina causes the red cells to tend to orientate parallel to the direction of flow at high shear rates. *Shear rate* is an important factor in haemodynamics: it is the change in fluid velocity per unit distance across the tube, i.e. the slope of the velocity profile in Figure 7.2a. Another effect of shear is that the cells are displaced a little towards the central axis ('*axial flow*'), leaving a thin cell-deficient layer of plasma at the margins. The *marginal layer* is only 2–4 μm thick, but it is important in arterioles, where it helps to ease the blood along (Section 7.5).

Turbulence in the circulation

If the pressure difference across a rigid tube is progressively raised, a point is reached where flow no longer rises linearly with driving pressure, as required by Darcy's law, but increases only as the square-root of pressure (see Figure 7.3). This is caused by a transition from smooth laminar flow to turbulent flow, in which swirling cross-currents dissipate part of the pressure-energy as heat. The conditions which create turbulence were explored in 1883, by the engineer Sir Osborne Reynolds, who visualized turbulence by injecting dye into water flowing down a tube. For a given smoothness of tube, turbulence is encouraged by a high fluid velocity (v), large tube diameter (D) and high fluid density (ρ, rho); these factors increase the fluid's momentum and thereby encourage any flow distortions to persist. Turbulence is discouraged by a high viscosity (η, eta) because this tends to damp out flow deviations. These factors can be

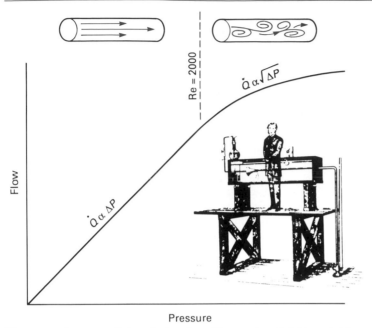

Figure 7.3 Pressure-flow relation for a Newtonian fluid in a rigid tube. Darcy's law, represented by the straight line though the origin, breaks down when turbulence develops. Inset shows Sir Osborne Reynolds' apparatus for studying the onset of turbulence; the flow-pattern (top) was visualized by injecting dye into the fluid

combined as a dimensionless ratio called the *Reynolds number* (Re):

$$Re = (v \cdot D \cdot \rho)/\eta$$

The critical Reynolds number at which turbulence sets in is around 2000 for steady flow down a rigid, straight uniform tube. The critical value is smaller in blood vessels because blood flow is pulsatile and the vessels are neither straight nor uniform. Even so a critical Re is not normally reached in most vessels; for example, Re is only approximately 0.5 in arterioles. Turbulence does occur in the ventricles where it helps to mix the blood and produce a uniform arterial gas content. Turbulence also occurs in the human aorta during peak flow*, and this sometimes creates an 'innocent' systolic ejection murmur audible over the aortic area (Section 2.5). Turbulence can also develop in leg arteries roughened by atheromatous plaques, and can cause a local 'bruit' (murmur) audible through the stethoscope. Normal laminar blood flow is of course silent.

Note: Re reaches approximately 4600 in the root of the adult human aorta. Peak velocity is approximately 70 cm/s. The diameter is 2.5 cm. Blood density is 1.06 g/cm^3. Blood viscosity is 4 milliPascal s (0.04 g cm^{-1} s^{-1}, Section 7.5).

Single-file flow in capillaries

The diameter of mammalian capillaries (5–6 µm) is less than the width of the human red cell (8 µm). Red cells are therefore compelled to proceed through capillaries in single file, and to deform into folded or parachute-like configurations even to enter the vessel, a fascinating and sometimes comic spectacle when seen through the microscope. Since the cell spans the full width of the tube, parabolic flow is impossible and the bolus of plasma trapped between the cells is compelled to move along at uniform velocity, albeit with some internal eddying ('bolus flow' or 'plug flow'). Bolus flow eliminates some of the internal friction associated with lamina sliding against lamina. Friction between the cell and capillary wall is thought to be minimized by a thin film of plasma or by the endothelial glycocalyx, a surface coating of mucopolysaccharides.

The efficiency of bolus flow depends critically on the deformability of the red cell, and this is impaired in many clinical conditions. The most dramatic of these is *sickle cell anaemia*, where the haemoglobin is abnormal, and in hypoxic situations this causes the red cell to adopt a rigid sickle shape which reduce capillary flow, impairs tissue nutrition and causes serious tissue damage (the 'sickling crisis'). Red cell flexibility is also impaired in *spherocytosis*, a condition in which many red cells are spherical rather than biconcave owing to a deficiency of spectrin, the fibrous 'skeleton' protein coating the inner surface of normal red cell membranes. Spheres cannot fold easily and tend to burst (haemolyze) when forced through narrow channels, causing a 'haemolytic anaemia'. The haemolysis is particularly severe as spherocytes pass through special narrow channels inside the spleen, so splenectomy is often helpful in such cases.

The *polymorphonuclear leucocyte* is rounder and much stiffer than the red cell, and it moves less freely along the microvessels, often creating a little 'traffic jam' of red cells behind it. If the leucocyte adheres to the wall, as happens in small venules during inflammation, the resistance to flow can increase markedly. This is an important factor impairing microvascular flow during inflammation, and also in ischaemia of the myocardium and severe haemorrhagic hypotension.

7.3 Measurement of blood flow

In anaesthetized animals, flow or velocity can be measured directly in large vessels by an *electromagnetic velocity* meter (Section 5.5), or by a heated wire in the bloodstream (*hot-wire anemometry*), the wire cooling rate being proportional to fluid velocity. A recent method is based on a brief intra-arterial injection of *radiolabelled microspheres* (typically of 15 μm diameter). The microspheres are washed into an organ in proportion to the flow it receives and lodge within its arterioles. The tissue can then be excised and its microsphere content determined by radioactive counting. The heterogeneity of flow within the tissue can be assessed by dicing the tissue into small pieces and counting each separately.

In man, less direct methods are used, as follows.

Doppler ultrasound velocity meter

This non-invasive method, described in Section 5.3, is used to assess femoral artery blood flow in patients with ischaemic limb disease, and to assess placental blood flow in late pregnancy. A variant, the laser-Doppler flowmeter, uses a laser light beam rather than ultrasound and is proving useful for estimating superficial skin blood flow.

Measurement of blood flow using Fick's principle

The Fick principle, described in Section 5.1, can be used to measure blood flow in several other organs besides the lungs. To measure renal blood flow, *para*-amino-hippuric acid (PAH) is injected intravenously and its rate of appearance in urine is measured. PAH is almost completely cleared from renal blood, so renal venous concentration is taken as zero, and the arteriovenous concentration difference is equated with the arterial concentration which is easily measured. Blood flow is then calculated as PAH excretion rate in urine (mg/min) divided by arterial concentration (mg/ml). The Fick principle has also been used to determine coronary and cerebral blood flow from the uptake of inhaled nitrous oxide gas from blood into these organs.

Venous occlusion plethysmography

This is used to measure blood flow in a limb, foot or digit. To measure forearm

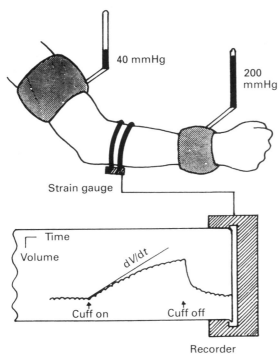

Figure 7.4 Venous occlusion plethysmography using a mercury-in-rubber strain gauge to record forearm circumference, from which volume can be calculated. Individual pulsations can be seen. The 'congesting cuff' (upper arm) occludes venous return and the wrist cuff eliminates hand blood flow from the measurement. The initial swelling rate (tangent to curve) measures forearm blood flow. Swelling rate tails off as venous back-pressure rises. After a few minutes (not shown here; see Figure 9.3) forearm blood volume stabilizes because venous pressure exceeded cuff pressure and venous outflow resumes

blood flow (see Figure 7.4), an inflatable cuff is placed over the brachial vein and inflated quickly to 40 mmHg. This arrests the venous drainage but not the arterial inflow, so the forearm begins to swell with blood. The initial swelling rate equals arterial blood flow. The swelling rate is measured by a mercury-in-rubber strain gauge, which is a mercury-filled elastic tube wrapped around the limb. As limb circumference increases, the mercury column is stretched and narrowed, and its electrical resistance increases; this is measured with a Wheatstone bridge circuit and converted to change in forearm circumference by a calibration factor. (In past years, swelling was measured by the displacement of air or water out of a rigid jacket around the limb, called a plethysmograph ('fullness record'): this is how the method's gargantuan name arose.) Another use of plethysmography, namely to study plasma ultrafiltration across the capillary wall, is illustrated in Figure 9.3.

Kety's tissue-clearance method

This method measures microvascular blood flow in a small, local region of tissue. A rapidly diffusing solute is injected into the tissue as a local depot (see Figure 7.5), from where it gradually diffuses across the capillary walls into the bloodstream and is washed away. If diffusion is fast enough the depot and incoming blood equilibrate during the blood's short residence in the capillary, and the rate of removal of solute is then directly proportional to capillary blood flow. This is called 'flow-limited exchange' and is explained more fully in Section 8.10. Fast-diffusing, lipid-soluble radioisotopes like xenon-133 and krypton-85 are suitable for this method, and their rate of removal is easily recorded by a gamma-counter over the depot. Kety pointed out in 1949 that, given the conditions outlined above, the isotope concentration declines exponentially with time, so a plot of the logarithm of concentration (C_t) against time (t) is linear (see Figure 7.5). The slope of the log. plot (k), is called the removal rate constant and is

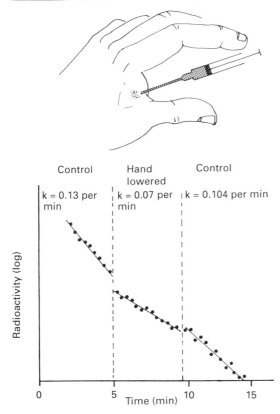

Control — k = 0.13 per min

Hand lowered — k = 0.07 per min

Control — k = 0.104 per min

Radioactivity (log)

Time (min)

Figure 7.5 Tissue clearance method, using a cutaneous injection of Xenon-133. With the hand at heart level the washout slope, k, was 13% per min (half-life 5.3 min; half-life is 0.693/k). The blood flow calculated from this slope (see text) was $9\,ml/min^{-1}\,100\,ml^{-1}$. The reduced slope on lowering the hand 40 cm below heart level revealed a fall in blood flow to $5\,ml/min^{-1}$ $100\,ml^{-1}$; this is due to local arteriolar constriction, probably induced by a local venoarteriolar reflex (see Chapter 11). (From Lassen, N. A., Henriksen, U. and Sejrsen, P. (1983) In *Handbook of Physiology, Cardiovascular System, Vol. 3, Part I, Peripheral Circulation*, (eds. J. T. Shepherd and F. M. Abboud), American Physiological Society, Bethesda, pp. 21–64)

given by the expression:

$$\log_e C_t = \log_e C_o - k \cdot t = \log_e C_o - (\dot{Q}/V\lambda) \cdot t$$

where C_o is initial concentration, V is the solute's distribution volume and λ (lambda) is the solute partition coefficient between blood and tissue. Local blood flow per unit volume of tissue ($\dot{Q}/V$) is calculated from the slope (k). One problem in practice is that the tissue composition and blood flow may not uniform, leading to complex multi-exponential clearance curves.

7.4 Haemodynamics in arteries

Waveform of the arterial pressure pulse

Arterial pressure oscillates and the size of the oscillation, or 'pulse pressure', depends on the volume and speed of the ventricular ejection, the rate of runoff through the peripheral resistance vessels, and the distensibility of the artery wall (see Section 5.4 and Figure 5.4). As ventricular ejection begins, input into the arterial system is much faster than runoff and arterial volume increases even though 20–33% of the ejected blood flows away through the peripheral resistance during this phase. The arterial pressure rises steeply, forming the 'anacrotic limb' of the pulse (see Figure 7.6). Pressure reaches its maximum when the declining ejection rate transiently equals the runoff through the resistance vessels. Pressure then declines as ventricular systole weakens and runoff exceeds ejection rate. A notch on the descending limb (dicrotic notch) marks aortic valve closure, and as the valve cusps check the backflow they create a small secondary rise in pressure, the 'dicrotic wave'. As runoff continues, pressure gradually declines to the diastolic value.

Lesions of the aortic valve cause characteristic abnormalities of the aortic waveform. If the aortic valve is narrowed by fibrosis (*aortic stenosis*), arterial pressure rises more sluggishly than normal during ejection and has an abnormal plateau (see Figure 7.6, inset). In *aortic incompetence*, pressure decays abnormally quickly in diastole owing to reflux into the ventricle. As a result, pulse pressure may be twice as big as normal, and may even cause relaxed limbs to jerk in time with the throbbing pulse.

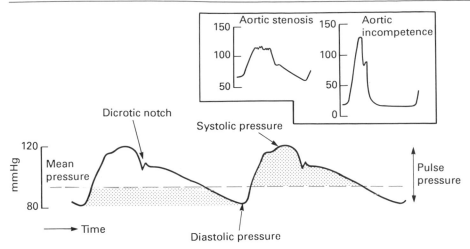

Figure 7.6 Pressure wave in the human subclavian artery over two cycles recorded by an electronic pressure transducer. The mean pressure, averaged over time, is the pressure at which the area above the mean (grey area, $\int P.dt$) equals the area below the mean. Inset shows abnormal waveform in aortic valve stenosis (slow rise, prolonged plateau) and aortic incompetence (excessive pulse pressure, low diastolic pressure (After Mills, C. J., Gale, I. T., Gault, J. H. *et al.* (1970) Cardiovascular Research, **4**, 405, omitting variable minor waves on the descending limb caused by reflections)

Mean arterial pressure ($\overline{P}_a$)

Mean arterial pressure is not a simple arithmetic average of systolic and diastolic pressures, because the blood spends relatively longer near the diastolic level than near systolic level. The true, time-weighted average is nearer the diastolic value, and can be worked out by dividing the area under the pressure wave ($\int P.dt$) by time (see Figure 7.6). An easier, working approximation, however, is to add a third of the pulse pressure to diastolic pressure:

$$\overline{P}_a = P_{diast} + (P_{sys} - P_{diast})/3 \qquad (7.4)$$

This approximation is valid for the brachial artery waveform, where human pressure is normally measured. If for example brachial pressure is 110/80 mmHg, the pulse pressure is 30 mmHg and the mean pressure is 90 mmHg.

As explained earlier, Darcy's law can be written in a form which defines the two

factors governing mean arterial pressure:

$$\overline{P}_a = CO \times TPR \qquad (7.5)$$

This important expression, the '*mean blood pressure equation*', shows that mean arterial pressure is set by the size of the cardiac output and the total peripheral resistance.

Measurement of arterial pressure

Direct methods

Arterial pressure was first measured by the vicar of Teddington, Stephen Hales, who in 1773 connected a vertical 9-foot glass tube to the carotid artery of a horse via a goose trachea, and noted the height to which the blood rose in the tube. A century later the French medical physicist J. L. M. Poiseuille developed the smaller *mercury manometer*, which is still in use today. The principle behind manometry is that the vertical column of manometer fluid exerts a downward pressure which opposes the blood pressure as in Figure 7.1. When the column reaches a stable height (h), the blood pressure must be equal to the pressure at the bottom of the column, namely ρgh (fluid density ρ × force of gravity g × h). Since mercury is very dense (ρ = 13.6 g/ml), a column about 100 mm high suffices to balance blood pressure. Using his new mercury manometer, Poiseuille proved that there is little change in mean pressure along the arterial system, and therefore little resistance to flow along arteries.

The mercury manometer is a good way of measuring mean blood pressure but it cannot follow fast changes in pressure, owing to the inertia of the mercury. To record the pressure waveform, a fast-responding *electronic pressure transducer* is needed, and this was developed by the American scientists Lambert and Wood as a spin-off from aviation research during World War II. The transducer contains a metal diaphragm which deforms slightly when subjected to arterial pressure via a linking catheter. The deformation of the diaphragm alters the resistance of a wire connected to it and the resistance is recorded. This produces a very fast response to pressure, but the transducer still has to be calibrated by a fluid column. For this reason blood pressure is normally reported in

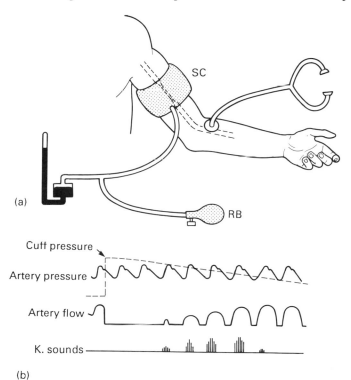

(a)

(b)

Figure 7.7 Measurement of human blood pressure. (a) Dashed line represents compressed brachial artery under the sphygmomanometer cuff (SC). Cuff pressure is controlled by the rubber bulb (RB) and measured by the mercury column. (b) Korotkoff sounds begin when cuff pressure is just below systolic pressure and diminish when cuff pressure is close to diastolic pressure

mmHg or cmH_2O rather than standard international units (Pascals; see Appendix for conversion factors).

Indirect measurement by sphygmomanometry

The mercury manometer is used in medical practice throughout the world to measure human blood pressure, by an indirect method called sphygmomanometry (see Figure 7.7). A Riva-Rocci cuff, which is an inflatable rubber sac within a cotton sleeve, is wound around the upper arm, with the inflatable sac located medially over the brachial artery at heart level. The cuff is inflated initially to around 180 mmHg, the applied pressure being measured by a mercury manometer. The applied pressure is transmitted through the tissues of the arm and occludes the artery. Auscultation of the brachial artery at the antecubital fossa (inner aspect of elbow) with a stethoscope therefore reveals no sound at this stage. Cuff pressure is then gradually lowered, and a sequence of sounds is heard.

1. When cuff pressure is just below systolic pressure, the artery opens briefly during each systole. The transient spurt of blood vibrates the artery wall downstream and creates a dull tapping noise called a Korotkoff sound. The pressure at which the Korotkoff sound first appears is conventionally accepted as systolic pressure, though it is actually about 10 mmHg less than the systolic pressure measured directly.
2. As cuff pressure is lowered further the Korotkoff sounds grow louder, because the intermittent spurts of blood grow stronger.
3. When cuff pressure is close to diastolic pressure, the artery remains patent for most of the cardiac cycle and the vibration of the vessel wall abruptly diminishes. This causes a sudden diminuendo of the Korotkoff sounds, and the cuff pressure at which this happens is accepted as the diastolic pressure

(though it is about 8 mmHg higher than diastolic pressure measure directly). A faint Korotkoff sound persists after the sudden diminuendo, and complete silence is often not attained until 8–10 mmHg below the true diastolic pressure.

What is the 'normal' blood pressure?

That mythological polymath 'every schoolboy' knows that 'normal' human blood pressure is 120/80 mmHg. For an adult male under certain conditions he would be right but it is quite wrong to adopt 120/80 mmHg as the normal standard for a resting child, a pregnant women in midterm or an elderly man. The lability of blood pressure is illustrated by the 24-h record shown in Figure 7.8, and some of the factors affecting blood pressure are as follows.

Age

Mean pressure increases progressively with age. In 3000 subjects studied in Wales, blood pressure averaged 100/65 mmHg at 6 years of age, 125/80 mmHg (women) to 130/85 mmHg (men) at 30 years of age, and 180/90 mmHg at 70 years of age. The increase in pulse pressure, which is especially striking, is caused by reduced arterial compliance, as illustrated in Figure 5.4. Reduced compliance is due to arteriosclerosis (hardening of the arteries by fibrosis and calcinosis), and is a virtually universal accompaniment to ageing. As a very rough rule, systolic pressure equals 100 mmHg plus age in years.

Sleep and exercise

Blood pressure can fall below 80/50 mmHg during sleep (see Figure 7.8). In exercise, mean blood pressure may either rise or fall, depending on the balance between increased cardiac output and reduced

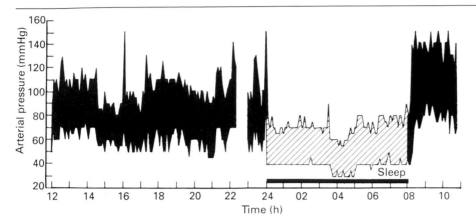

Figure 7.8 Arterial pressure in a normal subject recorded continuously for 24 h. Sleep (cross-hatched period) lowered the pressure. A painful stimulus at 16.00 h and sexual intercourse at 24.00 h markedly raised pressure. (From Bevan, A. T., Honour A. J. and Stott, F. H. (1969) *Clinical Science*, **36**, 329, by permission)

peripheral vascular resistance. In gentle dynamic exercise pressure can fall slightly, and even in heavy dynamic exercise, where cardiac output increases fourfold or more, the mean pressure increases by only 10–40 mmHg. In heavy static exercise such as weight-lifting, an 'exercise pressor reflex' can elevate pressure by approximately 60 mmHg (see Section 14.3).

Gravity: direct effect

Pressure increases in arteries below heart level owing to the weight of the column of blood between the heart and artery. In a foot 115 cm below heart level, arterial pressure will increase by 115 × 1.06/13.6 cmHg (1.06 is the relative density of blood and 13.6 the relative density of mercury): this is 90 mmHg, so arterial pressure in the foot is increased to approximately 180 mmHg above atmospheric pressure. Conversely, pressure is reduced in the arteries above heart level and is only 60 mmHg or so in the human brain during standing. (Our problems are slight, though, compared with the giraffe's. To ensure cerebral perfusion, the giraffe has to generate an aortic pressure of approximately 200 mmHg.)

Gravity: indirect effect

Upon moving from lying to standing, arterial pressure at heart level changes too due to changes in cardiac output and peripheral resistance. A transient fall in pressure (which can produce a passing dizziness) is followed by a small but sustained reflex rise (see Section 14.1).

Emotion and stress

Anger, apprehension, fear, stress and sexual excitement are all potent 'pressor' stimuli, i.e. they elevate blood pressure (see Figure 7.8). Since a visit to the doctor is stressful for many patients, a solitary high pressure measurement is not in itself proof of the disease 'hypertension'; the measurement needs to be repeated with the patient relaxed. The pressor effect of stress is particularly harmful to patients with ischaemic heart disease, as the cautionary history of John Hunter illustrates (see Section 6.10).

Other factors

Arterial pressure fluctuates with *respiration*, rising by 15–20 mmHg during each inspiration. The *Valsalva manoeuvre*, a forced expiration against a closed or narrowed glottis, causes a complex sequence of pressure changes which are described in Section 14.2. In *pregnancy* blood pressure gradually falls and reaches a minimum at approximately 6 months, so in obstetrical practice a pressure of 130/90 mmHg would cause grave concern, even though it is merely close to the upper limit of normal for a non-pregnant woman. Many *pathological processes* also alter arterial pressure, such as dehydration, haemorrhage, shock, syncope (fainting), chronic hypertension, acute heart failure and valvular lesions like aortic incompetence. Some of these conditions are described in Chapter 15.

It will be clear from this brief survey that the yardstick used to assess a subject's blood pressure must be matched to age, sex, physiological and psychological condition.

Pulsatile flow in arteries

Flow along the aorta and major arteries is pulsatile, and virtually ceases during diastole (see Figure 7.9). This flow pattern arises from moment-to-moment changes in the pressure gradient along the arterial system. Pressure rises first in the proximal aorta, where the stroke volume is initially accommodated, so there is at first a pressure gradient from the proximal aorta to the peripheral arteries, which causes the blood already in the system to accelerate. The pressure pulse then spreads along the arterial tree (see later), taking an appreciable time to do so; in the human radial artery, for example, the pressure rises about 0.1 s later than in the aorta. Consequently, there comes a period when distal pressure is transiently higher than proximal pressure (see Figure 7.9) and the pressure gradient is reversed. The reversed pressure gradient does not instantly reverse the flow because the blood has acquired forward momentum by now, but it does steadily decelerate the flow. Thus, flow in the major arteries first accelerates and then decelerates over the initial third of the cardiac cycle. During the

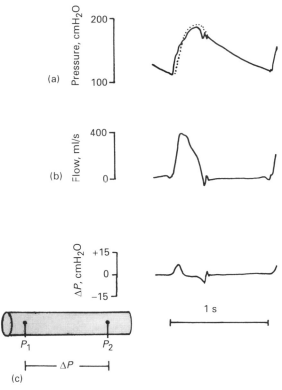

Figure 7.9 Pressure, flow and pressure gradient (ΔP over a distance of 5 cm) measured simultaneously in the ascending aorta of a normal human. The solid pressure line represents P_1, and P_2 has been reconstructed as the dotted line. Flow is virtually confined to the ejection phase and terminates with a brief backflow. Acceleration (not shown) has the same waveform as the pressure gradient trace. (After Snell, R. E., Clements, J. M., Patel, D. J., Fry, D. L. and Luchsinger, P. C. (1965) *Journal of Applied Physiology*, **20**, 691)

next two-thirds of the cycle flow is virtually zero. The instantaneous flow is not governed by Darcy's law (which, as pointed out earlier, is a steady-state expression) but is governed by Newton's second law of motion (acceleration = force/mass).

The period of near zero flow gradually shortens as blood enters smaller arteries, and in the smallest arteries flow becomes continuous, albeit still pulsatile.

Transmission of the pressure wave

If arteries had rigid walls, pressure would rise virtually instantaneously throughout the arterial system, but they do not, and the pressure wave takes a finite time to pass along the arterial tree. The pressure pulse travels at around 4 m/s in young people and 10 m/s in the elderly. This is an order of magnitude faster than blood travels, mean blood velocity being approximately 0.2 m/s in the ascending aorta. One is reminded at this point of the White Queen's suggestion that Alice should practise believing at least six 'impossible' things before breakfast. The difference between the pressure transmission velocity and blood velocity is, however, merely difficult to understand, not impossible (see Figure 7.10). Since blood is essentially incompressible, the blood entering the proximal aorta during ejection has to create space for itself. This it does partly by distending the proximal aorta (which raises the pressure) and partly by pushing ahead the blood previously occupying the required space. As the displaced blood moves forward, it too must make space for itself, partly by distending the wall downstream (which raises the pressure there) and partly by displacing the blood ahead. This 'shunting' sequence repeats itself, very rapidly, along the arterial tree. The pulse is thus transmitted by a wave of wall distension (at 4–10 m/s) while the ejected blood itself advances only 20 cm (the 'stroke distance') in 1 s.

Since the wall deforms as it propagates the pulse, the transmission velocity is affected by wall stiffness: it increases as wall stiffness increases. Arterial stiffness is greater at high blood pressures and in elderly subjects, so pulse transmission is faster in elderly subjects. The measurement of transmission velocity, by timing the central and peripheral pulse, forms a convenient way of assessing arterial distensibility in human subjects.

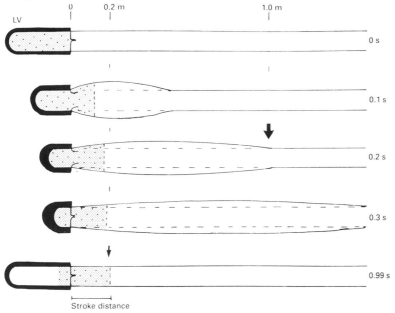

Figure 7.10 Sketch to illustrate transmission of the pressure pulse at 5 m/s along the arterial system. The left ventricle (LV) is shown ejecting 100 cm³ blood into an aorta of cross-section 5 cm²; the ejected blood travels only 20 cm per beat (small arrow, 1 s). This is the 'stroke distance'. The distension of the artery wall (the pulse) travels 1 metre in only 0.2 s (large arrow)

Change in the pressure waveform along the arterial tree

The shape of the pressure wave changes strikingly as it travels out to the periphery (see Figure 7.11). The pulse pressure, far from damping out as one might imagine, actually grows taller and steeper for some distance. This 'peaking' of the wave by around 50% is attributed partly to the tapering shape and increasing stiffness of the distal arterial tree and partly to the variation in transmission velocity with wall stiffness. The pulse pressure continues to increase as far as the third generation of arteries (e.g. femoral artery) but beyond this it becomes progressively damped out by the viscous properties of the blood and artery wall (see Figure 1.6). As the oscillations in pressure and flow dwindle the blood reaches the resistance vessels (arterioles).

7.5 Vascular resistance

Resistance to flow in tubes: Poiseuille's law

The resistance to laminar flow arises exclusively from the internal friction between adjacent laminae of fluid and has nothing to do with friction between tube and fluid – which is zero because of the zero-slip condition (see Figure 7.2). Nevertheless, resistance is greatly affected by tube geometry because the radius of the tube affects the rate of shear (sliding) of the laminae. If the same flow is forced through a narrow tube and a wide tube, the velocities are greater in the narrower tube (since mean velocity equals flow/cross-sectional area), so the shear rates are higher in the narrower tube; and high shear rates produce more internal friction.

The properties which determine resistance were elucidated in 1840 by the Parisian physician, Jean Leonard Marie Poiseuille, in an extraordinarily meticulous study of water flowing through glass capillary tubes. Poiseuille established that the resistance (R) to the steady laminar flow of a Newtonian fluid such as water (or plasma) along a straight cylindrical tube is proportional to tube length (L) and fluid viscosity (η); and is inversely proportional to tube radius raised to the fourth power (r^4):

$$R = 8\eta . L/\pi r^4 \qquad (7.6)$$

Combining this definition of resistance with Darcy's law (Equation 7.1a) we get an

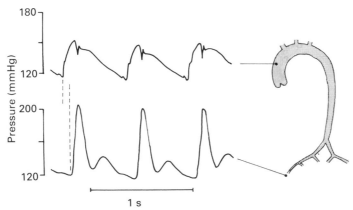

Figure 7.11 Simultaneous records of the pressure-wave in the canine ascending aorta and femoral artery. The dashed lines show the time required to propagate the pulse, which is 86 ms foot-to-foot. Note the change in waveform. (After Noble M. I. M. (1979) *The Cardiac Cycle*, Blackwell, Oxford, by permission)

expression for flow through a tube, called Poiseuille's law:

$$\dot{Q} = (P_1 - P_1) \cdot \frac{\pi \cdot r^4}{8\eta L} \qquad (7.7)$$

Poiseuille's law describes flow along a single tube. If several tubes are arranged in *series*, the overall resistance is the sum of the individual resistances. However, if the tubes are connected in *parallel* (e.g. capillaries), the same driving pressure will obviously produce more flow, because it is now the conducting capacities of the tubes that summate. If there are N tubes of equal conductance K, the net conductance is NK; and since resistance is the reciprocal of conductance, the net resistance is $1/NK$.

Armed with Poiseuille's law, we can now consider the total systemic resistance which, as explained in Section 1.6, is sited chiefly in the arterioles.

Tube geometry: importance of arteriolar radius

Owing to the fourth power term in Poiseuille's law, resistance is equisitively sensitive to vessel radius. The fall in radius from approximately 1 cm in the human aorta to 0.01 cm in an arteriole will in itself increase resistance 100 000 000 times, and this is essentially why arterioles are the main site of resistance. Of course, the radius of a capillary is even smaller (approximately 0.0003 cm), so why does the capillary network not offer an even greater resistance than the arteriolar network? The pressure drop per unit length of vessel (gradient dP/dx, a measure of the vessel's intrinsic resistance) is in fact five times greater in capillaries than arterioles but the pressure drop across the whole capillary bed, approximately 30 mmHg, is smaller than that across the arteriolar bed (40–50 mmHg) because of (1) the huge number of capillaries in parallel arrangement, (2) the shortness of capillaries (approximately 0.5 mm), and (c) the occurrence of bolus flow rather than laminar flow in capillaries.

The radius of an arteriole is actively controlled by the smooth muscle in its wall; contraction narrows the lumen (vasoconstriction) and relaxation produces vasodilatation. Owing to the fourth-power effect on resistance, active changes in radius constitute an extremely powerful mechanism for regulating both the *local blood flow* to a tissue and the *central arterial pressure* (Equation 7.5). A mere 16% reduction in arteriolar radius will in theory halve the blood flow to an organ (although this is somewhat mitigated by viscosity changes, described later). It is worthwhile, therefore, considering the relation between active wall tension and vessel radius a little more closely.

Wall mechanics during vasoconstriction and vasodilatation

The radius of a vessel depends partly on the active tension exerted by vascular smooth muscle, partly on the passive elastic properties of the wall and partly on blood pressure. The net tension in the wall (T) is the sum of the tension in active smooth muscle cells (T_a) and tension in passive structures like the collagen and elastin networks (T_p; see Figure 7.12). The net wall tension resists the distending force exerted by the blood pressure, which is the internal pressure (P_i) times the area it acts on in one direction, namely $2R_i \times 1$ for a tube of internal radius (r_i) and unit length: the internal distending force is therefore $2P_i r_i$. We must also take into account the pressure outside the wall (P_o) which tends to compress the vessels with a force equal to $2P_o r_o$, where r_0 is the outer radius of the vessel. At mechanical equilibrium, the wall tension on each side of the vessel ($2T$) must counteract the net distending force, and since the factor 2 cancels out, this gives:

$$T = P_i r_i - P_o r_o \qquad (7.8)$$

This the fundamental equation for mechanical equilibrium in any cylinder.

If the wall is thin relative to the radius,

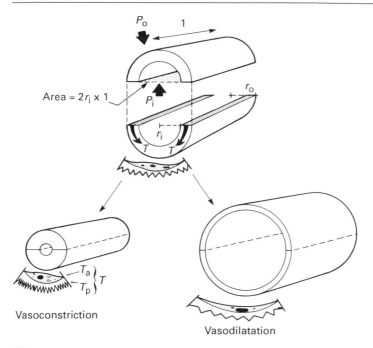

Figure 7.12 Wall mechanics in a blood vessel. The internal pressure (P_i) acting over area $2r_i$ is tending to push the two halves of the cylinder apart. This is opposed by the external pressure (P_o) and by the tension in the wall on each side of the cylinder (T). The tension is the sum of the force exerted by active smooth muscle (T_a) and the force exerted by passive elastic elements (T_p) represented here as compressed springs. T_a and T_p change in opposite directions during vasoconstriction and vasodilatation to produce mechanical stability. (After Azuma, T. and Oka, S. (1971) *American Journal of Physiology*, **221**, 1301–1308)

the internal radius and external radius are nearly equal and the expression simplifies to $T = \Delta P . r$, a relation known as Laplace's law for a tube. This simplified version tells us that the wall of a thin pressurized vessel is under tension, something we all know from our everyday experience of balloons, footballs etc. The mechanical situation in arterioles, however, is quite unlike this and is rather surprising. Because the ratio of wall thickness to internal radius is large in arterioles (up to 1.0; see Figure 1.8), the internal pressure acts upon a relatively small area while the external pressure (atmospheric) acts on a relatively large area.

As a result, the compressive force ($P_o r_o$) is larger than the distending force ($P_i r_i$) and the wall is, overall, in a state of compaction or 'negative tension' rather than positive tension. When the smooth muscle cells contract, producing a positive tension (T_a) within themselves, this reduces the vessel's internal radius but increases the state of compression of the passive elements in the wall, and the latter opposes further reduction of radius. In this way mechanical stability is achieved. The key point here is that without the opposing effect of the passive elastic elements, vasoconstriction would be a highly unstable process.

Viscosity of blood

The word viscosity stems from 'viscum', the Latin for mistletoe, because mistletoe berries contain a thick glutinous fluid. Viscosity was defined by Isaac Newton as 'defectus lubricitatis' or lack of slipperiness, meaning that it is a measure of the internal friction within a moving fluid, analogous to friction between two moving solid surfaces.

Viscosity is defined formally as the ratio of shear stress to shear rate. *Shear stress* is the shearing or sliding force applied per unit area of contact between two laminae (N/m^2), and it depends on the axial pressure gradient. *Shear rate* is the change in velocity per unit distance radially (see Figure 7.2), so its units are (m/s)/m, i.e. s^{-1}. The viscosity of water (shear stress/shear rate) is 0.001 Newton-s/m^2 at 20°C, which is usually expressed as 1 milliPascal-second or 1 centiPoise. The viscosity of a fluid relative to water is easily and accurately measured with a simple capillary viscometer. The time taken for the test fluid to flow between two marks in a glass capillary tube is simply divided by the time taken by water under the same pressure head. With a simple fluid like water or plasma, the viscosity is independent of the tube radius or shear rate, and such a fluid is called Newtonian. Whole blood by contrast has anomalous, non-Newtonian properties (see later).

Viscosity of plasma

The viscosity of water decreases as temperature rises, falling to 0.69 mPa s at body temperature (37°C). In plasma, the presence of voluminous protein molecules (albumin and globulins) raises the viscosity by 70% so the viscosity of plasma is 1.2 mPa s at 37°C. In myeloma, a cancer of globulin-secreting cells, the globulin concentration and viscosity rise to pathological levels. Moreover the globulins are themselves abnormal and some can cause red cells to agglutinate under cool conditions. In cold fingers, the viscosity can increase so dramatically that perfusion is badly impaired and necrosis of the fingertips ensues.

Importance of haematocrit

The addition of red cells to plasma greatly increases the amount of internal friction during flow, so blood viscosity is dominated by haematocrit (see Figure 7.13). Haematocrit is the red cell volume as a fraction or percentage of the blood volume. At a haematocrit of 47% the relative viscosity of human blood is approximately 4, as measured in a wide-bore viscometer at high shear rates (the significance of these conditions will be explained shortly). At this haematocrit, there are frequent collisions between the red cells during flow, and were it not for the great flexibility of the red cell the viscosity would be even higher. If the cells are hardened by glutaraldehyde, the relative viscosity rises to 100. Achieving the optimal haematocrit for oxygen delivery is a

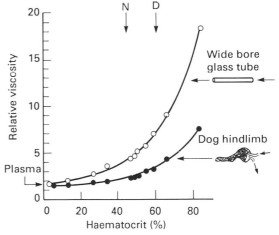

Figure 7.13 Effect of haematocrit on the viscosity of blood relative to water. Open circles; viscosity measured in a high-velocity glass viscometer. Closed circles; smaller effective viscosity in the vasculature of an isolated dog hindlimb. The lower viscosity in the hindlimb vasculature is due to the Fahreus–Lindquist effect- see text. Normal human or canine haematocrit (N). Haematocrit at which cells are packed so tightly that they begin to deform even at rest (D). (From the classic experiment of Whittaker, S. R. F. and Winton, F. R. (1933) *Journal of Physiology*, **78**, 339–369)

delicate balancing act: on the one hand a high haematocrit increases the oxygen-carrying capacity of the blood but on the other it also raises viscosity, which either reduces the flow or increases arterial pressure and cardiac work. Each species has an optimal haematocrit in this respect. In the camel, for example, the red cells are less flexible than human cells, and the camel's haematocrit is only 27%.

Abnormalities of haematocrit can have serious haemodynamic effects. *Polycythaemia*, a raised haematocrit, can develop either as a physiological adaptation to chronic hypoxia (e.g. high altitude) or as a pathological condition called polycythaemia rubra vera, in which an overproduction of red cells by the bone marrow raises the haematocrit as high as 70%. At haematocrits above 63% the red cells are so closely packed that they are deformed even at rest, doubling the viscosity and raising the resistance to flow. Polycythaemia rubra vera thus causes hypertension and a sluggish blood flow, which in turn predisposes to cerebral thrombosis and coronary thrombosis (strokes and heart attacks). Conversely, *anaemia* reduces viscosity and resistance, and in order to maintain blood pressure, cardiac output has to increase. If this situation is prolonged, a form of cardiac failure called high-output failure can develop. These examples illustrate the importance of the homeostasis of viscosity for normal cardiovascular functioning.

Non-Newtonian viscosity of blood. I-Effect of tube radius

In 1933 Whittaker and Winton made the remarkable observation that the effective viscosity of blood in the peripheral circulation is only half that of the same blood measured in a wide-bore viscometer (see Figure 7.13). Just before this, Fahreus and Lindqvist (1931) made a related finding – the apparent viscosity of blood in a viscometer depends on the radius of the glass tube through which the blood is driven (see Figure 7.14a), which is not the case for

plasma or water. In tubes of diameter greater than 1 mm (small arteries) the viscosity of blood is independent of bore, but below 1 mm diameter the apparent

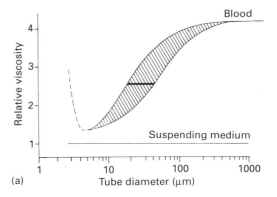

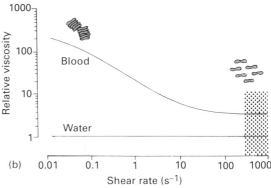

Figure 7.14 The anomolous viscous properties of blood flowing through glass tubing. (a) Fahreus–Lindquist effect: viscosity decreases with tube diameter. Hatched area indicates spread of data. The effective viscosity of blood in the intact circulation is approximately 2.5 (black bar), implying that the functional diameter of the resistance vessels is approximately 30 μm (arterioles). At diameters smaller than a blood capillary, viscosity rises again (dashed line). (b) Effect of shear rate on viscosity at a fixed haematocrit of 45%. Dotted region shows typical shear rates *in vivo*. Sketches show red cell aggregation into rouleaux at low shear rates and disaggregation at high rates. ((a) adapted from Gaetghens, P. (1981) In *The Rheology of Blood, Blood Vessels and Associated Tissues* (eds. D. R. Gross and N. H. C. Wang), Sijthoff & Noordhoff, Amsterdam; and (b) from Chien, S., Usami, S. and Skalak, R. (1984) see Further reading list)

viscosity decreases. In tubes of diameter 30–40 µm, which is the size of many arterioles, the relative viscosity is only approximately 2.5, and in capillary-size tubes (approximately 6 µm), blood viscosity reaches a minimum value almost as low as that of plasma. This constitutes a remarkable and important device for minimizing the pressure needed to perfuse the microcirculation, and explains the low viscosity of blood *in vivo* shown in Figure 7.13. The *Fahraeus–Lindquist effect*, as it is known, arises from several factors. In capillaries, the single-file pattern of flow lowers the viscosity. In arterioles, the viscosity is reduced by the peripheral plasma stream produced by axial flow (see Figure 7.2). Since shear rates are highest peripherally, a reduction in friction at that location has a particularly marked effect on the overall viscosity. This effect declines in wider tubes because the thickness of the marginal layer becomes insignificant relative to tube radius.

Fahreus described another curious consequence of axial flow; the concentration of red cells in blood flowing along a narrow tube (the dynamic or tube haematocrit) is lower than the central haematocrit in the feeding and draining vessel. In a tube of radius 15 µm, for example, fed from a central reservoir of haematocrit 40%, the dynamic haematocrit is only approximately 24%. This is a little hard to digest at first acquaintance. The explanation lies in the difference between axial and marginal stream velocities (see Figure 7.2). Let us consider, as a simple example, arterial blood of haematocrit 50% feeding an arteriole in which the cells have twice the velocity of plasma owing to their more axial location. If the haematocrit in the parent artery and vein is to remain at 50% (which it must, since the circulation is in a steady state) equal volumes of plasma and red cells must pass through the arteriole in a given time. Since the cell velocity is twice the plasma velocity in our example, equal volume flows are only possible if the concentration of red cells in the arteriole is half that in the parent blood. This is achieved by the red cells

speeding away from the plasma at the tube entrance, thinning out rather like traffic entering a fast road from a congested slip road.

Non-Newtonian viscosity. II-Effect of shear rate

The viscosity of blood varies not only with tube diameter but with velocity too, or more accurately with shear rate (see Figure 7.14b). Shear rate, it will be recalled, is the change in fluid velocity per unit distance normal to the direction of flow. At normal, physiological flows, the average shear rates are of the order 1000/s, and the viscosity is at a minimum. But at low flows the viscosity increases steeply, because low shear rates allow the red cells to adhere to each other and form aggregates; these resemble stacks of coins and are called rouleaux. This probably occurs in veins *in vivo* if blood flow is sluggish.

7.6 Pressure-flow curves for entire vascular beds

The conditions under which Poiseuille's law is strictly valid include a steady flow (cf. pulsatile flow *in vivo*) of a Newtonian fluid (cf. blood) along a long straight vessel (cf. branched, curved and tapering vasculature) with rigid walls (cf. distensible blood vessels). So one has to be cautious in applying Poiseuille's law to the circulation. In the lungs and in perfused hindlimbs with little vascular muscle tone, the pressure-flow relation is relatively linear at physiological pressure: but there is a positive pressure intercept at zero flow, and a curved relation at low pressures (see curve labelled 'blood' in Figure 7.15). The latter indicates that resistance decreases as pressure rises. This is caused by (1) the elasticity of the arterioles, which permits radius to increase and therefore resistance to decrease as pressure rises, and (2) alterations in blood viscosity at low shear rates due to rouleaux formation. The major role played by the anomalous viscous behaviour of blood becomes clear upon perfusing a dog's hindlimb with saline: a more linear pressure-flow relation is obtained, extrapolating virtually through the origin (see Figure 7.15).

In circulations with good arteriolar tone, a curious and physiologically-important pressure-flow relation exists, called an *autoregulation curve*. Flow increases with pressure up to a certain point, but the slope of the relation then flattens and thereafter flow changes relatively little with pressure, until pressure exceeds about 180 mmHg. A nearly constant flow in the face of a rising pressure indicates that resistance is rising in proportion to the pressure. This phenomenon, called autoregulation, is caused by active vasoconstriction and is described further in Section 11.3). Autoregulation occurs in most organs except for the lung.

7.7 Haemodynamics in veins

Venous distension curve

Peripheral venules and veins are thin-walled, voluminous vessels which contain roughly two-thirds of the circulating blood (see Figure 1.9). They act as a variable reservoir of blood for the thoracic compartment and thereby influence the cardiac

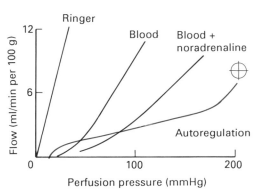

Figure 7.15 Pressure-flow curves for dog skeletal muscle. When the arterioles are in good physiological condition, autoregulation is present. When this is abolished, steeper curvilinear relations are seen: the curvature is caused by changes in resistance with pressure. Noradrenaline causes vasoconstriction and increases resistance. Perfusion with mammalian Ringer's solution (a physiological salt solution) produces a steeper line due to its low viscosity, and the line is almost straight because the anomolous viscous effects of blood are removed. (After Pappenheimer, J. R. and Maes, J. P. (1942) *American Journal of Physiology*, **137**, 187–199; and Stainsby, W. N. and Renkin, E. M. (1961) *American Journal of Physiology*, **201**, 117–122, by permission)

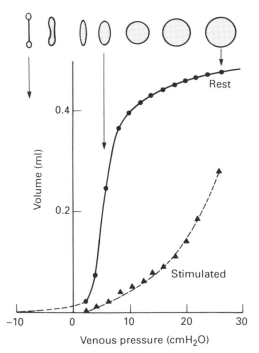

Figure 7.16 Distensibility of a vein in relaxed state (filled circles) and at maximum contraction (triangles). The change in cross-sectional profile of the vein in the relaxed state is indicated schematically above. (Data points for isolated canine saphenous vein from Vanhoutte, P. M. and Leusen, I. (1969) *Pfluger's Archiv*, **306**, 341–353, by permission)

filling pressure. The volume of blood in a peripheral vein depends on the venous blood pressure and on active wall tension, as illustrated in Figure 7.16.

The effect of pressure on venous volume is particularly steep between zero pressure and 10 mmHg because the thin-walled vein deforms easily. At a transmural pressure of 1 mmHg, the vein is almost collapsed and has a narrow elliptical profile. As pressure rises towards 10 mmHg, the elliptical profile becomes progressively rounder, enabling the vein to accommodate large volume changes with just a few mmHg change in pressure. The maximum distensibility, which occurs at approximately 4 mmHg, is estimated to be approximately 100 ml/mmHg for the human venous system – over 50 times greater than the compliance of the arterial system. Below zero transmural pressure, the vein collapses into a dumb-bell shape and any flow is confined to the marginal channels. Above 10–15 mmHg the profile is fully circular and since the stretched collagen in the wall is relatively inextensible the volume is less sensitive to pressures over 15 mmHg.

The other factor influencing venous volume is the degree of active tension in the smooth muscle of the venous tunica media. In the gastrointestinal, hepatic, renal and cutaneous circulations, the vein wall is innervated by sympathetic vasomotor nerves. Sympathetically-excited venoconstriction reduces the capacity of these peripheral veins (see Figure 7.16, lower curve) and displaces blood into the thoracic comparment. This provides an important mechanism by which the nervous system can regulate the filling pressure of the heart (see Sections 6.5 and 11.10).

Venous pressure and the measurement of human CVP

Blood enters the venules at a pressure of approximately 12–20 mmHg and by the time it reaches named veins like the femoral vein, pressure has fallen to approximately 8–10 mmHg. The subsequent venous resistance is very small (except in collapsed vessels) so the 8–10 mmHg pressure head suffices to drive the cardiac output from the periphery into the central veins and right ventricle, where the diastolic pressure is 0–6 mmHg.

In intensive care units, central venous pressure (CVP) is often monitored directly via a catheter advanced into the subclavian vein or superior vena cava. In the routine clinical examination of the cardiovascular system, however, the CVP is assessed indirectly by inspection of the neck veins (see Figure 7.17). The external jugular vein runs over the sternomastoid muscle and the internal jugular vein runs deep to it. With the subject in a semi-supine position the lower part of the external jugular vein is normally distended while the upper part is collapsed (an effect of gravity, see later). From the venous pressure-volume curve, we know that transmural pressure is zero at

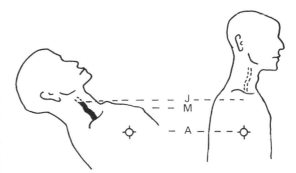

Figure 7.17 Estimation of central venous pressure (CVP) in a semi-supine human subject. CVP is the pressure at the point where the venae cavae enter the right atrium. This is equal to the vertical distance between point of collapse of the jugular vein (J) and right atrium (A). (The atria serve as a standard reference level when measuring circulatory pressures.) Since point A cannot be seen, the height of J above the manubriosternal angle (M) is measured instead; M-A is approximately 5 cm. CVP = (J-M) + 5 (cmH$_2$O). In the upright position the jugular vein is normally collapsed

the point of collapse of a vein. CVP equals the pressure at the point of collapse (zero) plus the pressure exerted by the vertical column of blood between this point and the right atrium. If, for example, the point of collapse (zero pressure) is 7 cm vertical distance above the right atrial midpoint, CVP is 7 cm of blood (7.4 cmH$_2$O). The position of the atrium cannot of course be observed directly but from anatomical studies it is known that the right atrial midpoint is approximately 5 cm lower than the manubriosternal angle, which is readily located. Thus by measuring the vertical distance between the point of collapse and the manubriosternal angle, and adding 5 cm, CVP can be estimated. Although the accuracy is only of the order ±2 cm, this is sufficient to detect the grossly elevated CVP which characterizes right ventricular failure (see Chapter 15). The normal subject has to be semi-supine because in the upright position the point of collapse is hidden below the clavicle: in right ventricular failure by contrast the CVP may be so high that the venous pulse is visible in the neck even when the patient is upright.

Effect of posture and altered gravity

The adoption of a standing position (orthostasis) increases the pressure in any blood vessel below heart level and reduces it in any vessel above heart level because gravity is then pulling on a vertical column of fluid between the heart and the vessel. This is particularly important in veins because their volume is so sensitive to transmural pressure.

Veins below heart level

Upon tilting a human subject upright, a transitory closure of the venous valves in the limbs prevents any significant backflow of blood away from the heart. Pressure in the dependent veins then rises steadily over approximately 30–60 s because blood continues to flow into the veins from the arterial system, and as it does so it opens the venous valves and re-establishes an uninterrupted column of blood between the heart and the feet. The weight of this fluid column raises venous pressure in the feet tenfold, from approximately 10 mmHg

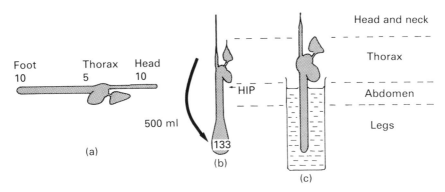

Figure 7.18 Changes in venous blood volume (shaded area) on moving from supine position (a) to standing (b). The thoracic compartment includes the central veins, heart and pulmonary blood; the lungs are shown disproportionately small here. Numbers are typical pressures in cmH$_2$O. The hydrostatic indifferent point (HIP) is the point where pressure is unaltered by tilting. (c) shows how immersion in water increases CVP. (After Gauer, O. H. and Thron, H. L. (1963) Handbook of Physiology, Circulation, Vol. 3 (eds. W. F. Hamilton and P. Dow, American Physiological Society, Bethesda, pp. 2409–2440)

supine to nearly 100 mmHg upright (see Figures 7.18 and 7.21). There is no counterbalancing rise in extramural pressure (unless the subject is immersed in water), so the veins distend: this is plainly visible in one's own hand on lowering it below heart level. In a human adult, about 500 ml of extra blood accumulate over approximately 45 s in the distended veins of the lower limbs. This is usually called venous 'pooling', but the phrase is misleading in that a pool is static whereas the venous blood is of course flowing continuously. Most of the additional blood comes ultimately from the intrathoracic compartment, so the CVP falls, impairing stroke volume by the Frank–Starling mechanism and provoking a transitory arterial hypotension and, sometimes, dizziness (postural hypotension; Section 14.1). A handcount amongst medical students indicates that most healthy individuals occasionally experience this orthostatic dizziness, especially when warm and venodilated.

*Caveat: flow in a syphon or 'How gravity does **not** work'*

Gravity acts equally on venous and arterial fluid columns; consequently, the pressure difference between the arteries and veins at any vertical level is not directly altered by orthostasis (see Figure 9.4), and nor therefore is blood flow. The assumption of some students that venous flow decreases in the dependent leg 'because blood has to go uphill against gravity' is utterly spurious, for it ignores the fact that gravity has an equal effect on the arterial column. The circulation through the limb or brain in fact resembles flow through a U-tube syphon, and flow through a rigid syphon is the same whether it is vertical, horizontal or upside down (see Figure 7.19). Indeed, if blood vessels were completely rigid, gravity would have no overall effect on the circulation. Limb blood flow does in fact decline with dependency (see Figure 7.5) but this is not because 'blood has to go uphill'; it is because (1) the orthostatically-induced fall

in cardiac volume and output elicits a reflex increase in vasoconstrictor nerve activity to limb arterioles, and (2) there is also a local

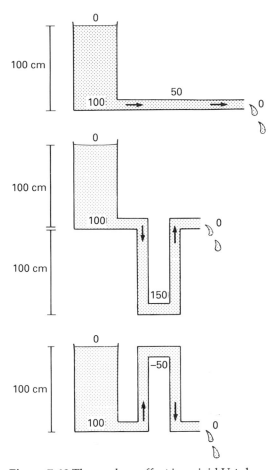

Figure 7.19 The syphon effect in a rigid U-tube. The feed-tank produces a pressure head of 100 cmH$_2$O to drive the flow. Flow passes through either a straight tube or a U-tube of equal resistance. The net pressure difference driving flow is the same in each case, so the flow is identical in all three situations (syphon principle). Numbers refer to pressure in cmH$_2$O at various points. In the horizontal tube, pressure has halved (50 cmH$_2$O) at the half-way point. In the dependent U-tube of limb-length 100 cm, pressure is (100 + 50) cmH$_2$O at the midway point, owing to the effect of gravity. In the inverted U-tube it is (−100 + 50) cmH$_2$O at the midway point

arteriolar constriction in response to the rise in local transmural pressure (see Chapter 11).

Veins above heart level

In vessels above heart level, the effect of gravity is to reduce blood pressure, and when the transmural pressure falls to zero or less the unsupported superficial veins collapse (see Figure 7.16). Deeper veins are better supported and do not collapse completely. Veins within the rigid cranial cavity are a special case and do not collapse because gravity reduces the pressure of the cerebrospinal fluid around them too, so the transmural pressure hardly changes.

When gravity alters

Air pilots experience altered g forces during aerobatic manoeuvres. A pilot pulling out of a steep dive can experience +3g to +4g along the body axis, and venous pooling in the lower body is so severe that the stroke volume falls rapidly and the pilot experiences a 'blackout' due to cerebral hypoperfusion. To prevent this, an anti-gravity suit is worn; bags inflate automatically around the legs to raise extramural pressure and minimize venous distension during turns. Conversely, in an inverted loop-the-loop manoeuvre a high negative g force is experienced, i.e. gravity is directed towards the head. This distends the retinal vessels and causes a 'redout' of vision. In space travel, the circulation is subjected to zero gravity for long periods but as this is not dissimilar to a supine posture or to floating in water, it presents no special problem for the cardiovascular system until the return to positive gravity (Section 14.1).

Oscillations in venous pressure and flow

Pressure is pulsatile in veins close to the right atrium, such as the jugular veins. The

pulse pressure is a few mmHg, just sufficient to move the overlying skin and render itself visible but too small to be palpable, unlike the arterial pulse. The waveform of the venous pulse was described fully in Section 2.3. Figure 7.20 relates the pressure waves to the oscillations in venous flow. Blood flow in central veins displays two spurts per cycle. Peak flow occurs during the x descent of the pressure wave and is caused by atrial relaxation. This inflow may be boosted by ventricular systole because the ballistic effect of firing out a mass of blood propels the ventricle downwards (Newton's law of action and reaction), stretching the atria and helping to suck

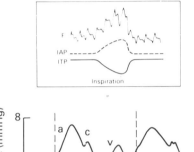

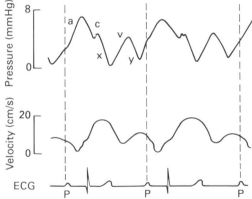

Figure 7.20 Pressure and flow in human superior vena cava over two cycles. Inset shows effect of breathing on venous return. F = flow in thoracic inferior vena cava; IAP = intra-abdominal pressure; ITP = intra-thoracic pressure. (After Brecher, G. A. (1956) *Venous Return*, Grune & Stratton, New York and Wexler, L., Bergel, D. L., Gabe, T. *et al.* (1968) *Circulation Research*, **23**, 349–359)

blood into them. The second flow-spurt, during the Y descent, is due to the tricuspid valve opening in diastole. This second spurt is boosted by the elastic recoil of the ventricular walls when end-systolic volume is low (e.g. exercise). Thus flow through the great veins, whilst primarily driven by the upstream pressure of about 8 mmHg (pressure from behind or *'vis a tergo'*), is aided by two transient reductions in downstream pressure due to the motion of the heart (suction from in front or *'vis a fronte'*). Venous flow can also be assisted by two non-cardiac factors: the skeletal muscle pump and the effect of breathing.

Skeletal muscle pump

When a skeletal muscle contracts it compresses the veins within, expelling their blood into the central veins. Venous valves prevent retrograde flow and ensure that the emptied segments refill from the periphery during muscle relaxation (see Figure 7.21). Rhythmic exercise thus has a pumping effect with several beneficial consequences. (1) The pump redistributes venous blood from the periphery into the central veins, and this prevents CVP from falling during exercise (cf. Figure 6.11). The muscle pump may even increase CVP slightly and move

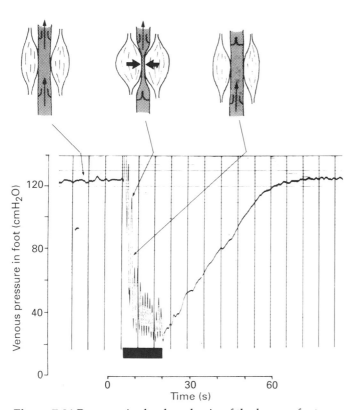

Figure 7.21 Pressure in the dorsal vein of the human foot during quiet standing, interrupted by rhythmic contraction of the calf muscles for a short period (black bar). Insets illustrate how the muscle pump operates. The subject, a physiologist, evidently has competent venous valves (as is of course to be expected of a physiologist!). (Unpublished data of Levick, J. R. and Michel, C. C.)

the ventricle up the Starling curve. (2) Like any pump, the muscle pump lowers pressure in the feed line: distal venous pressure falls because as the muscle relaxes blood drains rapidly from the distal veins into the empty muscle veins. At the same time, closure of the proximal valves interrupts the vertical column of blood between limb and heart. This reduces venous pressure in the foot and calf from around 90–100 mmHg in immobile orthostasis to 20–40 mmHg during walking, running, cycling etc. As a result, the arteriovenous pressure difference driving blood flow through the calf muscle increases by 50–60%. (3) The muscle pump reduces capillary filtration pressure in the legs too since capillary pressure is closer to venous pressure than to arterial pressure, and this reduces the tendency of a dependent limb to swell during exercise.

If the venous valves become *incompetent*, the muscle pump becomes ineffective. Since the vertical blood column can no longer be effectively broken up, the veins are subjected to a chronically-raised pressure load which leads to permanent distension (*varicose veins*). In addition the distal tissues swell due to the unrelieved, high capillary filtration pressure (see Chapter 9), and this can provoke trophic skin changes and ulceration.

Respiratory pump

Flow in the vena cava increases during inspiration (see Figure 7.20, inset), because the fall in intrathoracic pressure expands the intrathoracic veins. At the same time, the diaphragm compresses the abdominal contents, raising the abdominal venous pressure and enhancing venous flow from abdomen to thorax. Conversely, vena caval flow slows during expiration, especially during forced expiration or a Valsalva manoeuvre (see Chapter 14). Coughing for

example can elevate intrathoracic presure to 400 mmHg, and paroxysmal coughing can impede venous inflow to such a degree that fainting results.

The respiration-related oscillations in venous return evoke oscillations in stroke volume. Right ventricle stroke volume rises during inspiration owing to increased right ventricular filling. Left ventricular stroke volume on the other hand falls because the stretched pulmonary vessels have a greater capacitance and this reduces the left side filling pressure. The situation reverses during expiration, so the output of the two ventricles are regularly out of phase, though equal when averaged over the respiratory cycle.

Further reading

Blomqvist, C. G. and Stone, H. L. (1983) Cardiovascular adjustments to gravitational stress. *Handbook of Physiology, Cardiovascular System, Vol. 3, Part 1, Peripheral Circulation* (eds J. T. Shepherd and F. M. Abboud), American Physiological Society, Bethesda, pp. 1025–1063

Caro, C. G., Pedley, T. J., Schroter, R. C. and Seed, W. A. (1978) *The Mechanics of the Circulation*, Oxford University Press, Oxford

Chien, S., Usami, S. and Skalak, R. (1984) Blood flow in small tubes. In *Handbook of Physiology, Cardiovascular System, Vol. 4, Microcirculation* (eds E. M. Renkin and C. C. Michel), American Physiological Society, Bethesda, pp. 217–250

McDonald, D. A. (1974) *Blood Flow in Arteries*, Edward Arnold, London

Noble, M. I. M. (1979) *The Cardiac Cycle*, Blackwell, Oxford

Papppenheimer, J. R. (1984) Contributions to microvascular research of Jean Leonard Marie Poiseille. *Handbook of Physiology, Cardiovascular System, Vol. 4, The Microcirculation* (eds E. M. Renkin and C. C. Michel), American Physiological Society, Bethesda, pp. 1–10

Schmid-Schonbein, G. W. (1988) Granulocyte: friend or foe? *News in Physiological Science*, **3**, 144–147

Zweifach, B. W. and Lipowsky, H. H. (1984) Pressure-flow relations in blood and lymph microcirculations. In *Handbook of Physiology, Cardiovascular System, Vol. 4, Microcirculation* (eds E. M. Renkin and C. C. Michel), American Physiological Society, Bethesda, pp. 251–307

Chapter 8
Solute transport between blood and tissue

The heart and vasculature exist for one fundamental purpose: the delivery of metabolic substrate to the cells of the organism. This delivery takes place across the thin walls of capillaries, which thus subserve the ultimate function of the cardiovascular system. The capillary wall is also the site of fluid exchange between plasma and interstitial fluid, and thereby influences the volume of each compartment (see Chapter 9). Active metabolic functions are also being increasingly documented for endothelial cells.

8.1 Microvessel heterogeneity and density

Types of microvessel

The smallest arteries branch into first-order arterioles with muscular walls innervated by sympathetic nerves. These branch into second and third order arterioles (see Figure 1.9) and finally into the terminal arterioles whose walls contain smooth muscle but few vasomotor nerves, control being dominated at this level by local metabolites (see Chapter 11). The terminal arteriole gives rise to a cluster or module of capillaries (see Figure 8.1) and the smooth muscle tone in the terminal arteriole determines whether the capillary module is well perfused with blood ('open' capillaries) or not ('closed' capillaries). A few tissues, such as mesentery, have a ring of smooth muscle at the capillary entrance, the 'precapillary sphincter', governing capillary perfusion but most tissues lack these. Another specialized structure is the arteriovenous anasto-

mosis, a broad muscular vessel bypassing the capillary network in the skin of the extremities (fingers, nose, ear); it is involved in temperature regulation (see Chapter 12). The capillaries themselves are of microscopic size (diameter 5–8 µm; 'capillary' means hair-like), and were first proved to exist by Malpighi, a pioneer of microscopy who observed capillaries in a frog lung in 1661. The venous ends of capillaries unite to form pericytic venules (postcapillary venule) whose walls contain pericytes but no smooth muscle; these too are exchange vessels. Smooth muscle reappears in the walls of venules of 30–50 µm diameter.

Heterogeneity of length and blood flow

The non-uniformity within a microcirculation is striking. Capillaries vary in length (typically 500–1000 µm), in blood flow and in dynamic haematocrit even within the same tissue at the same time. Blood flow waxes and wanes ever 15 s or so in some capillary modules and can stop briefly due

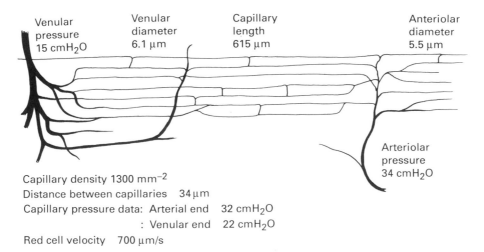

Venular pressure 15 cmH₂O — Venular diameter 6.1 µm — Capillary length 615 µm — Anteriolar diameter 5.5 µm — Arteriolar pressure 34 cmH₂O

Capillary density 1300 mm^{-2}
Distance between capillaries   34 µm
Capillary pressure data: Arterial end   32 cmH₂O
                    : Venular end   22 cmH₂O
Red cell velocity   700 µm/s

Figure 8.1 Capillary bed in relaxed cremaster muscle of a rat, with a terminal arteriole feeding a module of 14 capillaries. Numbers are means of observations. (From Smaje, L. H., Zweifach, B. W. and Intaglietta, M. (1970) *Microvascular Research*, **2**, 96–110, by permission)

to spontaneous rhythmic contractions in the terminal arterioles (*vasomotion*). This influences both solute exchange and fluid exchange (see Chapter 9). In well-perfused capillaries, however, the blood velocity is typically 300–1000 µm/s and the transit time is 0.5–2 s; this is the time available for plasma to unload oxygen, glucose etc. and load up with carbon dioxide etc. Mean transit time can fall to approximately 0.25 s in exercise.

Capillary density and its functional importance

Skeletal muscle contains roughly 1 capillary per muscle fibre or 300–1000 capillaries per mm^2 of muscle. The number of capillaries packed into a tissue is functionally important because it determines: (1) the total area of capillary wall available for exchange between blood and tissue (approximately $100 cm^2$/g in muscle), and (2) the intercapillary spacing and therefore the maximum blood-to-cell distance, which has a large effect on diffusion time (see Table 1.1). In the myocardium and brain, where oxygen consumption is high and sustained, the capillary density is even greater than in skeletal muscle, and provides around $500 cm^2$ surface per gram tissue. In the lung, capillary area is very high indeed (approximately $3500 cm^2$/g).

8.2 Structure of exchange vessels

The term 'exchange vessel' actually embraces both sides of the anatomical capillary bed because some oxygen diffuses through the walls of terminal arterioles and some fluid crosses the walls of pericytic venules. True anatomical capillaries are of three ultrastructural types: continuous, fenestrated and discontinuous (in order of increasing permeability to water).

Continuous capillary

Continuous capillaries are found in muscle, skin, lung, fat, connective tissue and the nervous system. In section, the circumference is formed by one to three flattened endothelial cells resting on a basement membrane (see Figure 8.2). The wall is only one cell thick so the diffusion distance is very small (approximately 0.5 µm). Pericytes, or 'Rouget cells', partly envelop the

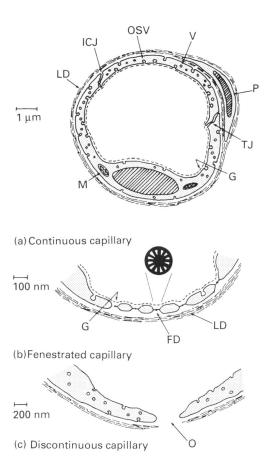

(a) Continuous capillary

(b) Fenestrated capillary

(c) Discontinuous capillary

Figure 8.2 Sketches of capillary wall in transverse section, based on electron micrographs. FD, fenestral diaphragm; inset shows diagram en face; G, glycocalyx; ICJ, intercellular junction; LD, lamina densa of basal lamina; M, mitochondrion; O, open intercellular gap; OSV, open surface vesicle; P, pericyte; TJ, tight part of junction; V, vesicle. Scale only approximate

capillary and there is some evidence that these can contract, though the significance of this is unclear. The endothelial cell contains mitochondria, endoplasmic reticulum, a Golgi apparatus and filaments of actin and myosin. The latter form distinct stress fibres in spleen endothelium, and splenic capillaries appear to be actively contractile. Most capillaries however, are through to be non-contractile under physiological conditions, although contraction of venular endothelial cells can occur in acute inflammation (Section 9.11).

Certain endothelial features are particularly important for solute transfer, namely the intercellular junction, vesicle system and surface coat (glycocalyx). Their structure is as follows.

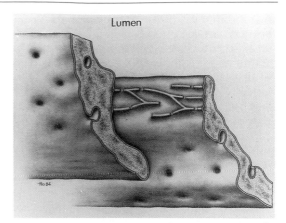

Figure 8.3 Three-dimensional reconstruction of tight junction between two endothelial cells, based on serial thin sections. The cleft is typically 500–1000 nm long from lumen to outside and 15–20 nm wide. Small lipophobic solutes and water can circumvent the junctional strands or pass through discrete interruptions in them. (From Bundgaard (1984) *Journal of Ultrastructural Research*, **88**, 1–17, by permission)

Intercellular junctions

These are parallel-sided clefts whose entrance occupies 0.1–0.3% of the capillary surface. They are almost certainly the transcapillary route taken by most fluid and by metabolites like glucose. In cross-section the gap is 15–20 nm wide for most of its length, and this is much wider than the diameter of a glucose molecule (0.9 nm) or even albumin (7.1 nm). At one to three points along the cleft, the adjacent cell membranes touch and form 'tight junctions', but the junctions do not provide a continuous seal around the cell perimeter. Freeze-fracture electron microscopy, which displays the junctional membrane *en face*, reveals the tight junctions as lines of membrane particles called junctional strands (see Figure 8.3). These run round the cell perimeter but are interrupted by two kinds of break. (1) Gaps of 5–11 nm occur between individual particles, and probably account for the 'open junctions' occasionally seen in transverse sections. (2) The junctional strands sometimes come to an abrupt end leaving a tortuous but open route through the intercellular cleft. The overlap of other strands prevents this tortuous bypass from being seen in a single transverse section but serial transverse sections confirm its existence. In pericytic venules, which are actually more permeable than capillaries, the overlap of the junctional strands is less marked. Conversely, in brain capillaries, which have a very low permeability to fluid, the strands are numerous and complex and extend without interruption around the perimeter to form a true seal (zonula occludens), as in tight epithelia.

Endothelial vesicles and vesicular transport

About a quarter of the cytoplasmic volume is occupied by vesicles of diameter 60 nm, which are thought to be involved in transporting macromolecules into the cell (endocytosis) and, more controversially, across it (transcytosis). Some vesicles open directly onto the cell surface by a stalk 20 nm wide while others look as though they are floating free in the cytoplasm. This led to the idea that vesicles might ferry plasma proteins across the cell, loading up at the surface, detaching and diffusing across the cell to release their load at the abluminal side. Recent studies of ultrathin serial sections, however, reveal that free-floating vesicles are mostly an illusion; 99% of seemingly 'free' vesicles are connected, out of the plane of section, to one or other surface via surface vesicles (see Figure 8.4). The vesicular system is now thought to be a racemose invagination of the plasmalemma resembling a bunch of grapes. Nevertheless, these structures might still play a role in macromolecular transport by a process involving transient fusion and exchange of contents between the luminal and abluminal systems (see Figure 8.4). Very rarely, two or three vesicles are seen to be fused to create a continuous channel across the endothelial cell, the 'multivesicular channel', and these two may contribute to

plasma protein passage (see Figure 8.4b, left).

Glycocalyx

The endothelial surface is coated with a thin layer of negatively-charged material called the glycocalyx, as revealed by cationic probes (positive charged probes; see Figure 8.5). The glycocalyx permeates the intercellular cleft and lines the caveoli too. It consists of a meshwork of fibrous molecules with protein cores and sugar-based side chains (namely sialoglycoproteins and glycosaminoglycans such as heparan sulphate), and there is growing evidence that this meshwork may act as a macromolecular sieve in the capillary wall (Section 8.8).

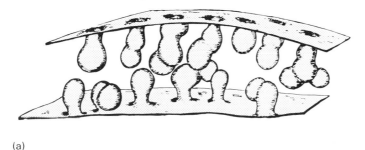

(a)

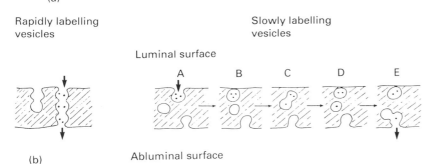

(b)

Rapidly labelling vesicles

Slowly labelling vesicles

Luminal surface

Abluminal surface

Figure 8.4 Diagrams of the endothelial vesicle system. (a) Reconstruction of serial sections showing that the vesicles are a racemose invagination of the surface plasmalemma, with hardly any truly free-floating vesicles. (b) Frames A–E show how a slow transcytosis of protein molecules (black dots) might occur from the plasma to interstitium by means of a process involving the fusion and mixing of contents (frame C) between luminal-linked and abluminal-linked vesicles. A few abluminal vesicles label unusually rapidly, perhaps because they are part of a multivesicular transendothelial channel (left). ((a) From Frokjaer-Jensen, J. (1983) *Progress in Applied Microcirculation*, **1**, 17–34. (b) From Clough, G. and Michel, C. C. (1981) *Journal of Physiology*, **315**, 127–142, by permission)

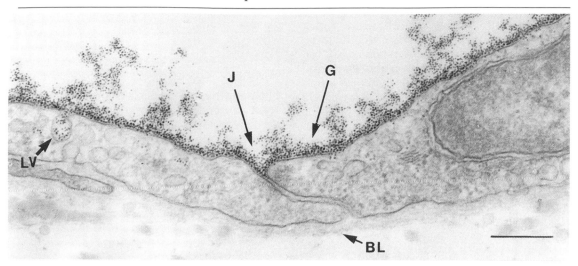

Figure 8.5 Electron micrograph of frog mesenteric capillary perfused with a solution of cationized ferritin (diameter 11 nm). Lumen at the top. The cationic molecules (black dots) bind to and delineate the glycocalyx (G). On the left, some ferritin molecules are seen labelling a vesicle (LV). The intercellular junction in the centre (J) appears impermeable to them. The cationized ferritin was found to reduce hydraulic conductance (see 'the protein effect', Section 8.8), whereas native ferritin, which does not bind to the glycocalyx, does not affect conductance. BL, basal lamina. Bar = 0.2 μm. (From Turner, M. R., Clough, G. and Michel, C. C. (1983) *Microvascular Research*, **25**, 205–222, by permission)

Basal lamina

The basal lamina (often called basement membrane) is 50–100 nm thick and consists of a dense region (lamina densa) separated from the cell by a lighter region (lamina rara). The lamina densa consists of a network of type IV collagen molecules and negatively-charged heparan sulphate proteoglycan. It is attached to the cell by a cross-shaped glycoprotein called laminin. The membrane endows the capillary with sufficient strength to withstand blood pressure, but the stresses involved are actually quite small owing to the small radius of curvature (see Laplace's law, Equation 7.8). The basal lamina retards but does not prevent the passage of protein molecules, except in renal glomerular capillaries where a double lamina densa holds back macromolecules such as ferritin.

Fenestrated capillary

Fenestrated capillaries are an order of magnitude more permeable to water and small hydrophilic solutes than most continuous capillaries. They occur in tissues specializing in fluid exchange (renal glomerulus and tubules, exocrine glands, intestinal mucosa, ciliary body, choroid plexus, synovial lining of joints) and also in endocrine glands. The endothelium is perforated by small circular windows, the fenestrae (diameter 50–60 nm), which allow plasma within 0.1 μm of the extravascular space. Fenestrae are the major route by which water and metabolites cross fenestrated capillaries. The fenestrae are mostly not open holes but are bridged by an extremely thin membrane, the fenestral diaphragm (thickness 4–5 nm) which is sandwiched between the glycocalyx and basement membrane. Viewed *en face* the diaphragm resembles a cartwheel, being perforated by about 14 wedge-shaped apertures of arc-length 5.5 nm (Figure 8.2b). In renal glomerular capillaries, however, diaphragms are totally absent.

Discontinuous capillary

Discontinuous or sinusoidal capillaries possess some intercellular gaps over 100 nm wide and a discontinuity in the underlying basal lamina. As a result, these capillaries are permeable even to plasma proteins. They occur wherever red cells need to migrate between blood and tissue, i.e. bone marrow, spleen and liver.

8.3 Transport processes: diffusion, reflection and convection

The passage of water and solutes across the capillary wall is a passive process requiring no energy expenditure by the endothelium. *Fluid movement* across the wall is a process of hydraulic flow down pressure gradients set up by the heart. (Water molecules also diffuse extremely rapidly across the wall but this diffusion is bidirectional and no net transport results.) *Metabolite exchange* is mainly a process of passive diffusion down concentration gradients set up by tissue metabolism. To a minor extent, metabolites are also swept along in the transcapillary stream of fluid ('convective transport'), but the stream is relatively slow and can often be neglected. For proof, see footnote to Table 8.2. *Macromolecules*, however, diffuse more slowly than metabolites like glucose, so convective transport across the capillary wall is relatively more important in their transport.

To understand capillary exchange, we must therefore consider the nature of diffusion and convection across a porous membrane and equip ourselves with a basic kit of simple transport expressions.

Free diffusion and Fick's law

Diffusion was studied by Adolf Fick, Professor of Physiology at Wurzburg, and Fick's first law of diffusion (1855) describes the rate of diffusion, i.e. the mass of solute transferred by diffusion per unit time (J_S) – see Figure 8.6a. Fick's law can be summarized thus:

$$J_S = -DS\frac{\Delta C}{\Delta x} \tag{8.1}$$

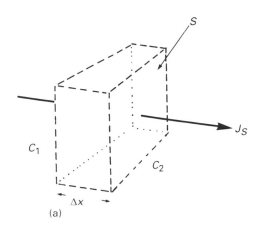

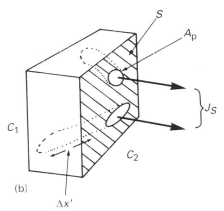

Figure 8.6 (a) Free diffusion in bulk solution. The solute traverses an unimpeded layer of fluid of thickness Δx and surface area S, driven by concentration difference $C_1 - C_2$. J_S is the diffusion rate (moles/s). (b) Membrane reduces J_S by confining solute to pores of total area A_p. The pathlength through the pore, $\Delta x'$, can exceed the membrane thickness Δx

Diffusion rate depends on the concentration difference driving diffusion (ΔC), the distance across which diffusion is occurring (Δx) – the ratio $\Delta C/\Delta x$ being the concentration gradient, surface area (S), and the diffusion coefficient (D), which represents the intrinsic velocity of the solute. The negative sign indicates that solute flux occurs down the concentration gradient. The diffusion coefficient (D) depends on temperature, solvent viscosity and solute size. For most solutes, D is inversely proportional to the cube root of the molecular weight, so small molecules diffuse faster than big ones. The diffusion coefficient is often used to calculate the *radius* of the molecule, idealized as a sphere (the Stokes–Einstein radius or diffusion radius; see Table 8.1 and Appendix).

Effect of a membrane on diffusion, and the meaning of 'permeability'

Diffusion through a large volume of solvent is called free diffusion. When a solute diffuses across a membrane, however, several factors slow the diffusional process.

Available area   If the solute is confined to solvent-filled pores penetrating the membrane, the area available for diffusion is reduced from the total surface (S) to the pore area (A_p) (see Figure 8.6b). Moreover, for a molecule of radius 'a' the centre of the molecule can get no closer to the pore rim than distance 'a', so only a fraction of the pore space is available to the molecule. If the pore happens to be a cylinder of radius r, the available fraction is $(r - a)^2/r^2$ (see Figure 8.7). This geometrical effect is called *steric exclusion*. Owing to steric exclusion, the concentration of solute in pore water is less than the concentration in bulk solution and the ratio of the two concentrations at equilibrium is called the equilibrium partition coefficient ϕ (phi). For a neutral cylindrical pore ϕ equals the fractional available space, $(r - a)^2/r^2$. Thus steric exclusion reduces the pore area available for diffusion to $A_p \times \phi$.

Pathlength   If the pores run obliquely through the membrane, as endothelial junctions mostly do, the diffusion distance ($\Delta x'$) is greater than the membrane thickness (Δx) as illustrated in Figure 8.6b.

Intrapore diffusion   Whenever a solute molecule moves through water it experiences frictional resistance, called hydrodynamic drag; this is what governs the diffusion coefficient. The hydrodynamic drag on a solute inside a pore rises progressively as the solute radius approaches pore radius in size, because the water slips less easily past the solute molecule in the confined space of a pore (Figure 8.7b). As a result, the solute's diffusion coefficient within the pore (D') is smaller than its free diffusion coefficient in bulk solution (D; *'restricted diffusion'*). The restriction to diffusion increases rapidly when solute/pore size exceeds one-tenth; for example, the restricted diffusion coefficient is half the free value when solute radius is 15% of pore radius.

Because of the above effects, Fick's law is modified by the presence of a membrane and becomes:

$$J_S = - D'\frac{A_p}{\Delta x'}\phi\,\Delta C \qquad (8.2)$$

This expression tells us the factors responsible for a membrane's 'permeability'. The permeability of a membrane (P) is, by definition, the rate of diffusion of solute across unit area of membrane per unit concentration difference, or in symbols:

$$J_S = - P\,S\,\Delta C \qquad (8.3)$$

Comparing Equation 8.3 with Equation 8.2, we find that permeability is governed by the restricted diffusion coefficient (D'), the pore length ($\Delta x'$) and the fractional pore area available to the solute ($A_p\,\phi/S$). The study of capillary permeability therefore has the potential to reveal a great deal about the porosity of the capillary wall.

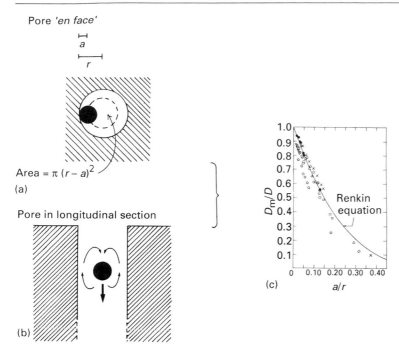

Figure 8.7 A spherical molecule of radius 'a' diffusing through a water-filled cylindrical pore of radius r. (a) End-on or 'bird's eye' view of pore, showing steric exclusion of the molecule centre from an annulus of fluid (shaded). (b) Longitudinal section through the pore. Arrows indicate how solute movement necessitates solvent flow, giving rise to enhanced hydrodynamic drag within the narrow confines of the pore. (c) Plot of reduced diffusion across a membrane (D_m) relative to the free diffusion coefficient (D) as the ratio of molecular size to pore size increases (a/r). The membranes were mica sheets containing pores of known, uniform radius produced by bombardment with uranium fission fragments. The 'Renkin equation' fitted to the data is D_m/D = steric exclusion × hydrodynamic drag (restricted diffusion) = $(1 - a/r)^2 \times (1 - 2.1\,a/r + 2.09(a/r)^3 - 0.95(a/r)^5)$. (From Beck, R. B. and Schultz, J. S. (1972) *Biochimica Biophysica Acta*, **255**, 273–303, by permission)

Reflection and reflection coefficients

Two additional parameters are needed to characterize exchange across capillaries besides permeability, namely the 'reflection coefficient' and 'hydraulic conductance'. The reflection coefficient (σ; sigma), is an index of the membrane's molecular selectivity (see Figure 8.8). It is defined as the

osmotic pressure that a given difference in solute concentration exerts across the test membrane divided by the full osmotic pressure that the same concentration difference exerts across a perfect semi-permeable membrane; thus σ equals $\pi_{\text{effective}}/\pi_{\text{ideal}}$. If the solute passes through the membrane as freely as water, it exerts no osmotic pressure and $\sigma = 0$ (see Figure 8.8c). If the solute is totally reflected by the pores, it exerts its full

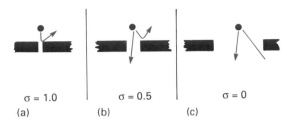

σ = 1.0   σ = 0.5   σ = 0

(a)   (b)   (c)

Figure 8.8 Sketch illustrating how the osmotic reflection coefficient (σ) relates to the molecular selectivity of a membrane. For explanation, see text

osmotic potential and σ = 1 (see Figure 8.8a). For most hydrophilic molecules at the capillary wall, reflection is partial and σ is between 0 and 1 (see Figure 8.8b). The size of σ is related to the partition coefficient φ:

$$\sigma = (1 - \phi)^2 \tag{8.4}$$

and as explained earlier φ (and therefore σ) depends on a/r, the ratio of solute radius to pore radius. Reflection coefficients are thus a very useful guide to equivalent pore size.

Convection and hydraulic conductance (L_p)

The rate of fluid movement across a membrane depends on the pressure gradients and on the *hydraulic conductance* of the membrane. The latter is defined as the filtration rate (J_V) produced by unit pressure difference acting across unit area of membrane, and is usually symbolized by L_p or K (see Darcy's law, Equation 7.1). Hydraulic conductance, like the other properties considered above, depends on the porosity of the membrane, A_p/Δ_x.

The reflection coefficient influences not only osmotic pressure but also solute transport by *convection* (wash along or 'solvent drag') because only the fraction of solute that is not reflected, namely (1 − σ), can be washed into a pore. The rate at which solute is dragged across a porous membrane by solvent thus depends on the

solvent flow, the concentration of solute and the 'admitted fraction', 1 − σ. This relation is important for macromolecule transport (see Section 8.7 and 9.4).

8.4 Measurement of capillary permeability

The above outline shows that all three membrane parameters, P, σ and L_p contain information about a membrane's porosity. Partly for this reason, and partly because of its intrinsic physiological importance, capillary permeability has been intensively investigated. To determine capillary permeability, three sets of measurement are required (see Equation 8.3): (1) the rate of solute transfer, (2) the membrane area, and (3) the concentration difference across the wall.

1. *Solute diffusion rate* across the capillary wall can be measured by direct optical methods in individual, cannulated capillaries perfused with dyes or fluorescent markers (see Figure 8.9). Alternatively, in whole organs the transfer of rapidly diffusing solutes (e.g. glucose) across the entire microvascular bed can be determined by the Fick principle (Section 5.1), transfer rate being the product of blood flow and arteriovenous concentration difference. With slowly-exchanging solutes like albumin, the difference between arterial and venous concentrations is small and unreliable, but net transfer by all processes (diffusion, convection and vesicles) can be calculated as prenodal lymph flow × concentration in lymph.

2. *The anatomical area of the capillary walls (S)* can be measured in tissue sections, but the value *in vivo* is less certain because perfusion of some capillaries is only intermittent in certain organs.

3. *The average concentration difference across the capillary wall (ΔC)* is particularly difficult to measure in whole organs

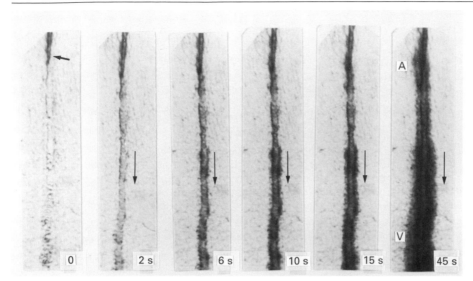

Figure 8.9 Photomicrographs of a single frog mesenteric capillary illustrating transcapillary diffusion of Evans blue, a lipid-insoluble molecule of Stokes–Einstein radius 1.3 nm. The tip of the perfusing micropipette is just visible in the arterial end of the capillary (thick arrow). Frame numbers are time in seconds after beginning perfusion; thin arrows show direction of flow. Transcapillary passage is visible within 6 s and permeability can be determined from the change in optical density with time. By 15–45 s an arteriovenous gradient of permeability is clearly visible (A–V); venous capillaries and pericytic venules are normally more permeable than arterial capillaries, probably due to their less well-developed junctional strands. (From Levick, J. R. (1972) Doctoral thesis, Oxford)

because the average plasma concentration is not simply the arithmetic average of venous and arterial concentrations. This arises from the fact that the concentration falls non-linearly along a capillary, declining most steeply at the inlet where the transmural concentration gradient and therefore efflux is highest (see Figure 8.10). In the 'osmotic transient method' this problem is circumvented; a solute such as glucose is injected intra-arterially and the average concentration difference across the capillary walls is calculated from the osmotic pressure which it transiently exerts there. This requires the use of an isolated, artificially perfused tissue.

The other main method available, the multiple indicator diffusion technique, can often be applied to organs *in situ* and relies on a simple theoretical analysis of solute exchange along a capillary. The method is presented below because it illustrates some important relationships.

Multiple-tracer indicator diffusion technique

A solution containing a diffusible test solute and a non-exchanging reference solute (usually radiolabelled albumin) is injected rapidly into the organ's main artery. A series of venous blood samples are collected over the next few seconds and analyzed. The concentration of test solute in the venous effluent is found to fall below that of the reference solute because some test solute has diffused out of the capillaries (see Figure 8.11). The reference concentration indicates what the test solute concentration would have been if no exchange had taken place. From these concentrations, the fraction of the test solute that is extracted

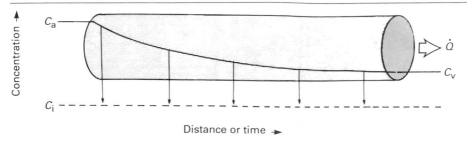

Figure 8.10 The concentration of a rapidly diffusing solute falls non-linearly along a capillary from arterial level (C_a) to venous level (C_v). Arrows indicate size and direction of diffusional flux. If interstitial concentration (C_i) is constant (or zero) and permeability is uniform, the concentration profile is exponential, and the mean plasma concentration equals $C_i + (C_a - C_v)/\ln((C_a - C_i)/(C_v - C_i))$. $\dot{Q}$ is blood flow

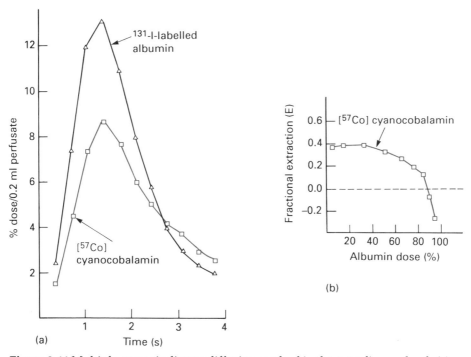

Figure 8.11 Multiple-tracer indicator diffusion method in the cat salivary gland. (a) Concentration of test solute, radiolabelled vitamin B12 (cyanocobalamin, radius 0.84 nm) and reference solute, radioiodinated serum albumin (radius 3.6 nm), in successive 0.2 ml samples of venous effluent. The marker content is expressed as a percentage of the mass injected intra-arterially in a brief bolus. The lines cross at 3 s because B12 that has reached the interstitium begins to diffuse back into the plasma as plasma concentration falls below interstititial concentration. (b) Extraction of B12, calculated as (arterial concentration − venous concentration)/arterial concentration, for successive venous samples. Extraction falls as interstitital concentration rises and becomes negative when the concentration gradient reverses. (After Mann, G. E., Smaje, L. H. and Yudilevich, D. L. (1979) *Journal of Physiology*, **297**, 335–354, by permission)

during its passage through the capillary bed (extraction, *E*) can be calculated for each successive venous sample. The extraction must be measured at several, increasing blood flows ($\dot{Q}$) in order to check that exchange is not being limited by delivery rate, rather than by capillary permeability; the former situation causes permeability to be underestimated (Section 8.10). The permeability surface area product (PS) for the whole capillary bed can then be calculated from the extraction and plasma flow ($\dot{Q}$) using the Renkin–Crone expression:

$$E = 1 - \exp.(-PS/\dot{Q}) \qquad (8.5)$$

which is based on an exponential concentration profile along the capillary (see Figure 8.10). The Renkin–Crone expression is an important one and will be met later in Section 8.10, for it tells us how blood flow affects solute exchange.

8.5 Classes of solute. I: Lipid-soluble molecules

The lipid-solubility of a molecule dramatically influences its permeation through the capillary wall. Capillary permeability to the

lipid-soluble molecule oxygen, for example, is many thousand times greater than permeability to the lipid-insoluble molecule glucose. The second main factor affecting permeation is the solute's molecular size, and capillary permeability to glucose (M_w 180) is nearly 1000 times greater than to albumin (M_w 69 000; see Table 8.1). Solutes thus fall into three main classes: lipid-soluble molecules, small lipid-insoluble molecules and large lipid-insoluble molecules (macromolecules).

Taking the lipophilic molecules first, capillary permeability increases in proportion to the solute's oil:water partition coefficient, indicating that these molecules traverse the capillary wall by dissolving in the lipid cell membrane. Virtually the entire capillary surface is therefore available for diffusion and this explains the high permeability. The oil:water partition coefficient is approximately 5 for oxygen and 1.6 for carbon dioxide, so capillaries are extremely permeable to respiratory gases: indeed, the permeability to oxygen is so high that some gas exchange even occurs in arterioles, and the haemoglobin saturation can fall to approximately 80% even before the blood enters the true capillaries. Anaesthetic

Table 8.1 Capillary permeability to various solutes

Solute	M_w	D (cm²/s, × 10⁵)	a (nm)	Membrane	Permeability (cm/s, × 10⁶)
Oxygen	32	2.11	0.16	continuous capillary	~100 000
Urea	60	1.90	0.26	continuous capillary	26–28
Glucose	180	0.91	0.36	continuous capillary	9–13
Sucrose	342	0.72	0.47	continuous capillary	6–9
				cerebral capillary	0.1
				fenestrated capillary	>270
Albumin	69000	0.085	3.55	continuous capillary	0.03–0.01
				fenestrated capillary	0.04

M_w, molecular weight in Daltons. D, free diffusion coefficient in water at 37°C. a, Stokes–Einstein diffusion radius. Continuous capillaries of cat leg, dog heart and human forearm, fenestrated capillaries of cat salivary gland (After Renkin, E. M. (1977) *Circulation Research*, **41**, 735–743; Clough, G. E. and Smaje, L. H. (1984) *Journal of Physiology*, **354**, 445–455; Landis, E. M. and Pappenheimer, J. R. (1963) *Handbook of Physiology, Cardiovascular System, Circulation, Vol. II*, (eds W. F. Hamilton and P. Dow), American Physiological Society, Bethesda, pp. 961–1034)

agents and the flow-tracer xenon also fall into this class of solute.

8.6 Classes of solute. II: Small lipid-insoluble molecules and the small pore concept

Hydrophilic solutes like electrolytes, glucose, lactate, amino acids, vitamin B12, insulin and many drugs cannot easily penetrate the endothelial cell membrane but they can cross the capillary wall via paracellular water-filled channels. The existence of these channels explains why the permeability of capillaries to water and small lipid-insoluble molecules is two to three orders of magnitude higher than the permeability of most cell membranes. The permeability of a protein-retaining dialysis membrane on the other hand is an order of magnitude higher than capillary permeability. This comparison, along with direct calculations of the pore area needed to explain permeability data, led Pappenheimer, Renkin and Borrero to propose, in 1951, that the capillary wall is penetrated by small aqueous channels occupying only a tiny fraction of the capillary surface (less than 0.1% in dog hindlimb capillaries).

Restricted diffusion and equivalent pore size

In their seminal paper of 1951, Pappenheimer and co-workers described a phenomenon that firmly established the 'small pore theory' of capillary permeability. They found that the permeability of hindlimb capillaries to hydrophilic solutes declined as molecular size increased, and fell off more steeply than the free diffusion coefficient (see Figure 8.12). This happens in artificial porous membranes too as the solute diameter becomes a significant fraction of the channel width, and is caused by steric exclusion and restricted diffusion within narrow channels (see Figure 8.7). The existence of restricted diffusion in the capillary wall indicates that narrow aqueous channels perforate the wall. Moreover, the degree of restriction enables the equivalent radius of the channels to be calculated (see legend to Figure 8.7 for details), and it emerges that the channels in the walls of cardiac, skeletal muscle and intestinal capillaries restrict diffusion to the same degree as cylindrical pores of radius 4–5 nm. It must be emphasized, however, that nobody believes that the small pores are actually cylindrical tubes, only that some of their properties can be equated with those of tubes. The true nature of small pores is considered in Section 8.8.

Another way of estimating the size of the small pores is to measure the reflection coefficient. The capillary reflection coefficient for plasma albumin (radius 3.55 nm) is typically 0.8–0.9, from which an equivalent pore radius of 4.6–5.3 nm is easily calculated (see Equation 8.4).

Differences in permeability due to differences in pore density

The permeability of capillaries to small lipophobic solutes like potassium and sodium ions varies very widely for different organs; the permeability of frog mesenteric capillaries and salivary gland capillaries, for example, is over ten times larger than the permeability of muscle capillaries. The hydraulic conductance of the wall varies similarly, and increases in linear proportion to the solute permeability. If pore radius (r) were different in the various capillaries, hydraulic conductance would increase disproportionately more than permeability owing to the r^4 effect in Poiseuille's hydraulic law. It thus appears that channel width is the same in capillaries with a high or low permeability: the physiological variation in permeability between organs with continuous or fenestrated capillaries is not due to differences in pore size but is due to differences in the total area of the wall occupied by pores and/or the pore length.

By dividing the hydraulic conductance ($\propto r^4$) by the solute permeability ($\propto r^2$) it is

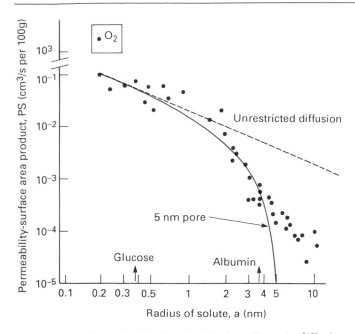

Figure 8.12 Effect of molecular size (Stokes–Einstein diffusion radius) on the permeability of continuous capillaries. All points refer to lipophobic solutes except that for oxygen. Dashed line shows effect of fall in free diffusion coefficient with molecular size; the plot is a log-log plot, so the dashed line has a slope of −1. Dotted line shows decline in permeability due to steric exclusion and hydrodynamic drag in cylindrical pores of radius 5 nm (see legend to Figure 8.7). Data for small solutes conform approximately to the theoretical line for 5 nm-radius pores but the permeability to solutes larger than albumin indicates the existence of additional larger pores. Data are from mammalian skeletal muscle and skin except for oxygen point (lung). The solutes range from urea to immunoglobulins, dextrans and polyvinylpyrrolidones. (Results of Pappenheimer *et al.*, 1951; Diana *et al.*, 1972; Garlick and Renkin, 1970; Carter *et al.*, 1974; Youlten, 1969, as reviewed by Renkin, E. M. and Curry, F. E. (1978) In *Membrane Transport in Biology*, Vol. IV (eds G. Giebisch, D. C. Tosteson and H. H. Ussing), Springer-Verlag, Berlin, pp. 1–45)

possible to work out the radius of the channels, if it is supposed that the channels are cylindrical tubes. The radius works out, however, to be 9 nm, which is incompatible with the estimates of 4–5.3 nm based on restricted diffusion and reflection (see earlier). This discrepancy indicates that the channels are not in reality cylindrical tubes. Even if the channels are assumed to be parallel-sided slits, like the endothelial clefts, a similar discrepancy still occurs. A solution to this puzzle is offered by the new fibre-matrix theory, which is described in Section 8.8.

8.7 Classes of solute. III: Large lipid-insoluble molecules and the large pore concept

Capillary permeability to plasma macromolecules can be assessed either by recording the extravascular accumulation of radiolabelled albumin after an intravascular injection or by collecting and analyzing prenodal lymph. Plasma protein is found in lymph at 20–70% of the plasma concentration, demonstrating that normal capillary beds have a finite low permeability to macromolecules. This should not be thought of as some lamentable 'leak', on the contrary, it is a functional necessity, both for interstitial defence by immunoglobulins and for the transfer of protein-bound substances like iron, copper, vitamin A, lipids, thyroxine, testosterone and oestradiol.

The permeability versus molecular size plot again

Permeability declines steeply as solute radius approaches 3.5 nm (albumin) because the solute radius is approaching the width of the equivalent small pores. Beyond this point, however, the permeability-size plot changes slope, and the permeability to larger macromolecules declines only slightly more than one would expect from the decline in free diffusion coefficient (see Figure 8.12). The persistent permeability to macromolecules that are too large to penetrate the small pore system led Grotte, in 1956, to propose the co-existence of a tiny population of large pores of radius 20–30 nm. More recently, measurements of the reflection coefficient have confirmed this: the 'admitted fraction' $(1 - \sigma)$ decreases rapidly with solute radius between 1 nm and 3.5 nm but does not fall to zero even for molecules as large as fibrinogen (Stokes-Einstein radius 10 nm).

Transport through the large pore system

Calculations suggest that there is typically around one large pore per 12 000 small pores. Consequently, the large pore system contributes very little to the transport of small solutes. As regards fluid transport, it is estimated that at a normal filtration rate, 73–94% of the fluid passes through the small pore system and only 6–27% through the large pores in dog hindpaws and cat intestine. In the discontinuous capillaries of liver, however, the large:small pore ratio is higher, approximately 1:100; conversely, renal glomerular capillaries and cerebral capillaries may lack a large pore system entirely.

Large pores, like small pores, are a functional concept and their actual physical nature is controversial (Section 8.8). This makes it difficult to be dogmatic about how macromolecular transport occurs. The processes that might be involved are diffusion, convection and vesicular transport (Section 8.2). A rise in filtration rate is known to increase the rate of transport of macromolecules across the capillary wall many fold, so convective transport (solvent drag) through hydraulically continuous large pores clearly dominates protein transport at moderate to high filtration rates; but diffusion or vesicular transport might become relatively more important at low filtration rates.

Effects of molecular size and charge

Large proteins like fibrinogen undergo more reflection by the capillary wall than do smaller proteins like albumin, so the larger plasma proteins are less abundant in interstitial fluid. This process is called 'molecular sieving' and is considered further in Section 9.4.

The permeation of a macromolecule is affected by its charge, as well as size. Many capillaries, such as glomerular vessels, are less permeable to net negatively-charged macromolecules like albumin than to neutral or positively-charged ones of similar size. This is due to repulsion of the molecule by the fixed negative charges of the glyco-calyx and basement membrane. The charge effect is most marked for the smaller macromolecules which are known to pass partly through the small pores, so the charge effect probably arises mainly in the small pores.

8.8 In search of the pores

Anatomical location of the small pores

The small pore system in continuous capil-laries can be defined as a set of channels of equivalent radius 4.0–5.5 nm occupying less than 0.1% of the capillary surface. The intercellular cleft is generally regarded as the region where the small pores are located because: (1) intercellular clefts occupy only 0.1–0.3% of the surface, and (2) electron micrographs show that small lipid-insoluble molecules like ferrocyanide and micro-peroxidase can penetrate the cleft. The tight junctions in the cleft are an obstacle in this pathway but they can be circumvented by the tortuous route around the ends of junctional strands and/or via the inter-particle gaps (see Figure 8.3). The differ-ences in permeability between organs appear to be due principally to differences in the fraction of the cleft's surface area that is functionally 'open' (i.e. not sealed). In muscle capillaries, for example, only approx-imately 10% of the intercellular cleft needs to be open to account for the wall's permeability; in frog mesenteric capillaries it is about 50% and in cerebral capillaries less than 0.1%. The fenestra, when present, is a second site for small pores as indicated by the fact that capillary permeability increases in proportion to fenestral area.

Fibre matrix theory

The location of the small pores is reasonably certain but their physical nature remains controversial because the dimensions of the clefts and fenestrae do not correspond adequately with those deduced for small pores: the cleft is too wide (15 nm) and the fenestral diaphragm too thin. One long-standing view is that the small pores are actually narrowings of the cleft where the membranes approach to within 8 nm with-out fusing (open junctions), but it is difficult to reconcile the shortness of such regions with certain experimental data. In 1980, Michel and Curry proposed the 'fibre matrix theory' of permeability, which offers a new view of the nature of the small pore system (see Figure 8.13). They suggested that the glycocalyx, a fibrous material which fills the intercellular junction and covers the fenes-tra, may be dense enough to function like small pores, i.e. the spaces between the molecular chains may be narrow enough to restrict solute diffusion, increase hydraulic resistance and reflect macromolecules. The uniformity of the glycocalyx would explain why the albumin reflection coefficient is about the same (0.8–0.9) in both highly permeable capillaries and less permeable ones. Also, while cylindrical and slit-pore models predict unreasonable values for small-pore radius from hydraulic data (the 'puzzle' mentioned in Section 8.5), the fibre matrix equations produce no such incon-sistency. This is because Poiseuille's law does not apply to irregular porosities with discontinuous walls; such channels offer less hydraulic resistance for a given reflec-tion coefficient.

An observation which highlights the importance of the glycocalyx is *the protein effect*. When albumin is washed out of a capillary, P and L_p increase and σ decreases. This indicates that albumin normally helps maintain the narrowness of the notional small pores, and albumin is known to bind to glycocalyx via its cationic arginine groups. Moreover cationized ferritin, which as Figure 8.5 shows also binds to the glycocalyx, has the same permeability-reducing effect. The protein effect is thought to be due to the adherent protein molecules changing the density and spacing of the matrix (see Figure 8.13). An exciting aspect of this hypothesis is that it opens up

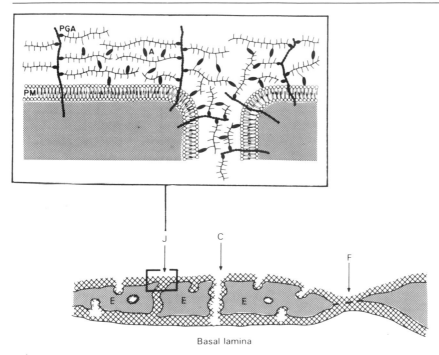

Figure 8.13 Fibre matrix model of capillary permeability. Glycocalyx covers the endothelium (E) and fenestra (F) and fills the intercellular junction (J). It also lines the vesicles. The multivesicular transcellular channels (C) is shown as a clear route, representing the large pore system; these are very few in number. (Inset) Putative details of matrix structure. Proteoglycan fibres (PGA) of radius 0.6 nm are thought to occupy approximately 2.5% of glycocalyx volume. Albumin (A) binds via its positively-charged arginine groups to form 8.5% of the matrix by volume. Molecular sieving is governed by the mesh size while the anatomical width and number of channels (J,F,C) influences total hydraulic and diffusional permeability. PM, plasmalemma membrane. (After Michel, C. C. (1980) *Journal of Physiology*, **309**, 341–355 and Curry, F. E. (1986) *Circulation Research*, **59**, 367–380, by permission)

the possibility of modifying capillary permeability therapeutically by changing the density and pattern of the fibre matrix.

Large pores

In discontinuous capillaries there are open intercellular gaps which are undoubtedly the large pores (see Figure 8.2). In healthy continuous and fenestrated capillaries the nature of the large pore is more controversial, there being three main candidates (see Figure 8.17). (1) The occasional *intercellular cleft* might open wider than normal and loosen its fibre matrix. Gaps of 50–500 nm have indeed been seen occasionally in venules but they are so rare that the investigator is never sure whether they are normal or artefact. (2) The rare *transendothelial channels* formed by fused vesicles (see Figures 8.4 and 8.17) are about the right size to serve as large pores, being 15–35 nm in radius. (3) Some form of transport by the invaginated *vesicular system* is conceivable. Macromolecules like gold-labelled albumin and ferritin (radius 5.5 nm) undoubtedly enter the luminal vesicles (see Figure 8.5) and some minutes later appear in abluminal vesicles. The mechanism might be a transient fusion of luminal and abluminal systems with an exchange of contents, but how much this contributes to net protein transport is not clear. Protein transport is certainly not an active process for it is not abolished by metabolic poisons or tissue cooling. Moreover, the increase in net protein transport when filtration pressure is increased implies that there exists a hydraulically conductive 'large pore', i.e. a continuous channel. At normal low filtration rates, however, the observed protein flux is greater than expected for conductive channels alone, so it may turn out that both conductive channels and some form of vesicular transport are important at normal capillary pressures.

8.9 Carrier-mediated transport across cerebral capillaries

Although cerebral capillaries are highly permeable to oxygen and carbon dioxide they are exceptionally impermeable to small lipophobic molecules like K^+, L-glucose, sucrose, mannitol, polar dyes, catecholamines and plasma proteins. This gives rise to the concept of a '*blood-brain barrier*' which protects the delicate neuronal circuits from interference by plasma solutes. The barrier is created by the dense complex junctional strands, which form a continuous seal around the endothelial cell perimeter (zonula occludens), and by the scantiness of the vesicular system. These structural specializations are probably induced by the surrounding astrocytes whose 'feet' cover over 80% of the abluminal surface of the cerebral capillary.

The brain's chief energy source is glucose in its natural dextro-rotatory form (D-glucose, dextrose). D-glucose, unlike the stereo-isomer L-glucose, rapidly crosses the blood-brain barrier by *facilitated diffusion*. It binds reversibly to a specific carrier protein in the endothelial cell membrane, very like the glucose carrier in red cell membranes, and this renders the capillary wall selectively permeable to D-glucose. The movement of the glucose is nevertheless a passive diffusion down the concentration gradient set up by neuronal activity, and is not an active transport. Because receptor sites are involved, glucose transport displays the classic phenomena of saturation at high concentrations, stereospecificity (D-glucose is transported but L-glucose is not) and competitive inhibition by analogues like deoxyglucose and galactose. Carriers also exist for the metabolic acids lactate and pyruvate and for adenosine. There are three distinct carriers for amino acids: one for large neutral amino acids like phenylalanine, one for anionic amino acids like glutamate and one for cationic amino acids like arginine. There is evidence also for facilitated diffusion of adenosine across the coronary endothelium.

The cerebral endothelium also seems capable of active ionic transport. If brain interstitial K^+ concentration rises due to neuronal electrical activity, K^+ is pumped out across the abluminal membrane of the endothelial cell, where a Na^+-K^+ ATPase is located (rather as in a tight epithelium). This stabilizes the level of K^+ in brain interstitium and probably explains why cerebral endothelium has five to six times as many mitochondria as muscle endothelium.

8.10 Effect of blood flow on exchange rate

To illustrate how blood flow influences solute exchange, let us consider a simple and rather artificial situation in which the pericapillary concentration is held constant.

This is in fact the situation during the first few seconds of an indicator diffusion experiment (see Figure 8.11), since the pericapillary concentration of the test solute is zero at the early times. Under these conditions the limiting factor for solute exchange can be either the rate at which plasma delivers solute (flow-limited exchange) or the rate at which the wall allows solute across (diffusion-limited exchange).

Flow-limited exchange

The permeability of a capillary to lipophilic solutes and very small lipophobic solutes is so high that the solute equilibrates with the pericapillary fluid within the plasma transit time. Consequently, exchange involves only the initial segment of the capillary (see Figure 8.14, curve F). If flow is increased, an equilibrium may still be reached before the end of the capillary, albeit a little further

downstream, and the arteriovenous concentration difference is unaltered (curve F′). From the Fick principle (Section 5.1), we know that solute transfer is direct proportional to blood flow if the arteriovenous difference is constant, and this is indeed how lipid-soluble molecules like antipyrine behave (see Figure 8.14b). A rise in blood flow is thus an important way of increasing the transfer of lipid-soluble substances, like oxygen, in an active tissue. Permeability cannot be measured when exchange is flow-limited, because the fraction of capillary wall involved in exchange is unknown. Instead, the solute transfer rate is a measure of blood flow (see Figure 8.14b, antipyrine line; and Figure 7.5).

Diffusion-limited exchange

Bigger lipophobic molecules like inulin and cyanocobalamin (vitamin B12) cross the

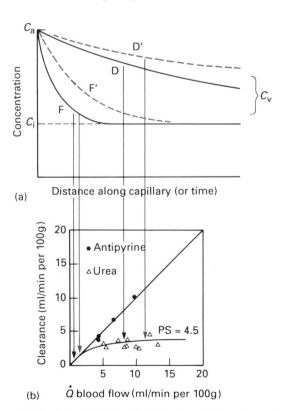

(a)

(b)

Figure 8.14 (a) Plasma concentration of a diffusible solute during a single transit along a capillary. If plasma equilibrates with interstitium (C_i) before the end of the capillary, exchange is flow limited (curve F). Raising blood flow causes more of the downstream capillary wall to participate in exchange (curve F′) and increases the transfer rate. If plasma fails to equilibrate with interstitium, exchange is limited by the diffusion barrier (curve D). Raising blood flow now limits the time for diffusion so extraction falls and venous concentration rises (curve D′): this offsets the effect of blood flow, and solute transfer barely increases. (b) Effect of blood flow on rate of solute clearance (solute transfer rate from plasma to interstitium per unit arterial concentration) in skeletal muscle. The lipid-soluble solute antipyrine undergoes flow-limited exchange; so too does urea at low flows. At flows $>10\,\text{ml/min}^{-1}\,100\,\text{g}^{-1}$, urea undergoes diffusion-limited exchange and the plateau equals PS for urea. ((a) After Michel, C. C. (1972) In *Cardiovascular Fluid Dynamics* (ed. D. Bergel), Academic Press, London; (b) From Renkin, E. M. (1967) In *International Symposium on Coronary Circulation* (1966) (eds G. Marchetti and B. Taccardi)Karger, Basel, pp. 18–30, by permission)

capillary wall too slowly to equilibrate with pericapillary fluid during the normal transit time (Figure 8.14a, curve D). The same is true for smaller molecules like glucose if the blood transit time is shortened sufficiently by raising the blood velocity. The exchange rate is then limited not by flow but by the diffusional resistance of the capillary wall. Diffusion-limited exchange is relatively insensitive to blood flow because raising the blood flow reduces the time spent in the capillary; extraction falls and venous concentration rises (curve D'). By applying the Fick principle, we find that the effect of the rise in blood flow is largely offset by the fall in Ca-Cv, and solute exchange increases relatively little. This is illustrated by the urea curve in Figure 8.14b.

The above examples show that a crucial factor in solute transfer is the ratio of the diffusing capacity of the wall (PS) to the blood flow ($\dot{Q}$). If the diffusing capacity is high relative to flow, exchange is flow-limited; conversely if flow is high relative to diffusing capacity the exchange is diffusion-limited. The concepts of flow-limited exchange and diffusion-limited exchange are actually extremes of a continuous spectrum; the whole spectrum is described by the Renkin–Crone expression (Equation 8.5), which tells us how extraction varies with the $PS/\dot{Q}$ ratio. As a rough rule exchange may be regarded as flow-limited if $PS/\dot{Q}$ is 5 or more and diffusion-limited if it is less than 1. For nutrients like glucose, $PS/\dot{Q}$ can vary between about 5 at rest (low blood flow; flow-limited exchange) and 1 during exercise (high blood flow; diffusion-limited exchange).

8.11 Physiological regulation of exchange rate

In a steady state, the rate of transfer of oxygen and nutrients across the capillary wall must keep pace with the rate of consumption by the tissue. In exercising muscle, for example, the oxygen transfer rate can increase 20–40-fold, and this is achieved by three mechanisms: capillary recruitment, an increased concentration gradient across the wall and increased blood flow.

Capillary recruitment

In skeletal muscle each capillary supplies a surrounding envelope of muscle, called a Krogh cylinder (see Figure 8.15). The radius of the Krogh cylinder depends on capillary density, which is increased by physical training or living at high altitude. A low capillary density or, equivalently, perfusion of only a fraction of the existing capillaries produces a broad Krogh cylinder and a poor oxygen tension at its periphery. In resting skeletal muscle, half to three-quarters of the capillaries are either not perfused or perfused only sluggishly at any moment,

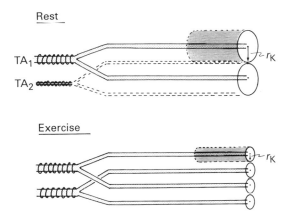

Figure 8.15 Diagram illustrating the Krogh cylinder concept and capillary recruitment in exercising skeletal muscle. At rest, contraction of terminal arteriole 2 (TA$_2$) stops the perfusion of one capillary module (dashed lines) so the Krogh cylinder radius is large (r_K). Metabolic vasodilatation during exercise dilates terminal arteriole 2, increasing the perfused capillary area and reducing the maximum diffusion distance r_K. Strictly, the Krogh 'cylinder' is really a hexagonal column

owing to vasomotion in the terminal arterioles (Section 8.1). During exercise the metabolic dilatation of terminal arterioles increases the number of well-perfused capillaries, and this not only increases the *surface area* for exchange but also reduces the *diffusion distance*, i.e. radius of the Krogh cylinder (see Figure 8.15). Diffusion distance is especially important for gas exchange, because the main diffusional resistance in the overall blood-to-tissue pathway is not at the capillary wall but in the tissues, due purely and simply to the greater length of the extravascular pathway (>10 μm). The main fall in oxygen concentration is thus not from plasma to pericapillary space but from pericapillary space to cell interior. The intracellular P_{O_2} is only approximately 5 mmHg in resting muscle, despite a blood P_{O_2} of 100 mmHg (arterial) to 40 mmHg (venous).

Tissue concentration gradient

An increased cellular metabolic rate lowers the intracellular concentration of glucose, oxygen etc. and thereby increases the concentration difference between plasma and the cell. In combination with the shortened cell-to-capillary distance, this raises the concentration gradient ($\Delta C / \Delta x$) driving solute from the plasma to the cell. For glucose, for example, the mean concentration difference across the capillary wall increases from approximately 0.3 mM in resting muscle to approximately 3 mM during heavy exercise (see Table 8.2).

Blood flow

Blood flow usually increases in proportion to an organ's metabolic rate (see Figure 8.16), and if exchange is flow-limited (e.g. oxygen transfer), the flux into the pericapillary space increases in proportion to blood flow for a given pericapillary concentration. Metabolites such as glucose and urea are flow-limited at resting blood flows but become diffusion-limited at the high flows

occurring during exercise. Once the exchange process has become diffusion-limited, further rises in blood flow have only a small benefit.

The various factors that increase glucose delivery to exercising muscle fibres are brought together in Table 8.2 and Figure 8.16.

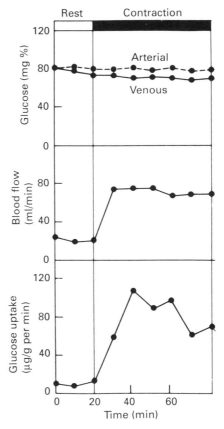

Figure 8.16 Glucose transfer rate from plasma to dog skeletal muscle (lower panel) calculated by the Fick principle from the data in the upper panels. Uptake rate = blood flow × arteriovenous concentration difference. Increased exchange in moderate exercise is achieved partly by increased blood flow and partly by increased extraction, i.e. widening of the arteriovenous difference. (From Chapler, C. K. and Stainsby, W. N. (1968) *American Journal of Physiology*, **215**, 995–1004)

Table 8.2 Transport of glucose from blood to 100 g skeletal muscle *in vivo*

	Rest	Heavy exercise	Fractional change (exercise/rest)
Glucose consumption rate (J_s)†	1.4 μmol/min	60 μmol/min	43 x
Arterial concentration (C_a)	5 mM	5 mM	–
Venous concentration (C_v)	4.44 mM	4 mM	0.9 x
Extraction (E)	11.2%	20%	1.8 x
Blood flow	2.5 ml/min	60 ml/min	24 x
Perfused capillary density	250/mm^2	1000/mm^2	} 4 x
Diffusion capacity (PS)	5 cm^3/min	20 cm^3/min	
Mean concentration difference across capillary wall (ΔC)*	0.3 mM	3 mM	10 x
Mean pericapillary concentration (C_i)	4.7 mM	2 mM	0.4 x
Krogh cylinder radius	36 μm	18 μm	0.5 x

* ΔC is calculated as J_s/PS – see Equation 8.3.
† Diffusion, and not fluid filtration, is the dominant transcapillary transport process for glucose and other small solutes. This is clear from a simple calculation based on the consumption by resting skeletal muscle (1.4 μmol/min^{-1} 100g^{-1}). Net transcapillary fluid flow is 0.005 ml/min^{-1} 100 g^{-1} and since plasma contains 5 μmol glucose per ml, the maximum convective transport of glucose is 0.025 μmol/min. This is a mere 2% of the total glucose transfer. The major process driving a metabolite across the capillary wall is thus diffusion.
(After Crone, C. and Levitt, D. G. (1984) In *Handbook of Physiology, Cardiovascular System Vol. IV, Microcirculation* (ed Renkin and Michel), American Physiological Society, Bethesda, pp. 431)

8.12 Active functions of endothelium

In addition to its passive functions as a porous membrane, the capillary wall has many active metabolic functions. (1) It secretes *structural components*: the glycocalyx and basal lamina. (2) The endothelial cell produces several *vasoactive substances*, for example, prostacyclin (PGI₂), a vasodilator and anti-platelet aggregating factor. Arterial endothelium secretes endothelium-derived relaxing factors and endothelin (see Chapter 11). An endothelial surface enzyme converts circulating angiotensin I into the vasoactive form angiotensin II, and degrades the circulating vasoactive substances bradykinin and serotonin. (3) There is evidence that *carbonic anhydrase*, an enzyme catalysing the conversion of plasma HCO_3^- to carbon dioxide, occurs as another surface enzyme in lung microvessels. (4) Another cluster of activity concerns the *clotting system*, endothelium being a producer of thromboxane and von Willebrand factor (factor VIII-related substance). Von Willebrand disease is a genetically-determined failure of endothelial cells to synthesize the haemostatic factor, resulting in a prolonged bleeding time. (5) Endothelium is also involved in the *defence against pathogens*. Venular endothelium interacts with polymorphs and lymphocytes during inflammation as the first step in white cell emigration. Moreover, endothelial cells are themselves capable of actively phagocytosing bacteria, although perhaps not of killing them.

8.13 Summary

Although the active aspects of endothelial cells are exciting great current interest, we must not lose sight of the fact that the single

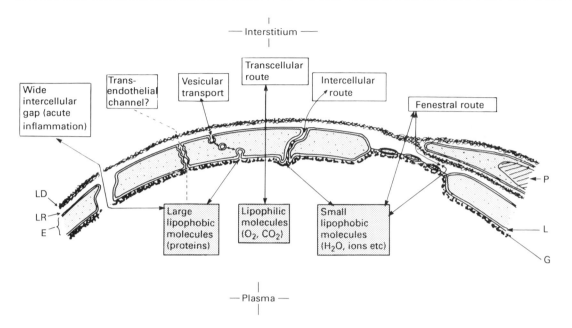

Figure 8.17 Main transport pathways across the capillary wall. The flux of small proteins through the small pore system is not shown. While water passes mainly through the small pore system as shown, some also passes directly through the cell membrane and some through the large pore system. E, endothelium; G, glycocalyx; L, lipid plasma membrane; LR, LD, lamina rara and lamina densa of basement membrane; P, pericyte. (After Levick, J. R. (1983) In *Studies in Joint Disease*, 2nd edn (eds A. Maroudas and E. J. Holborow), Pitman, London, pp. 153–240, by permission)

most important function of the endothelium is to determine the permeability of the capillary wall. The transfer of material across the capillary wall is summarized in Figure 8.17 but the concepts of John Pappenheimer and Gene Renkin, originators of pore theory, and Charles Michel and Roy Curry, developers of fibre matrix theory, deserve a less prosaic summary.

The Pore
The Pappenheimer pore's so small
You cannot make it out at all,
Though many sanguine doctors hope
To see one through the microscope –
Rectangular or round in shape
With many nanometres gape.
Some say the pore contains a fluff
Of glyco-proteinaceous stuff,
Whose fibres subdivide the space
Constructing there a random lace
'Til albumin, a protein, lands
And tidies up those tangled strands,
Through which there flows dilute sal-ine.
All this has never yet been seen,
But Scientists, who ought to know,
Assure us that it must be so.
Oh let us never, never doubt
What nobody is sure about.

(With apologies to Hilaire Belloc)

Further reading

Bradbury, M. W. B. (1985) The blood-brain barrier: Transport across the cerebral endothelium. *Circulation Research*, **57**, 213–222

Crone, C. (1984) The function of capillaries. In *Recent Advances in Physiology 10* (ed. P. F. Baker), Churchill Livingstone, London, pp. 125–162

Curry, F. E. (1986) Determinants of capillary permeability; a review of mechanisms based on single capillary studies in the frog. *Circulation Research*, **59**, 367–380

Hudlicka, O., Egginton, S. and Brown, M. D. (1988) Capillary diffusion distances – their importance for cardiac and skeletal muscle performance. *News in Physiological Science*, **3**, 134–138

Michel, C. C. (1988) Capillary permeability and how it may change. *Journal of Physiology*, **404**, 1–29

Renkin, E. M. (1985) Capillary transport of macromolecules: pores and other endothelial pathways. *Journal of Applied Physiology*, **58**, 315–325

Rippe, B. and Haraldsson, B. (1987) How are macromolecules transported across the capillary wall? *News in Physiological Science*, **2**, 135–138

Ryan, U. S., Ryan, J. W. and Crutchly, D. J. (1985) The pulmonary endothelial surface. *Federal Proceedings*, **44**, 2603–2609

Simionescu, M., Ghitescu, L., Fixman, A. and Simionescu, N. (1987) How plasma macromolecules cross the endothelium. *News in Physiological Science*, **2**, 97–100

Wissig, S. L. and Charonis, A. S. (1984) Capillary ultrastructure. In *Edema* (eds N. C. Staub and A. E. Taylor), Raven Press, New York, pp. 117–142

Chapter 9
Circulation of fluid between plasma, interstitium and lymph

Impelled by the pressure within the capillaries, fluid filters slowly across the capillary wall, passes through the interstitial space and returns to the bloodstream via the lymphatic system. The entire plasma volume (except the protein) circulates in this fashion in under a day, so the maintenance of normal plasma and interstitial volume depends directly on capillary and lymphatic function. Abnormalities of these vessels can give rise to inflammatory swelling and lymphoedema respectively, while abnormalities of the filtration forces give rise to other forms of clinical oedema.

9.1 Starling's principle of fluid exchange

Fluid movement across the capillary wall is a passive process driven by the pressure on either side of the wall. Capillary blood pressure determines filtration into the tissue, as pointed out by Carl Ludwig in 1861; and the osmotic 'suction' pressure of the plasma proteins determines absorption from the tissue, as Ernest Starling realized

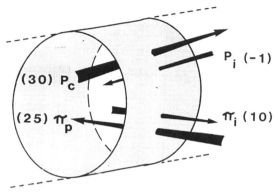

Figure 9.1 The Starling principle. The four pressures governing fluid exchange are capillary pressure (P_c), pericapillary interstitial pressure (P_i), plasma protein osmotic pressure (π_p) and pericapillary interstitial protein osmotic pressure (π_i). Numbers are pressures in mmHg above atmospheric, e.g. -1 mmHg is really $760 - 1 = 759$ mmHg; so the P_i arrow points into the capillary. Intravascular values from warm human skin at heart level. (From Levick, J. R. and Michel, C. C. (1978) *Journal of Physiology*, **274**, 97–109). Interstitial values from human subcutaneous tissue. (From Aukland, K. (1987) *Advances in Microcirculation*, **13**, 110–123)

in 1896 (see Figure 9.1). A simple experiment led Starling to advance the hypothesis that the capillary wall is a semipermeable membrane across which plasma proteins exert an osmotic pressure. He injected saline into the tissue spaces of a dog's hindlimb and found that when blood was perfused through the leg the blood underwent haemodilution. This established that

saline can be absorbed from the tissue spaces into the capillaries, and Starling further showed that the osmotic pressure of the plasma 'colloids' (i.e. proteins) was large enough to produce absorption. Following Starling's usage, the osmotic pressure of plasma protein is still called the 'colloid osmotic pressure' (or oncotic pressure). Starling recognized that it is the colloid osmotic pressure that retains water within the circulation, and this discovery led to the use of colloid solutions as plasma volume expanders for the wounded in World War I, and subsequently to the development of modern therapeutic colloids like urea-linked gelatin (Haemaccel) – a striking example of the practical benefit that can accrue from 'pure' research.

The modern form of the 'Starling principle' may be stated as follows. The net rate and direction of fluid movement (J_V) depends on the net filtration pressure across the capillary wall; the net filtration pressure is the difference between the hydraulic pressure drop and colloid osmotic pressure drop across the wall (see Figure 9.1). The drop in hydraulic pressure is the capillary pressure (P_c) minus interstitial pressure immediately outside the wall (P_i), and the osmotic pressure difference is plasma colloid osmotic pressure (π_p) minus the colloid osmotic pressure of interstitial fluid immediately outside the wall (π_i). In other words:

filtration rate α

{(hydraulic drive) − (osmotic suction)}

or in symbols

$$J_V \alpha \{ (P_c - P_i) - (\pi_p - \pi_i) \}$$

The proportionality factor here depends on the surface area of wall (S) and its hydraulic conductance (L_p; Section 8.3). Writing in the proportionality factors, we get the following equation:

$$J_V = L_p . S . \{(P_c - P_i) - (\pi_p - \pi_i)\} \quad (9.1)$$

The above expression is not quite complete, however, in that the capillary wall is not a perfect semipermeable membrane, being slightly permeable to plasma proteins. As explained in Section 8.3, the potential osmotic pressure of a solution is not fully exerted across a leaky membrane, and the reduced osmotic pressure, expressed as a fraction of the full osmotic pressure exerted by the same concentration difference across a perfect membrane, is called the reflection coefficient (σ);

$$\sigma = \Delta\pi_{\text{effective}}/\Delta\pi_{\text{ideal}} \qquad (9.2)$$

For plasma proteins, σ is typically 0.75–0.95, meaning that only 75–95% of the potential osmotic pressure difference across the wall is actually exerted. It must be stressed that this effect is not caused by the presence of protein in the interstitium, which is another important but quite distinct phenomenon. There is a reduction of the osmotic pressure exerted by the protein concentration difference whatever that difference is. Thus the osmotic pressures in the filtration equation are reduced by factor σ, and the correct expression for fluid movement is:

$$J_V = L_p S\{(P_c - P_i) - \sigma(\pi_p - \pi_i)\} \qquad (9.3)$$

This final form is called the *Starling equation* and it is central to the understanding of clinical oedema (see later). The Starling equation applies to each consecutive small segment of the capillary wall, where P_c etc. can be regarded as uniform. Over the whole length of the vessel, however, the capillary pressure changes and the implications of this are considered in Section 9.6.

Proof of the Starling principle in single capillaries

The cannulation of individual capillaries was pioneered by an American medical student, Eugene Landis in 1926, and the method has proved a powerful tool for investigating transcapillary flow. Figure 9.2 shows a modern extension of Landis' method. A single frog mesenteric capillary is cannulated and perfused with red cells

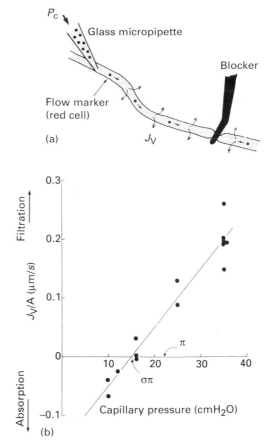

Figure 9.2 (a) Modified Landis red cell method for measuring fluid exchange in a single capillary (frog mesentery; see text). Filtration rate is calculated as red cell velocity × cross-section of filtering segment (πr^2). The process is repeated at several capillary pressures (P_c). (b) Initial filtration rate per unit wall area (J_V/A) as a function of capillary pressure. The pressure intercept at zero filtration equals the effective osmotic pressure ($\sigma\pi$, 15 cmH$_2$0) exerted *in vivo* by the albumin solution perfused from the micropipette; the perfusate COP *in vitro* (π) was 22 cmH$_2$O here. The slope equals hydraulic conductance (L_p) – 0.001 to 0.01 μm s^{-1} per mmHg^{-1} for mammalian continuous capillaries, and up to 0.1 μm s^{-1} mmHg^{-1} for fenestrated renal capillaries. (From Michel, C. C. (1980) *Journal of Physiology*, **309**, 341–355, by permission)

suspended in an albumin solution of known colloid osmotic pressure (COP). The vessel is then blocked downstream by a glass rod. If ultrafiltration occurs out of the blocked segment, red cells creep towards the block as the lost fluid is replaced from the pipette. Conversely, if absorption of interstitial fluid occurs, the red cells are pushed back towards the pipette, slowing down as haemodilution occurs. Transcapillary flow is calculated from the initial cell velocity after blockage and pressure is measured through the micropipette. Key observations are as follows. (1) The initial filtration rate is linearly proportional to capillary pressure and the steepness of the relation represents wall conductance (L_p). (2) Filtration rate is zero when capillary pressure is close to the perfusate's COP, and lowering the pressure beyond this point produces a transient absorption of interstitial fluid. This unequivocally proves Starling's hypothesis, namely, that the capillary wall acts like an osmometer membrane. (3) The capillary pressure that produces zero filtration is equal, on average, to 82% of the perfusate's COP. Since the interstitial pressures are probably negligible in the mesentery, the reflection coefficient for albumin is about 0.82 in these vessels.

Fluid exchange in whole organs

Changes in the weight or volume of a tissue are often used to assess filtration rate in whole organs, including human limbs (see Figure 9.3). If capillary pressure is suddenly raised (usually by congesting the venous outflow and waiting several minutes for blood volume to stabilize), the early filtration rate increases linearly with pressure. Conversely, if plasma COP is raised, the filtration rate decreases, in accordance with the Starling equation. (Care is needed, however, in applying the Starling equation to prolonged filtration states because prolonged filtration may alter the interstitial pressures P_i and π_i; Sections 9.6 and 9.10.) The slope relating initial tissue swelling rate

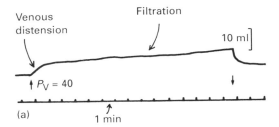

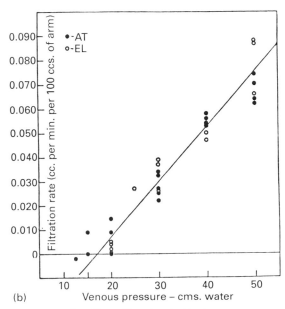

Figure 9.3 Capillary filtration in the human forearm. (a) Forearm volume measured by plethysmography (see Section 7.3). First arrow marks inflation of a venous congesting cuff around upper arm to 40 cmH$_2$O, and second arrow deflation. (b) Filtration rate (swelling rate after >2 min) at a series of venous pressures. The slope, 0.003 ml min^{-1} mmHg^{-1} venous pressure per 100 ml forearm, depends on the aggregate capillary filtration coefficient. (From the classic work of Krogh, A., Landis, E. M. and Turner, A. H. (1932) *Journal of Clinical Investigation*, **11**, 63–95)

to capillary pressure is called the capillary filtration coefficient (or capacity), and it represents the sum of the hydraulic permeabilities of all the exchange vessels within the tissue, i.e. the sum of their surface area × conductance values, $\Sigma(L_p S)$.

The capillary filtration capacity in 100 g of human forearm is 0.003–0.005 ml min^{-1} mmHg^{-1} rise in venous pressure (see Figure 9.3); in cat intestine, where the mucosal capillaries are fenestrated, it is 20 times higher.

The Starling principle relates primarily to a short segment of capillary wall over a brief period of time, since each pressure term is then invariant. For the whole capillary bed, however, capillary pressure changes with distance along the capillary, and can also vary with time, so we must next consider the pressure in a little more detail.

9.2 Capillary pressure and its control

Pressure in the exchange vessels is the most variable of the four Starling pressures, and is the only one under nervous control. It is influenced by distance along the capillary, arterial and venous pressures, vascular resistance and gravity.

Axial distance

Pressure falls by approximately 1.5 mmHg per 100 μm length of mammalian capillary owing to the vessel's hydraulic resistance. In human skin at heart level, for example, pressure falls from 32–36 mmHg at the arterial end of the capillary loop to 12–25 mmHg at the venous end; the values vary somewhat with skin temperature. The average pressure is lower in portal circulations (hepatic sinusoids approximately 6–7 mmHg, renal tubular capillaries approximately 14 mmHg) and in the pulmonary circulation (approximately 10 mmHg).

Control by the pre- to postcapillary resistance ratio

Capillary pressure must lie between arterial pressure and venous pressure but the precise value, whether closer to venous or arterial pressure, depends on the resistance of the precapillary vessels (R_a) and post-capillary vessels (R_v). If precapillary resistance is high, the capillary is well shielded from arterial pressure; the precapillary pressure drop is great and capillary pressure is close to venular pressure (see Figure 9.4). If postcapillary resistance is relatively high, the situation is somewhat like a hosepipe whose outlet is squeezed – the pressure rises until it nearly equals supply pressure, namely arterial pressure. Mean capillary pressure depends on the balance between these two effects, i.e. on the ratio of pre- to postcapillary resistance (R_a/R_v).*

The value of R_a/R_v is typically 4 or more in systemic organs, so capillary pressure is more sensitive to venous pressure than to arterial pressure. This is why venous congestion affects filtration rate so markedly (see Figure 9.3).

The pre- to postcapillary resistance ratio is actively controlled by central mechanisms (sympathetic vasoconstrictor nerves and circulating hormones) and by local mechanisms (myogenic response and tissue metabolites), as described in Chapter 11. An example of central regulation is provided by skin vasodilatation following a rise in core temperature; reduced sympathetic vasoconstrictor activity lowers R_a/R_v to about 2, which raises mean capillary pressure to over 25 mmHg (plasma COP) and enhances filtration. This is why fingers swell and rings feel tighter in hot weather. An example of local regulation is seen in the human foot during standing (see later).

Importance of gravity

Both arterial and venous pressures increase linearly with vertical distance below heart level, reaching 180 mmHg and 90 mmHg

---

* The exact relationship is easily derived, since blood flow from an artery to midcapillary point equals (P_a − P_c)/R_a by Darcy's law, and from capillary to vein it equals (P_c − P_v)/R_v. Setting these two expressions equal, we get the Pappenheimer – Soto–Rivera expression defining how capillary pressure depends on arterial and venous pressure and the pre- to postcapillary resistance ratio: mean $P_c = (P_a + P_v R_a/R_v)/(1 + R_a/R_v)$.

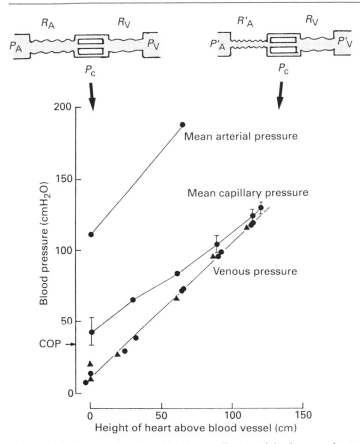

Figure 9.4 Pressure in nailfold skin capillaries of the human foot (P_c) measured by direct micropuncture with the foot at various distances below heart level. Popliteal artery pressure and dorsal foot vein pressure increase with distance below heart level in the expected fashion, but capillary pressure increases relatively less. Plasma colloid osmotic pressure (COP) is also indicated. Top inset illustrates the vasoconstrictor response that buffers the capillary pressure rise (see text). (After Levick, J. R. and Michel, C. C. (1978) *Journal of Physiology*, **274**, 97–109, by permission)

respectively in the feet of a standing man of average height (see Figure 9.4). Capillary pressure inevitably increase too, but does so by a smaller amount than either the arterial or venous pressure, because local vasoconstriction raises R_a/R_v to 20–30 and shifts the capillary pressure very close to the lower venous limit of its range. The mechanism of the local vasoconstriction may be partly myogenic and partly a local axon reflex involving sympathetic nerve terminals. Even so, capillary pressure reaches approximately 95 mmHg in the motionless dependent foot and exceeds plasma COP throughout most of the lower body in the upright position.

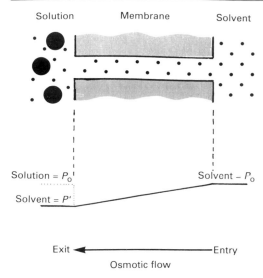

Figure 9.5 Osmotic flow between a solution (left) and solvent (right) exposed to equal pressure, P_o (atmospheric pressure). Owing to the presence of solute the 'partial pressure' P' of solvent within the solution (i.e. its energy level) is less than P_o (see text), and is thus lower than on the pure solvent side (P_o). This sets up a hydraulic pressure gradient within the pore which produces an osmotic flow of solvent into the solution. (After A. Mauro (1981) In *Water Transport across Epithelia* (eds Ussing, H. H., Bindslev, N., Lassen, L. A. and Sten-Knudsen, D.) Munksgaard, Copenhagen, pp. 107–110)

9.3 Colloid osmotic pressures of plasma

How does osmotic pressure arise?

The process of osmosis across a semipermeable membrane is illustrated in Figure 9.5. In a solution under atmospheric pressure, the pressure exerted on an imaginary plane within the liquid arises partly from random bombardment by solvent particles and partly from random bombardment by solute particles. Since these effects together add up to atmospheric pressure, it follows that the pressure exerted by the solvent is less than atmospheric pressure. (The situation is somewhat like that in a gas mixture where each gas has a partial pressure less than the total pressure.) The solute thus lowers the energy level of the solvent. Since the solute cannot enter the pore of a semipermeable membrane the pore contains pure solvent. If there is pure solvent on the opposite side of the membrane at atmospheric pressure, a pressure gradient exists between the pore exit (solution side, solvent at less than atmospheric pressure) and pore entry (solvent side, solvent at atmospheric pressure), and this drives solvent through the pore. The intrapore pressure gradient can be cancelled out by raising the solution's hydrostatic pressure or by lowering pressure on the solvent side. Flow then ceases, and the hydrostatic pressure required to produce this equilibrium is called 'osmotic pressure'. It should be noted that, contrary to the common belief that water moves through the membrane by diffusion during osmosis, it actually flows hydraulically along an intrapore pressure gradient: experiments confirm that osmotic flow obeys a hydraulic law, not Fick's law of diffusion.

Total osmotic pressure of plasma versus *colloid osmotic pressure*

Osmotic pressure is a 'colligative' property, like freezing point depression, which means that it depends on the number of particles in solution but not on their chemical identity. The osmotic pressure (π) of an 'ideal' dilute solution is described by van't Hoff's law, namely $\pi = RTC$, where R is the gas constant, T the absolute temperature and C the molar concentration of particles: RT is 25.4 atmospheres per mole/litre at 37°C. Since plasma contains about 0.3 moles of particles per litre, mostly in the form of sodium, chloride and bicarbonate ions, the van't Hoff osmotic pressure is enormous (7.6 atmospheres or 5800 mmHg). However, this *potential* osmotic pressure is simply not exerted across the capillary wall because (1) the high permeability of the small pore system to electrolytes (except in the brain) quickly establishes an electrolyte equilibrium between plasma and interstitium, and

(2) the average reflection coefficient of the capillary wall to electrolytes is only approximately 0.1. Thus it is only the plasma proteins, at a concentration of merely 0.001 moles/litre, that exert a sustained osmotic pressure across the capillary wall.

Non-ideal nature of plasma colloid osmotic pressure

Plasma COP is 21–29 mmHg in man, corresponding to 65–80 g protein/litre. In mammals like dog, rabbit and rat the COP is lower, approximately 20 mmHg, and in amphibia it is only approximately 9 mmHg. Albumin comprises only half of the plasma protein by weight but is responsible for two-thirds to three-quarters of the plasma COP because its molecular weight (69 000) is half that of gamma globulins (150 000) and hence its molar concentration is higher. Plasma COP thus depends on the albumin:globulin ratio as well as total protein concentration. The COP exerts a negative feedback on the rate of albumin synthesis by the liver, which accounts for the stable level of the plasma COP in a given species.

Protein COP is 'non-ideal', i.e. it considerably exceeds the COP predicted from van't Hoff's ideal law (see Figure 9.6). The excess COP is due partly to the space taken up by the voluminous protein molecules (0.7 ml/g), which increases their effective concentration, and partly to their charge. The albumin molecule carries a net negative charge of -17 at pH 7.4 and this attracts an excess of Na^+ ions into the albumin solution (the *Gibbs–Donnan effect*). These ions are confined electrostatically to the albumin side of the membrane, so they are osmotically effective and account for roughly a third of albumin's COP.

9.4 Colloid osmotic pressure of interstitial fluid

Over half the plasma protein mass in the body is actually in the interstitial compartment (16% body weight) rather than in the

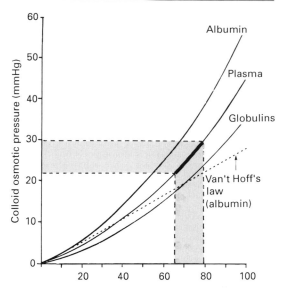

Figure 9.6 Osmotic pressure of human plasma proteins in isotonic saline (pH 7.4, 37°C). Shaded region shows the normal range for human plasma. Dashed line is the van't Hoff law prediction for albumin. The actual COP deviates considerably from van't Hoff, and is given, in mmHg, by polynomial equations:

albumin $\pi = 0.28C + 1.8.10^{-3}C^2 + 1.2.10^{-5}C^3$
plasma $\pi = 0.21C + 1.6.10^{-3}C^2 + 0.9.10^{-5}C^3$
globulins $\pi = 0.16C + 1.5.10^{-3}C^2 + 0.6.10^{-5}C^3$

where concentration C is g/l. (From Scatchard, *et al.* summarized by Landis, E. M. and Pappenheimer, J. R. (1963) In *Handbook of Physiology* 2, Circulation III, (eds W. F. Hamilton and P. Dow, American Physiological Society, Washington, pp. 961–1034)

smaller plasma compartment (4% body weight). The whole-body average interstitial concentration is estimated to be 20–30 g/litre (see Figure 9.14), corresponding to a COP of 5–8 mmHg, but the value varies greatly between tissues. To assess the protein levels in individual tissues, the interstitial fluid can be collected directly by inserting a nylon wick into the tissue and allowing the fluid to equilibrate with it. Alternatively, since interstitial fluid and

Table 9.1 Starling pressures in human subcutaneous tissue (mmHg)

	Normal subjects	Congestive cardiac failure	
Chest			
Plasma COP	26.8	23.3	
Interstitial fluid COP†	15.6	10.5	
Interstitial fluid pressure	–1.5	–1.4	
Ankle			
Plasma COP	26.8	23.3	
Interstitial fluid COP†	10.7	3.4	} mild
Interstitial fluid pressure	0.1	0.4	} oedema

COP = colloid osmotic pressure
Arterial plasma from the arm. Interstitial fluid from a soaked wick. Interstitial pressure by wick-in-needle method. (From Noddeland, H., Omvik, P., Lund-Johansen, P. *et al.* (1984) *Clinical Physiology*, **4**, 283–297.)
† These values for subcutaneous tissue are higher than those implied in the text by the protein concentration in skin lymph (15–20 g/litre, COP 3–5 mmHg). Whether these differences reflect difference is sampling site or methodology is unclear.

prenodal lymph probably have the same composition lymph duct can be cannulated for fluid collection. Human skin lymph contains 15–20 g/litre of plasma protein, skeletal muscle lymph 13–33 g/litre, intestinal lymph 30–40 g/litre and lung lymph 40–50 g/litre. These protein levels represent between 23% (skin) and 70% (lung) of the plasma concentration. Interstitial COP is therefore *far from negligible*, and substantially reduces the absorptive force into plasma ($\pi_p - \pi_i$: see Table 9.1). Moreover, there is some evidence that the protein concentration in the pericapillary region of venous capillaries is even higher than the average interstitial level, further impairing the absorptive force at the venous end of the microcirculation.

Effect of fluid filtration on interstitial protein concentration

Despite a continuous flux of plasma proteins across the capillary wall (Section 8.7), the concentration of plasma proteins in interstitial fluid is lower than in plasma; this is because there is a simultaneous input of water from the capillaries. In the steady state, the interstitial protein concentration (C_i), and therefore interstitial COP, depends on the rate of protein arrival (i.e. mass m per unit time t; flux J_S) relative to the rate of water arrival (volume V per unit time t; flow, J_V):

$$C_i = \frac{m/t}{V/t} = \frac{J_S}{J_V} \quad (9.4)$$

The concept that concentration is set by the ratio of two fluxes is illustrated in Figure 9.7a. If water filtration were to cease altogether, the interstitial concentration would gradually rise to plasma level due to diffusion (see initial point in Figure 9.7b). When capillary pressure and filtration rate are increased, interstitial protein concentration falls because the water flow increases more than the protein flux. (It is true that protein flux increase too, due to increased convective transport, but this increase is not as great as the rise in water flow owing to the reflection of protein by the small pore system.)

There is a limit to the decline in interstitial protein concentration and COP and this limit is set by the average reflection coeffi-

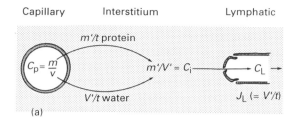

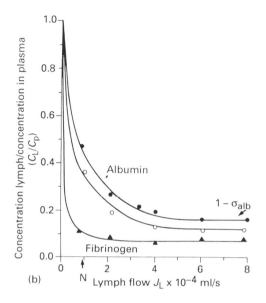

Figure 9.7 Effect of capillary fitration rate on interstitial protein concentration (and thus on interstitial COP). (a) Mass of protein transferred by all processes into the interstitium in a given time (m'/t or J_s) is diluted by the volume of capillary filtrate over the same period (V'/t or J_v) to form interstitial fluid (concentration C_i). This drains away as lymph ($C_L = C_i$). (b) Effect of net filtration rate (equal to lymph flow, J_L) on lymph/plasma concentration ratio (C_L/C_p) in dog paw. Filtration rate was varied by venous congestion. N is normal value. Curves are for albumin (●, radius 3.55 nm), γ-globulin (○, radius 5.6 nm) and fibrinogen (▲, radius 10 nm). At high flows C_L/C_p falls to a limit, namely 1-σ. (From Renkin, et al. (1977) plotted by Curry F. E. (1984) *Handbook of Physiology Cardiovascular System, Vol. IV, Part II Microcirculation*, (eds E. M. Renkin and C. C. Michel), American Physiological Society, Bethesda, pp. 309–374, by permission)

cient of the capillary wall to plasma proteins. At high filtration rates protein transport by diffusion or vesicles is relatively negligible and macromolecular sieving is maximal, so the ratio of interstitial concentration to plasma concentration (C_i/C_p) equals the non-reflected fraction of solute (1 − σ).*

If for example, the reflection coefficient is 0.9 (90%), then 10% of the solute is not reflected and the minimal value of C_i/C_p is 0.1. This approach is often used to assess σ.

We see, then, that interstitial COP is not only a determinant of filtration rate (Equation 9.3), but also a function of filtration rate (Equation 9.4). This has important physiological consequences, as explained in Sections 9.6 and 9.10.

9.5 Nature of the interstitial space and interstitial pressure

Fibrous interstitial meshwork

To understand the fourth Starling term, interstitial fluid pressure, we must recognize that the interstitial space is not simply a pool of liquid, but has a complex fibrous structure (see Figure 9.8). The space is intersected by periodic collagen fibrils of diameter 20–50 nm (collagen types I and III) and microfibrils of diameter approximately 10 nm (collagen type VI), and the interfibrillar spaces are themselves subdivided by a class of fibrous molecule called glycosaminoglycan (GAG). These are long-chain polymers of amino sugars, and the chief types are hyaluronate, keratan sulphate,

* This is easily proved. If transport in the filtration stream is increased to the point where diffusion and vesicular transport are by comparison insignificant, then the protein transport rate (J_s) equals filtration rate (J_v) × plasma concentration (C_p) × the non-reflected fraction (1 − σ). Substituting $J_v C_p$ (1 − σ) for J_s in Equation 9.4, we get $C_i/C_p = 1 − σ$.

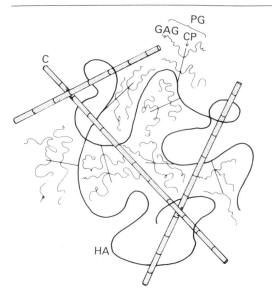

Figure 9.8 Organization of interstitial matrix. Fixed components include collagen fibrils (C, large rods), hyaluronate molecules (HA, long unbranched heavy line) and proteoglycans (PG), composed of a protein core (CP, thin straight lines) and sulphated glycosaminoglycan side chains (GAG, thin curly lines). Microfibrils and glycoproteins are not shown. In most tissues, the density of GAG is greater than shown here, so the inter-molecular spaces are smaller; these spaces are where the interstitial fluid is located. (From Granger, H. J. (1981) In *Tissue Fluid Pressure and Composition* (ed. A. R. Hargens), Williams & Wilkins, Baltimore, pp. 43–96, by permission)

dermatan sulphate, heparan sulphate and chondroitin sulphate. The sulphated GAGs are up to 40 nm long, whilst hyaluronate is several micrometres long. The sulphated GAGs are covalently bound at one end to a linear protein core, itself up to 400 nm long, to create an enormous brush-shaped molecule called a proteoglycan. The molecular weight of a proteoglycan can be up to 2.5 million in cartilage. The proteoglycans are immobilized by attachment to the long strands of hyaluronate (M_w 1–6 million) and by entanglement within the scaffolding of collagen fibrils; the whole system forms a

3-dimensional fibrous network. The sulphate and carboxyl groups of the GAGs represent fixed negative charges and this has an important effect on interstitial pressure (see later).

Near-immobilization of water

Interstitial fluid occupies the minute spaces within the meshwork of fibrous molecules, and the average effective width of these spaces ranges from a mere 3 nm (cartilage) to 30 nm (Wharton's jelly). The resistance to flow through such tiny spaces is very high and as a result the interstitium behaves much like a gel; Wharton's jelly in the umbilical cord is a classic example. Cells are not so much 'bathed' in interstitital fluid (a traditional but misleading metaphor) as 'set' in a gel. The gel is functionally important in (1) preventing a flow of interstitial water down the body under the drag of gravity, (2) impeding bacterial spread, and (3) influencing the pressure-volume curve of the interstitial space (Section 9.7).

Meaning of 'interstitial fluid pressure'

The term 'interstitial fluid pressure' is not quite as straightforward as it sounds because the energy level of the interstitial liquid depends not just on mechanical pressure but also on the influence of the interstitial GAGs, and both these effects are rolled into one in the Starling term 'interstitital pressure' (P_i). Wharton's jelly nicely illustrates the influence of the GAGs: if a slice of the jelly is brought into contact with saline at atmospheric pressure it imbibes the saline and swells. The swelling tendency is due to the osmotic pressure of the trapped GAGs, which in turn is caused mainly by their fixed negative charges exerting a Gibbs–Donnan effect (explained in Section 9.3). If a subatmospheric pressure is applied to the fluid in contact with the gel, imbibition can be halted, and the pressure which produces an equilibrium is called the 'gel swelling pressure'.

With the aid of this background information, we can understand what physiologists traditionally call the 'interstitial fluid pressure'. If we imagine a GAG-free fluid phase, such as plasma ultrafiltrate, brought into contact with the interstitium, the pressure which must be applied to the free fluid to prevent it either flowing into the interstitium or drawing fluid out of it equals the Starling term 'interstitial fluid pressure'. In other words, it is the pressure at which a free fluid phase has the same potential

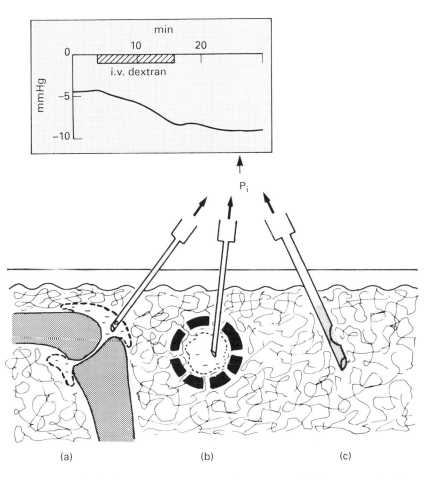

(a)   (b)   (c)

Figure 9.9 Methods for assessing interstitial pressure. (a) Where a pool of free interstitial fluid exists naturally, as in a joint cavity, pressure can be measured via a hypodermic needle. (b) Guyton's capsule, a chronic method (see text). (c) Wick-in-needle, an acute method developed by Fadnes, Reed and Auckland; saline between the nylon filaments in the tip equilibrates with interstitital fluid via the side hole. The multiple channels within the nylon wick facilitate pressure transmission. *Inset* illustrates how pressure in the giant interstitial cavity of a knee grows more subatmospheric when capillary absorption is induced by a high plasma COP (i.v. dextran infusion over hatched period; rabbit) (From Levick, J. R., (1983) In *Studies in Joint Disease 2*, (eds A. Maroudas and E. J. Holborow), Pitman, London, pp. 153–240)

energy as the fluid within the interstitial matrix.

Measurement of interstitial fluid pressure

Because fluid mobility is so low in the interstitium, the fluid is exceedingly slow to equilibrate with saline in an inserted hypodermic needle, making pressure measurement difficult. Guyton overcame this problem in 1960 by creating an artificial pool of free fluid, which was allowed to equilibrate with the surrounding interstitium for several weeks (see Figure 9.9). The artificial pool was formed by surgically implanting a hollow perforated capsule under the skin, and leaving it to fill with fluid. After several weeks interval, a hypodermic needle was inserted through one of the wall perforations to measure the pressure of the intra-capsular fluid, which was assumed to be in equilibrium with the surrounding interstitial fluid. Intra-capsular pressure was found to be subatmospheric (around −5 mmHg), and this discovery evoked a lively controversy since interstitial pressure was formerly regarded as slightly positive. Guyton pointed out, however, that sub-atmospheric pressures could explain why untethered skin hugs concave surfaces such as the anatomical snuffbox (base of first metacarpal) and ankles. Later, more rapid techniques such as the wick-in-needle method (see Figure 9.9 legend for details) have confirmed that the equilibrium pressure is indeed slightly subatmospheric in many tissues and in interstitial spaces like the joint space and epidural space (−1 to −3 mmHg). This accords with what we now know about the imbibition pressure exerted by the interstitial matrix. The maintenance of the subatmospheric pressure in the face of a continual input of fresh capillary ultrafiltrate is probably due to suction of fluid from the tissues by lymphatic vessels (see later). Pressure is above atmospheric, however, in encapsulated organs like the kidney (+1 to +10 mmHg), in certain muscles, myocardium, bone marrow and flexed joints.

9.6 Imbalance of Starling pressures: filtration and absorption

There is still much to be learned about the 'fine tuning' of the Starling forces *in vivo*. It is certain, however, that the balance varies greatly between tissues: contrast for example, the dependent foot where capillary pressure is far higher than plasma COP and the lung where pressure is much lower than plasma COP.

The filtration fraction

Virtually all tissues form lymph, including the lung, proving that there is normally a net filtration of fluid out of the microcirculation. The fraction of plasma filtered per transit (the filtration fraction) is actually very small in most tissues, around 0.2–0.3%, but since 4000 litres or so of plasma pass through an adult subject's microcirculation each day, a large volume of fresh interstitial fluid is generated over the course of a day, probably about 8–12 litres. Lymphatic drainage ensures that the interstitial volume remains normal despite this large input. In certain tissues, such as the renal glomerulus, salivary gland and dependent foot the filtration fraction is around a hundred times higher, being 20% in glomerular capillaries.

Critical capillary pressure for filtration/ absorption

Figure 9.10 shows some typical pressures along a systemic capillary; the values apply to mammalian muscle, mesentery or warm skin at heart level. Plasma colloid osmotic pressure does not change significantly because the filtration fraction is small, but capillary pressure falls progressively with distance. We can work out the critical capillary pressure at which filtration would

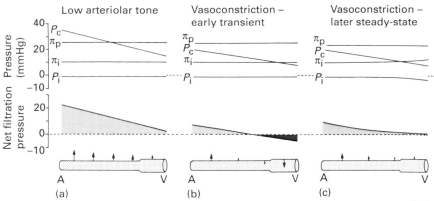

Figure 9.10 Change in pressures and fluid exchange with axial distance; symbols as in text. The shaded net filtration pressure equals $(P_c - P_i) - \sigma(\pi_p - \pi_i)$, and σ is taken to be 0.9. Bottom sketches represent the exchange vessels from arterial end of capillary (A) to pericytic venules (V) and arrows indicate direction of fluid exchange. Values are typical of skin, muscle and mammalian mesentery. (a) Well-perfused capillary. (b) Immediate effect of precapillary vasoconstriction, before interstitial forces have had time to change. (c) Later, during vasoconstriction. The transient absorption period has caused π_i to rise, P_i to fall and π_p to fall slightly, re-establishing a steady state in which there is again filtration throughout the vessel

cease (the nul-flux or isogravimetric pressure) by inserting values for π_p, π_i, P_i and σ into the Starling equation and setting J_V to zero. For the values in Figure 9.10a (and taking $\sigma = 0.9$), the nul-flux pressure works out to be 12.5 mmHg. Pressure does not actually fall as low as 12.5 mmHg until the larger venules are reached, however, and is well above this nul-flux value in the pericytic venules of many tissues. It therefore seems that well-perfused systemic capillaries are commonly in a state of filtration over virtually their whole length, contrary to earlier views.

The transience of fluid absorption

The flow of thoracic duct lymph is quite low, at most 4 litres per day in an adult, and this leads physiologists to believe that some of the capillary filtrate is re-absorbed directly into the blood stream. When and where this absorption takes place is, however, not

well understood. Textbooks have traditionally depicted fluid filtration out of the arterial half of the capillary and reabsorption into the venous half, because P_c falls below π_P in the venous segment (the Landis model of exchange). But as shown in Figure 9.10a, this view is untenable for well-perfused capillaries when modern measurements of interstitial COP and interstitial pressure are taken into account. Vasoconstriction, however, can reduce the capillary pressure sufficiently for the Landis filtration-absorption model to develop transiently (see Figure 9.10b) but even then the downstream absorption cannot be maintained because the absorption process raises the interstitial protein concentration and π_i, and lowers P_i, and these changes gradually abolish the net absorptive force (see Figure 9.10c); the absorption fades away with time. Both theory and experiment indicate that reabsorption cannot be sustained by capillaries with a finite permeability to plasma protein, (see Figure 9.11) except under

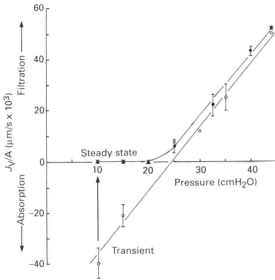

Figure 9.11 Proof of transience of fluid absorption when capillary pressure is lowered. Flow across unit wall area (J_V/A) is plotted against capillary pressure; modified Landis red cell method of Figure 9.2; perfusate COP = $32\,cmH_2O$. When the capillary is perfused with the high-COP test solution for only a few seconds prior to blockage, the COP gradient across the wall is close to that of the perfusate. Under these conditions, labelled 'transient' (open circles), a capillary pressure below $25\,cmH_2O$ produces absorption of fluid. When, however, the capillary was perfused with the high-COP solution for a long period (2–5 min) at a low pressure, to establish a new steady state prior to blockage, absorption was found to have ceased (filled circles, steady-state situation). This is because interstitial COP had increased during the absorption transient (see Figure 9.7). (From Michel, C. C. and Phillips, M. E. (1987) *Journal of Physiology*, **388**, 421–435, by permission)

special conditions (see later). The reason is that the normal low value of pericapillary protein concentration is entirely dependent on the filtration process as illustrated in Figure 9.7. If filtration ceases, protein accumulates outside the wall and progressively reduces the COP difference upon which absorption depends, until finally absorption ceases.

It is probable therefore that reabsorption occurs transiently in most tissues during periods of arteriolar vasoconstriction, and occurs chiefly at the venous end of the microcirculation. Fortunately (from the point of view of fluid balance), the absorption rate across the venular walls is enhanced by their greater hydraulical conductance, which is another facet of the arteriovenous gradient of permeability illustrated in Figure 8.9. Thus fluid exchange downstream may oscillate between periods of filtration and transient reabsorption. Another way of looking at this is to say that the functional *mean* capillary pressure, averaged over both time and axial distance, is much lower than the pressure recorded directly in well-perfused capillaries: indeed, the functional average is calculated to be very close to venous pressure, so there is only a very small net filtration force (perhaps as little as 0.5 mmHg) and a low lymph flow.

Fluid exchange in the low-pressure pulmonary circulation

The lung is an important example of the steady state reached when capillary pressure (approximately 10 mmHg) is less than plasma COP (25 mmHg). Although these values might suggest, superficially, that lung capillaries are absorbing fluid, the lung in fact produces lymph, showing that the capillaries have a net filtration pressure. The cause of the net filtration pressure is that lung interstitial fluid has a high protein concentration (approximately 70% plasma level) with a COP of 16–20 mmHg, which greatly reduces the net absorptive force. The flat part of the steady-state line shown in Figure 9.11 (from an experiment on mesenteric capillaries) is probably quite a good qualitative representation of the normal situation in the lung.

Tissues where sustained absorption occurs

Although reabsorption of filtrate from a 'closed' interstitium (one with no other fluid input than capillary filtrate) can only be transient, continuous absorption certainly occurs in renal tubular capillaries, in intestinal mucosal capillaries and probably in lymph node capillaries. This is possible because the interstitial space in these tissues has a second input of fluid, in the form of renal tubule absorbate, gut lumen absorbate or lymph respectively. This fluid flushes the interstitial space and prevents the accumulation of interstitial plasma protein. In the cat intestine, 80% of the intestinal absorbate is absorbed into the microcirculation while 20% acts as flushing solution and drains into the lymphatic vessels; in the kidney the corresponding figures are 99% and 1%.

9.7 Fluid and protein movement through the interstitial matrix

Ten to twelve litres of fluid occupy the interstitial compartment of a 70 kg man and act as a reservoir for the smaller plasma compartment (3 litres). If the plasma volume is reduced by a haemorrhage, some of the interstitial fluid is absorbed to top up the plasma; conversely, if plasma volume is increased by renal fluid retention or overinfusion, the excess fluid can 'spill over' into the interstitium, raising the interstitial volume and therefore pressure.

The interstitial pressure-volume curve and oedema

Figure 9.12 shows the effect of fluid volume on pressure in subcutaneous interstitium. This tissue is of special interest because it is where peripheral oedema accumulates chiefly. In normally hydrated interstitium, the fluid pressure is slightly subatmospheric, and small changes in volume affect the pressure markedly. The ratio of volume change to pressure change is called 'compliance', and normal interstitial compliance is rather small. This is because any removal

of water raises the GAG concentration, which makes the gel swelling pressure more negative and opposes further volume change. The same mechanism operates in reverse during fluid addition but its range of action is then more limited because a point is soon reached where the swelling pressure is negligible and interstitial pressure is close to atmospheric. Beyond this point, the accumulation of fluid is no longer opposed by changes in GAG swelling pressure: the only opposition comes from the stretching of the collagen and elastin network. Since the subcutaneous network is rather loose and the overlying skin highly distensible, this opposing force is small; moreover, it decays with time (a process called stress relaxation or delayed compliance). Interstitial compliance therefore increases rather abruptly just above atmospheric pressure, to around 20 times normal. Large volumes of fluid then accumulate with little opposing rise in pressure, and this creates pools of freely mobile liquid – a condition called oedema. In an oedematous leg, about 98% of the excess fluid is found in the subcutaneous plane and its pressure is only just above atmospheric (see Table 9.1). In tissues bounded by an inelastic fibrous capsule, such as the anterolateral muscle compartment of the leg, the pressure-volume curve is steeper and interstitial pressure can reach higher levels.

Interstitial conductivity at normal hydration and in oedema

Although water normally constitutes 65–99% of interstitium by weight (depending on the tissue), it is not easily displaced owing to the low hydraulic conductivity of the interstitial matrix (Section 9.5). The hydraulic conductivity of the interstitial matrix depends on the ratio of the fractional water content (the void volume fraction or 'porosity', ϵ) to the surface area of the fixed proteoglycan and collagen fibres (S), which are the source of hydraulic resistance. The ratio ϵ/S is called the mean hydraulic radius,

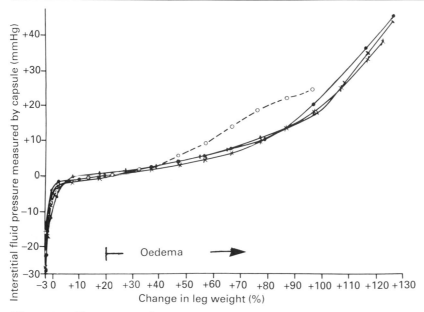

Figure 9.12 The interstitial compliance curve. Pressure was recorded in a subcutaneous capsule in four dog hindlimbs (mean − 6 mmHg). Changes in interstitital volume were assessed by changes in leg weight. Absorption was induced by hyperosmotic dextran solution i.v. and oedema by perfusing with saline at raised venous pressure. Clinical oedema was detected at +20% leg weight, corresponding to an estimated 300% rise in subcutaneous fluid volume. (From Guyton, A. C. (1965) *Circulation Research*, **16**, 452, by permission)

and it ranges from 30 nm in Wharton's jelly down to 3 nm in articular cartilage. At normal hydration, interstitial conductivity is between 10^{-11} and 10^{-13} cm^4/s per dyne (depending on tissue), and this is sufficient to allow capillary ultrafiltrate to percolate slowly through the matrix to the lymphatic vessels, under a small pressure gradient. It has also been suggested that some flow may pass preferentially along pathways of low GAG concentration, but this is a controversial matter (see later reference to 'gel chromatography' effect).

Interstitial conductivity increases dramatically with hydration especially if pools of free fluid form (oedema). This is illustrated by the clinical test for subcutaneous oede-ma, the *pitting test*, which is essentially a test of fluid mobility. When a finger is pressed firmly onto normal skin for a minute and then withdrawn, no impression is left behind because interstitial fluid mobility is very low and little fluid is displaced. In oedematous tissue, however, finger pressure leaves behind a distinct pit (see Figure 9.13), indicating that free, abnormally mobile fluid had accumulated in the interstitium.

Non-pitting forms of oedema also exist. The commonest of these is chronic lymphoedema (Section 9.10), in which the tissue reacts to high plasma protein levels in the oedema fluid by synthesizing collagen and fat. This produces a fibrotic kind of

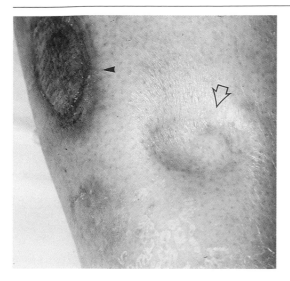

Figure 9.13 Photograph of back of calf (the ankle is off lower edge of picture), to show pitting oedema (arrow), in a patient with cardiac failure. The oedema was exacerbated by immobility and dependency. Note the skin damage (arrowhead, top left) after an oedema blister had stripped the epidermis from the dermis. (Courtesy of Dr P. Mortimer, Department of Dermatology, St. George's Hospital, London)

oedema that does not pit easily ('brawny' oedema).

Protein transport through the interstitium

Small solutes like glucose move easily through the interstitial proteoglycan meshwork, and the diffusional resistance to such solutes is simply related to pathlength (typically 10–20 µm). Larger solutes like albumin, however, experience a modest restriction and steric exclusion when diffusing through the interstitial matrix (see Section 8.3 for explanation of these terms). Ogston and others have shown that such phenomena also occur in GAG meshworks *in vitro* because the irregular interfibre spaces are small enough to partially exclude protein molecules and impede their diffusion. The greater the GAG concentration

the smaller the average interfibre distance and the greater these effects. Indeed, albumin is almost totally excluded from cartilage, the densest of all interstitia. In the less dense interstitia of subcutis and muscle, albumin is excluded from 20–50% of the water volume.

Convective transport (wash-along or 'solvent drag') by the stream of capillary filtrate contributes to protein transport through the interstitial space too, but the relative importance of diffusion and convection is not clear. If preferential flow pathways exist (see earlier), convection would predominate in such channels. Large molecules would diffuse laterally out of the channels into the adjacent matrix less rapidly than small molecules, so that large molecules should have a shorter transit time from capillary to lymph – rather as in a gel-chromatography column. Some laboratories have found evidence that this may occur *in vivo*.

9.8 Lymph and the lymphatic system

The anatomy and transport functions of the lymphatic system were explored in the seventeenth century by the Swedish biologist Rudbeck (after whom our yellow-flowered garden annual Rudbeckia is named) and others. The following functions are recognized.

1. Preservation of fluid balance. Lymph vessels return capillary ultrafiltrate and plasma proteins to the bloodstream at emptying points into the neck veins, and some fluid also returns to the blood in the lymph nodes. This completes the extravascular circulation of fluid and protein (see Figure 9.14) and secures the homeostasis of tissue volume. Since the plasma volume circulates in this fashion in less than 24 h, any impairment of lymphatic function leads to a severe protein-rich oedema. The brain and eye, which lack a normal lymphatic system,

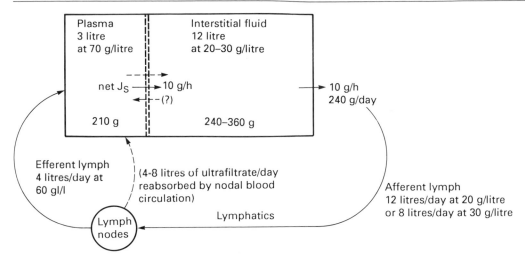

Figure 9.14 Estimate of extravascular circulation of fluid and plasma protein in a 65 kg human. (From Renkin, E. M. (1986) *American Journal of Physiology*, **250**, H706–710, by permission)

have unique fluid-draining systems, namely the arachnoid granulations and cribriform-nasal route for cerebrospinal fluid, and the canal of Schlemm for aqueous humor.

2. Nutritional function. Intestinal lymph vessels (lacteals) absorb digested fat in the form of tiny globules or 'chylomicra' and transport them to the plasma.

3. Defence function. Fluid draining out of the interstitial compartment carries with it foreign materials such as soluble antigens, bacteria, carbon particles etc. These are carried in afferent lymph to the lymph nodes scattered along the drainage route (see Figure 9.15), providing an effective and economical method for the immunosurveillance of virtually the whole body. Particulate matter is filtered out and phagocytosed in the nodes (hence the black mediastinal nodes of smokers and coal miners). Antigens stimulate a defensive lymphocyte response and the activated lymphocytes and plasma cells enter the efferent lymph (which has a much higher cell count than afferent lymph) for transport to the circulation.

Structure

Lymphatic capillaries

The lymphatic system begins as a set of lymphatic capillaries (terminal lymphatics, initial lymphatics) which are either blind terminal sacs, as in intestinal villi, or else an anastomosing network of tubes of diameter 10–50 μm (see Figure 9.15). The very thin wall consists of a single layer of endothelial cells resting on an incomplete basement membrane. Some of the cell junctions are 14 nm wide or more, so lymphatic capillaries are highly permeable to plasma proteins and even to particulate matter like carbon. The junctions run very obliquely and may function like flap valves, allowing fluid to enter readily but closing to prevent egress when lymph pressure rises above interstitial pressure (see Figure 9.16). The outer surface of the wall is tethered to the surrounding tissues by radiating fibrils, the 'anchoring filaments', which may help dilate the vessels in oedematous tissue.

Collecting vessels and afferent lymph trunks

Lymphatic capillaries unite to form collecting vessels which feed into the afferent

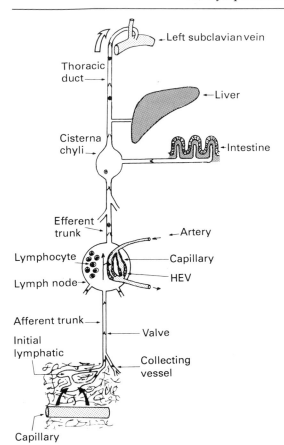

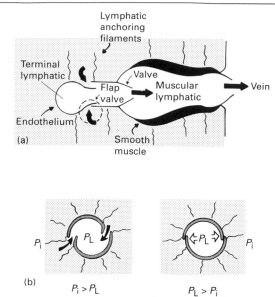

Figure 9.15 Sketch of principal elements of the lymphatic system (not to scale). HEV = high endothelial venule, where lymphocytes re-enter the node from blood. Arrows within node indicate absorption of some water by nodal capillaries

Figure 9.16 Simplified model for lymphatic transport. (a) Interstitial fluid enters the initial lymphatic down a pressure gradient. Each muscular segment or 'lymphangion' pumps lymph into the next one and ultimately into the venous system. (b) Proposed operation of endothelial junctions in the initial lymphatics as flap valves. P_i = interstitial pressure, P_L = lymph pressure. (From Granger, H. J. *et al.* (1984) In *Edema*, pp. 189–228, see Further reading list, by permission)

lymph trunks alongside major vascular bundles like the popliteal vessels. *Semilunar valves* direct the lymph centrally. From the collecting vessels onwards the lymphatics possess a coat of smooth muscle and connective tissue, the muscle element being abundant in man and ruminants but scanty in the dog and rabbit.

Lymph nodes

Several afferent vessels drain into the hilum of each lymph node. The node is a complex cellular body containing maturing lympho- cytes in germinal centres, plus many phago- cytic cells. The afferent lymph flows through sinuses, which are endothelial tubes with frequent gaps through which mature lymphocytes can enter the lymph. If the afferent lymph presents a foreign antigen to the node, the generation and release of lymphocytes into postnodal lymph increases dramatically and some lymphocytes develop into 'plasma cells', which secrete gamma-globulin antibodies. The node is supplied with nutrients by a network of continuous capillaries, and these drain into special high-endothelial venules. Lymphocytes in the bloodstream re-enter the node by penetrating the intercellular junctions of the high-endothelial venules, thus completing their own unique circula- tion.

The major lymphatic ducts

Lymphocyte-rich efferent lymph from the lower limbs and viscera flows into a large lymphatic trunk on the posterior abdominal wall. This possesses a saccular dilatation, the cisterna chyli, which acts as a temporary receptacle for chyle, the fatty lymph arriving from lacteals during absorption of a fatty meal. The ultimate lymphatic trunk, the thoracic duct, receives around three-quarters of the body's efferent lymph and empties into the left subclavian vein at its junction with the jugular vein. The small cervical and right lymphatic trunks have much smaller flows.

How is lymph formed and propelled?

The composition of prenodal lymph indicates that lymph is simply interstitial fluid drawn from the neighbourhood of the lymphatic capillary; the concentration of electrolytes, non-metabolized solutes and plasma protein in prenodal lymph matches that in samples of interstitial fluid. But what process drives the interstitial fluid into the lymphatic capillary, where the pressure is slightly higher than interstitial fluid pressure for much of the time? The likely answer is that first the lymphatic capillary empties proximally, either because it is compressed by the surrounding tissue during movement or because of contractions of the lymphatic wall (see Figure 9.17). Then, as the vessel re-expands elastically (probably helped by the anchoring filaments) the pressure inside falls transiently below interstitial fluid pressure, setting up a pressure gradient for filling. (The process may be not unlike the filling of a fountain pen or Pasteur pipette by first squeezing the rubber bulb empty and then allowing its recoil to suck in fluid.) The presence of lymphatic valves upstream ensures that backflow does not interfere with this process.

Once formed, lymph is moved along by both intrinsic and extrinsic mechanisms.

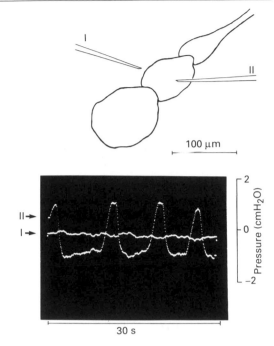

Figure 9.17 Evidence for the lymphatic suction theory of interstitial drainage in the bat wing. (a) Micropipettes were inserted into the interstitium and a contractile terminal lymphatic to assess the interstitium-to-lymph pressure gradient. (b) Interstitial pressure (I) exceeded lymph pressure (II) for 43% of the time, because lymph pressure fell to subatmospheric levels during relaxation (normalized data). Measurements of interstitial pressure at more distant sites revealed a gentle interstitial pressure gradient towards the lymphatic ($0.004\,cmH_2O$ per μm). To what extent less specialized tissues behave like the bat wing is still unclear. (From Hogan, R. D. (1981) In *Interstitial Fluid Pressure and Composition* (ed. Hargens, A. R.) Williams & Wilkins, pp. 155–163, by permission)

Intrinsic rhythmic contractions occur in lymph vessels with abundant smooth muscle, including those in the human leg, and the contraction rate is typically 10–15 per min. Animal studies show that each inter-valve segment, or 'lymphangion', functions as a rhythmic pump and the lymphangion resembles the heart in some of its characteristics: each has a pacemaker, a filling and ejection phase, a stroke volume, contractility that depends on extracellular Ca^{2+}, and even a sympathetic innervation in the larger vessels. Both the frequency and stroke volume of the lymphangion increase with lymph volume.

The chief *extrinsic propulsion* process is intermittent compression during movement. The flow of lymph from the leg of an anaesthetized dog is greatly increased by passive or active flexion, and the flow of mesenteric lymph is enhanced by intestinal peristalsis. Extrinsic propulsion is obviously important for non-contractile vessels. The lymph valves permit lymph pressure to rise stepwise in successive segments so that the lymph finally drains into venous blood at several mmHg above atmospheric pressure.

Fluid exchange in the lymph nodes

Lymph nodes like the popliteal and iliac nodes are known to modify the volume and protein concentration of lymph; postnodal lymph in the dog and sheep has up to twice the protein concentration of prenodal lymph, due mainly to absorption of water by the node's continuous capillaries. Presumably capillary pressure is less than plasma COP in such vessels and extravascular accumulation of protein is prevented by the stream of lymph. Thus postnodal lymph is not representative of either interstitial composition or formation rate. The proportion of human prenodal lymph that gets absorbed is uncertain; Figure 9.14 is based on a plausible estimate but the exact value probably varies with posture, since nodal capillary pressure must increase with dependency.

Regional differences in flow and composition of lymph

Postnodal lymph flow in the human thoracic duct averages 1–3 litres per day. The greatest producer of lymph is the liver which contributes 30–50% of thoracic duct flow (see Table 9.2), and owing to the discontinuities in hepatic capillaries, hepatic

Table 9.2 Postnodal lymph flow and composition in man

	Flow* (%)	L/P**
Thoracic duct	(1–3 litres/day)	0.66–0.69
Liver	30–49 %	0.66–0.89
Gastrointestinal	~37 %	0.50–0.62
Kidney	6–11%	0.47
Lungs	3–15 %	0.66–0.69
Limbs and cervical trunks	<10%	0.23–0.58

* expressed as percentage of total thoracic duct flow. The flows are approximate owing to the difficulty of collection and variability of flow.
** concentration of protein in postnodal lymph relative to plasma. (From Joffey, J. M. and Courtice, F. C. (1970) *Lymphatics, Lymph and the Lymphomyeloid Complex*, Academic Press, London)

lymph is particularly rich in plasma protein. Intestinal lymph flow is abundant after a meal and makes the second greatest contribution to thoracic duct flow. Renal and lung lymph flows are substantial too. The limbs contribute a variable quantity of lymph depending on the exercise level. The concentration of plasma protein in lymph varies from region to region and depends on the permeability and reflection coefficient of the tissue's exchange vessels, the molecular size and charge of the individual protein and the capillary filtration rate, as described earlier.

9.9 Regulation of fluid exchange: posture and exercise

Two physiological events which markedly disturb fluid exchange are the adoption of an upright posture and exercise.

Posture and swelling of the foot

The rise in capillary pressure below heart level, illustrated in Figure 9.4, increases the filtration rate in dependent tissues: the foot, for example, swells at an initial rate of around 30 ml/h. This is often noticed by people compelled to sit still for hours in aeroplanes and in the cinema, and many find it necessary to unlace their shoes to ease the expanded foot. Conversely, capillary pressure declines above heart level and causes a transient absorption of tissue fluid, which probably explains why earlobe thickness falls by 2–3 mm during the daytime, and why stubble growth apparently accelerates in the first few hours after rising. The overall effect of orthostasis, however, is increased filtration: plasma volume falls by 6–12% during a 40-min period of standing, with a corresponding rise in haematocrit and protein concentration. Plasma COP in students is found to increase from 25 mmHg to 29 mmHg after an 8-h day involving sitting through lectures, reading in the library etc.

The swelling of dependent tissues and decline in plasma volume would be very much worse but for several compensatory mechanisms.

1. Increase in R_a/R_v. Arteriolar contraction in the dependent tissue increases the pre- to postcapillary resistance ratio (R_a/R_v) and limits the rise in capillary pressure to about two-thirds of the rise in arterial and venous pressures (see Figure 9.4). This is a local response, not a baroreceptor reflex, and it involves the myogenic response and/or a local sympathetic axon reflex (see Chapter 11). Autonomic neuropathy in diabetic patients can abolish the local reflex and this probably explains why leg oedema is common in such patients.
2. The skeletal muscle pump (Section 7.7). Dynamic exercise reduces venous pressure to 30–40 mmHg in the calf during walking, cycling etc. and this reduces capillary pressure correspondingly. The muscle pump also enhances lymph transport.
3. Reduced blood flow. The dependent vasoconstriction referred to above re-duces not only capillary pressure but also blood flow, as shown in Figure 7.5. The low plasma flow, coupled with the increased filtration pressure, raises the filtration fraction enormously, namely to 20–27% in the foot, and the ensuing haemoconcentration raises the COP to 35–44 mmHg in the downstream capillaries. This helps to offset the increased blood pressure there. The low blood flow also causes temperature to fall, as is commonly experienced in dependent feet. There is no evidence that interstitial pressure changes significantly (see Table 9.1).

4. Reduced capillary filtration capacity. It is possible that the constriction of some terminal arterioles may stop flow completely through capillary modules for short periods, lowering the local filtration capacity. There is evidence of this in intestinal preparations, but rather conflicting evidence in limbs.

Exercise and the swelling of muscle

During exercise, local vasodilatation not only increases muscle blood flow but also increases capillary pressure by reducing R_a/R_v. At the same time, the number of perfused capillaries is increased. As a result the filtration rate rises in exercising muscle and the muscle can swell by 20% over the course of 15 min. Rock-climbers, for example, often notice a marked swelling of the forearm muscles after a 'fingery' ascent. An additional factor promoting swelling is the release of small solutes like lactate and K^+ by the active muscle fibres, which increases the interstitial osmolarity by 20–30 mOsm/litre (7–10%). Such molecules exert only a fraction of their potential osmotic pressure at the capillary wall because their reflection coefficient is only approximately 0.1, but nevertheless an osmotic pressure of 0.1×25 mOsm is 42 mmHg (van't Hoff's law, Section 9.3) and this is a substantial force enhancing filtration. The osmotic pressure of such small solutes is probably exerted mainly across the endothelial cell membrane which has a small but finite permeability to water, rather than across the small pore system.

When exercise involves the whole body, the increased transcapillary filtration can reduce the plasma volume by 16–20%. The decrease in plasma volume by up to 600 ml (man) is actually less than the increase in muscle volume, which can rise by 1100 ml during strenuous cycling. This difference is due to a compensatory absorption of interstitial fluid into plasma from non-exercising tissues, which minimizes the fall in blood volume.

9.10 Oedema

Oedema is an excess of interstitial fluid, and in clinical practice the two commonest sites for oedema are the subcutaneous plane (peripheral oedema) and the lungs (pulmonary oedema).

In *subcutaneous oedema*, the volume of fluid may be regarded as 'excessive' when it moves the tissue onto the flat, unstable part of the compliance curve (see Figure 9.12). Subcutaneous oedema is not detected clinically, however, until the interstitial volume has increased by about 100% (well along the compliance curve), which corresponds to a 10% increase in limb size. The nature of oedematous interstitium was described in Section 9.7. Peripheral oedema has unpleasant, albeit non-fatal, effects including impaired cell nutrition (due to increased diffusion distances), skin ulceration, blistering (see Figure 9.13) and deformity.

Pulmonary oedema is most commonly caused by left ventricular failure, which elevates the left-side filling pressure and therefore pulmonary venous pressure (whereas right ventricular failure causes peripheral, subcutaneous oedema). Pulmonary oedema has serious consequences, partly because the stiff oedematous lung is difficult to inflate, causing dyspnoea (difficulty in breathing), and partly because the gas-to-blood distance increases, slowing down gas exchange and causing hypoxia. If the interstitial oedema spills over into the alveolar spaces and floods them, pulmonary oedema can be fatal.

Causes of oedema

Oedema develops when the capillary filtration rate exceeds the lymphatic drainage rate for a sufficient period, i.e. the pathogenesis involves either a high filtration rate or a low lymph flow. Since the factors governing filtration are given in the Starling expression, Equation 9.3, the terms of this equation provide a logical classification for oedema.

Raised capillary pressure

Elevation of capillary pressure is usually secondary to chronic elevation of venous pressure caused by ventricular failure or fluid overload (as in overtransfusion and acute glomerulonephritis), or deep venous thrombosis (which raises postcapillary resistance and may later lead to venous valve incompetence). Pressures of 20–40 mmHg develop in the venous limbs of skin capillaries during right ventricular failure. The oedema fluid in such cases has a reduced protein level, namely 1–10 g/litre, due to the diluting effect of a high filtration rate (see Figure 9.7).

Reduced plasma COP

Hypoproteinaemia raises lymph flow as well as net capillary filtration rate (see Figure 9.18). At the same time the lymph protein concentration falls (1–6 g/litre), and these changes provide some protection against oedema formation (see later). Clinically, it is found that overt oedema only develops when the protein concentration in plasma falls below 30 g/litre. Hypoproteinaemia can be caused by malnutrition or malabsorption due to intestinal disease, by excessive loss of plasma protein either into urine (nephrotic syndrome) or into the gut lumen (protein-losing enteropathy), or

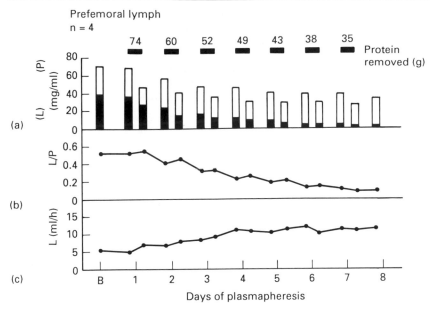

Figure 9.18 Experimental study of hypoproteinaemia in sheep. (a) Protein was removed by daily plasmapheresis (the removal of blood and replacement of only the cells, water and electrolytes). This caused protein concentration in the plasma to fall (P, white bars). The concentration of protein fell relatively further in postnodal leg lymph (L, black bars), so the L/P ratio fell (b). Lymph flow, an indication of capillary filtration rate, more than doubled (c). These data illustrate the operation of two 'buffers' or safety factors against oedema formation, namely dilution of interstitial protein (lowering the pericapillary COP) and increased lymphatic drainage. (From Kramer, *et al*. In Renkin, E. M. (1986) *American Journal of Physiology*, **250**, H706–710, by permission)

by hepatic failure, the liver being the site of synthesis of the plasma proteins albumin, fibrinogen, α-globulins and β-globulins. The commonest cause of hepatic failure is a fibrotic condition called cirrhosis, which gives rise to abdominal oedema (ascites) by raising portal vein pressure as well as lowering plasma COP. The nephrotic syndrome is characterized by albuminuria due to leakage of albumin through the glomerular membrane, often exceeding 20 g/day.

Changes in capillary permeability (L_p, σ, P)

In inflammation, the properties of the capillary wall itself change: hydraulic conductance and protein permeability increase and the reflection coefficient decreases. This causes a severe high-protein oedema. Inflammation is such a fundamental pathological process that it is described separately in Section 9.11.

Lymphatic insufficiency

Impairment of lymphatic drainage causes the accumulation of both fluid and protein since both enter the interstitial space in substantial amounts over a day (see Figure 9.14). Because lymph is the sole route for returning escaped protein to the plasma, lymphoedema fluid is rich in protein. In limb lymphoedema, the protein content is 30 g/litre or more and the lymph:plasma ratio is >0.4, in contrast to the dilute oedemas described above (<10 g/litre). The high interstitial COP exacerbates the oedema by raising the net filtration force across the capillary wall. Chronic exposure of the tissue to the highly proteinaceous oedema fluid evokes a fibrotic-fatty overgrowth, so long-standing lymphoedema does not pit easily. In Western countries lymphatic insufficiency is usually due either to poor formation of limb lymph trunks (idiopathic lymphoedema) or to damage to lymph nodes during cancer therapy (see Figure 9.19). The commonest cause world-wide, however, is filariasis, a nematode worm infestation transmitted by mosquitoes. The nematodes impair lymphatic function in the limbs and scrotum, causing a gross lymphoedema associated with hyperkeratotic elephant-like skin (elephantiasis).

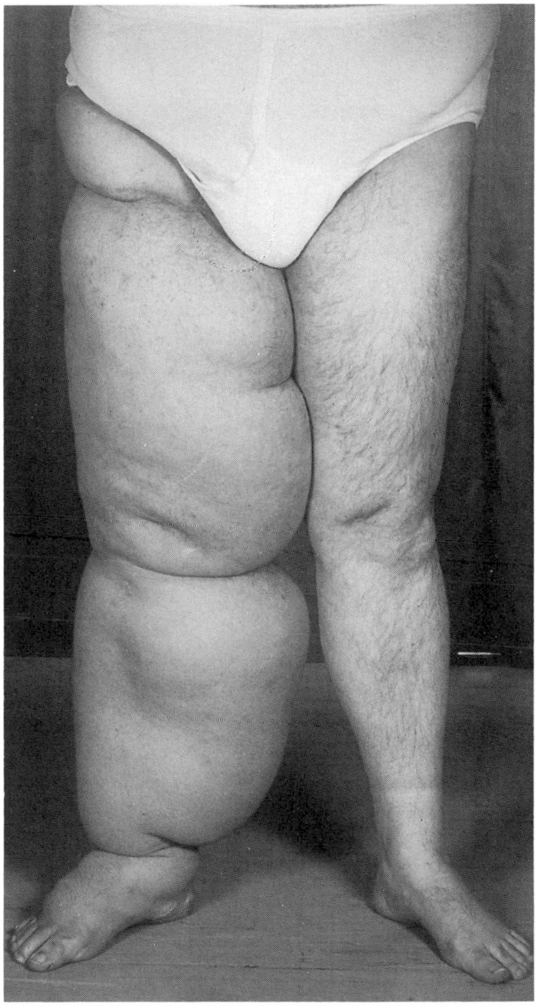

Figure 9.19 Lymphoedema caused by surgery of the groin to treat testicular cancer. (Courtesy of Dr P. Mortimer, Department of Dermatology, St. George's Hospital, London)

The safety margin against oedema

Clinicians have long recognized that clinical oedema does not develop unless plasma COP or venous pressure have changed by at least 15 mmHg. There is thus a margin of safety against oedema of 15 mmHg or thereabouts, and this is due to three buffering factors: changes in interstitial fluid pressure, interstitial COP and lymph flow (see Figure 9.20).

1. Changes in interstitial fluid pressure (P_i). When filtration rate is increased the interstitial pressure of normally-hydrated tissue rises markedly for only a very small rise in interstitial volume (see Figures 9.10 and 9.20) and this reduces the filtration pressure ($P_c - P_i$). If P_i is normally -2 mmHg and clinical oedema appears at, say, $+1$ mmHg, the change in P_i gives a safety margin of 3 mmHg. This mechanism fails above 1 to 2 mmHg in tissues where compliance increases sharply.

2. Changes in interstitial COP (π_i). An increase in filtration rate lowers the interstitial and lymph protein concentration. This lowers the interstitial COP and thereby increases the absorptive force ($\pi_p - \pi_i$), as shown by the COPs of ankle fluid in Table 9.1. Like mechanism 1, this buffer has a limited capacity, because the ratio of interstitial:plasma protein concentrations can fall no lower than $1 - \sigma$ (see Figure 9.7). Interstitial dilution is most effective as a buffer mechanism in tissues where the interstitial protein concentration is normally high, e.g. the lung. In the limbs, where interstitial COP

is 5–10 mmHg, interstitial dilution offers a safety factor of 4.5–9 mmHg.

3. Increased lymph flow. When interstitial volume and pressure increase, the lymph flow from the tissue increases too: in the cat intestine for example raising venous pressure to 30 mmHg causes a 20-fold rise in lymph flow. Some workers find, however, that the rise in lymph

flow reaches a limit, the maximum rise being equivalent to a 5 mmHg safety factor. The combined effects of the changes in P_i, π_i and lymph flow add up to a total safety margin of around 15 mmHg. The relative importance of P_i and π_i in this process depends on their starting levels but in most tissues π_i seems to be the major buffer. The lung in particular is well protected against oedema by its high interstitial COP.

9.11 Inflammatory swelling

The ancient definition of inflammation by Celsus (30BC–38AD) is hard to better: inflammation is a combination of redness (rubor), heat (calor), swelling (tumor) and pain (dolor), to which Galen (130–200AD) added a fifth criterion, loss of function. Some diverse examples include scalded skin, rheumatoid joints, infective peritonitis and disseminated cancer. The redness, heat and swelling all arise from microvascular changes. The redness and heat are due to vasodilatation caused by locally-released substances. These 'chemical mediators of inflammation' include histamine, bradykinin, prostaglandins, substance P, platelet-activating factor, superoxide radicals and others. Most of these substances, plus the

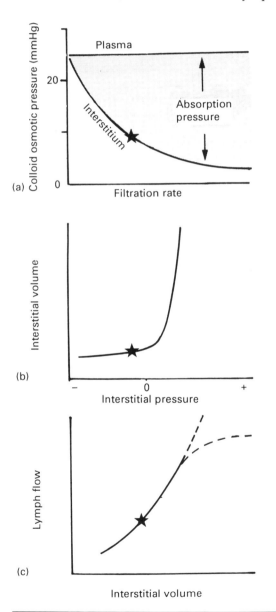

(a)

(b)

(c)

Figure 9.20 Three safety factors against oedema. Stars indicate normal state. (a) When capillary filtration rate increases, the COP gradient opposing filtration increases too, owing to the fall in interstitial protein concentration. (b) Change in interstitial pressure with volume in subcutaneous space; compliance is low at subatmospheric pressure so marked pressure changes oppose filtration. But compliance is very large above atmospheric pressure, so there is little further rise in pressure. This is essentially Figure 9.12 turned on its side. (c) Lymph flow increases with interstitial hydration, opposing oedema formation. Dashed lines indicate that lymph flow reaches a limit in stationary tissue but may not do so in moving flexed limbs. (After Taylor, A. E. and Townsley, M. I. (1987) *News in Physiological Science*, **2**, 48–52)

potent leukotrienes (Section 11.4) and cytokines also initiate a series of changes in the walls of pericytic venules. (Cytokines are mediators of the inflammatory response produced by monocytes, endothelial cells and fibroblasts. They include tumour necrosing factor and the interleukin series.) The changes are as follows:

1. The endothelial surface becomes attractive to polymorphonuclear leucocytes. These adhere to the wall (margination) and then push through the intercellular junctions to invade the inflamed tissue. They are followed more slowly by migrating lymphocytes.

2. Quite independently of leucocyte migration some endothelial junctions open up into gaps as much as $0.5\,\mu m$ wide (see Figure 9.21). Gap formation is thought to be due to the contraction of the actin and myosin filaments that exist within endothelial cells, mediated probably by a rise in intracellular Ca^{2+}. Gap formation in response to histamine can be reduced by the H_1-receptor antagonist mepyramine and by the β-adrenergic agonists isoprenaline and terbutaline.

The *inflammatory swelling* which ensues is caused partly by the gap formation and partly by changes in net filtration pressure. Capillary pressure rises due to arteriolar vasodilatation and a consequent fall in R_a/R_v. The hydraulic conductance of the wall increases many times, as can be seen from the increased slope in Figure 9.21. The gaps also raise the permeability to plasma protein: this speeds the movement of immunoglobulins into the tissue but at the same time raises the interstitial concentration of plasma proteins, reducing the gradient of COP that opposes filtration. The gaps also lower the protein reflection coefficient to around 0.4, which further reduces the effective COP across the wall (see Figure 9.21b, arrow). The ensuing capillary filtration rate is very rapid because the fall in the COP gradient is aggravated by the fall in σ, and the rise in net filtration force is further amplified by the rise in L_p

(Equation 9.3). Eventually the combination of a high filtration fraction and adhering leucocytes can lead to plugging of the capillary lumen by a packed column of red cells, a condition called vascular stasis.

Inflammation is a high-permeability disorder and the oedema fluid has a high protein concentration; it is sometimes called an 'exudate' by way of contrast with the low protein oedema or 'transudate' of venous hypertension and hypoproteinaemia. Large volumes of exudate can form in the peritoneal, pleural, pericardial and synovial

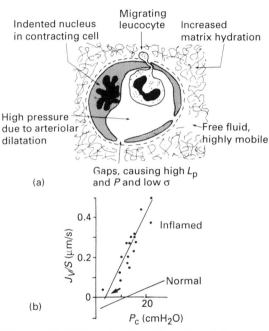

Figure 9.21 (a) Structural and functional changes during acute inflammation. (b) Filtration rate per unit surface area of frog mesenteric capillaries, injured by mercuric chloride or alcohol, studied by the red cell method. The perfusate was the frog's own plasma (COP approximately $13\,cmH_2O$). The sevenfold increase in slope shows that hydraulic conductance (L_P) increases in inflammation. The fall in intercept to $3\,cmH_2O$ (arrow) shows that the effective osmotic gradient across the capillary wall ($\sigma\Delta\pi$) is reduced by inflammation. (After the classic paper of Eugene M. Landis (1927) *American Journal of Physiology*, **82**, 217–238)

cavities during local inflammatory conditions (e.g. peritonitis, pleurisy, pericarditis, rheumatoid arthritis), and the exudate contains relatively high concentrations of fibrinogen owing to the loss of molecular selectivity at the capillary gaps. Fibrin adhesions can then develop and complicate the disease (e.g. intestinal adhesion after peritonitis).

Ischaemia-reperfusion injury is an interesting, recently-developed concept whereby, after a period of poor perfusion due to vascular obstruction, the restoration of blood flow delivers oxygen which is partly converted into toxic superoxide radicals by the tissue. These radicals damage the endothelium, leading to inflammation and impairment of tissue recovery, e.g. after myocardial infarction.

9.12 'Internal fluid transfusion' during hypovolaemia

The interstitial compartment represents a reservoir of fluid, some of which can be transferred to the plasma after a haemorrhage or any other form of hypovolaemia (low blood volume), such as dehydration. In hypovolaemia, a sympathetically-mediated arteriolar constriction raises R_a/R_v. This, coupled with a fall in arterial and venous pressures, reduces the mean capillary pressure and produces a net absorption force across the capillary wall. A transient absorption of interstitial fluid ensues (see Figure 9.10), partially restoring the plasma volume but reducing the haematocrit and plasma protein concentration. This 'internal transfusion' of fluid explains why most haemorrhage patients have a low haematocrit by the time they reach hospital. The amount of fluid absorbed is limited by changes in the other three Starling pressures: a reduction in plasma COP due to haemodilution, a rise in interstitial COP and a fall in interstitial pressure. Nevertheless as

much as half a litre is absorbed from the interstitial compartment during the first hour after a severe haemorrhage: much of it comes from skeletal muscle, since this is 40% of the body weight.

A second factor aiding the internal transfusion of fluid is post-haemorrhagic glycolysis in the liver, induced by sympathetico-adrenal stimulation and glucagon. The output of glucose by the liver rises sharply and raises the osmolarity of plasma and interstitial fluid by as much as 20 mOsm. The rise in interstitial osmolarity draws fluid osmotically from the huge intracellular compartment into the interstitial compartment and this replenishment of the interstitial compartment allows capillary absorption to continue for 30–60 min, much longer than would otherwise be possible. It is estimated that, overall, about half the internal transfusion comes ultimately from the intracellular compartment.

Further reading

Aukland, K. and Nicolaysen, G. (1981) Interstitial fluid volume: local regulatory mechanisms. *Physiological Reviews*, **61**, 556–643

Comper, W. C. and Laurent, T. C. (1978) Physiological function of connective tissue polysaccharide. *Physiological Reviews*, **58**, 255–315

Granger, H. J., Laine, G. A., Barnes, G. E. and Lewis, R. E. (1984) In *Edema* (eds N. C. Staub and A. E. Taylor), Raven Press, New York, pp. 189–228

Grega, G. J. (1986) (chairman) Role of endothelial cells in the regulation of microvascular permeability to molecules (Symposium). *Federal Proceedings*, **45**, 75–109

Michel, C. C. (1984) Fluid movement through capillary walls. In *Handbook of Physiology, Cardiovascular System, Vol. IV, Microcirculation, Part 1*, American Physiological Society, Baltimore, pp. 375–409

Olszewski, W. C. (1985) *Peripheral Lymph: Formation and Immune Function.* CRC Press, Florida

Renkin, E. M. (1986) Some consequences of capillary permeability to macromolecules: Starling's hypothesis reconsidered. *American Journal of Physiology*, **250**, H706–710

Staub, N. C., Hogg, J. C. and Hargens, A. R. (1987) (eds) Interstitial-lymphatic liquid and solute movement. *Advances in Microcirculation 13.* Karger, Basel

Staub, N. C. and Taylor, A. E. (1984) *Edema.* Raven Press, New York

Willoughby, D. A. (1987) Inflammation – mediators and mechanisms. *British Medical Bulletin*, **43**, 2

Chapter 10
Vascular smooth muscle

10.1	**Structure of the VSM cell**	**10.4**	**Neuromuscular excitation**
10.2	**Mechanism of contraction**	**10.5**	**Pharmacomechanical coupling**
10.3	**Ionic channels in the VSM cell membrane**	**10.6**	**Automaticity**

The tunica media of a blood vessel contains smooth muscle cells arranged in a helical pattern, and their degree of contraction controls the vessel radius and blood flow. Our knowledge of vascular smooth muscle (VSM) has expanded enormously in recent years, and it is now possible to discuss the ultrastructure, contractile mechanism and electrical behaviour of the VSM cell rather as we did for the myocyte in Chapter 3.

10.1 Structure of the VSM cell

The VSM cell is spindle-shaped, about $20–60\,\mu m$ long and $4\,\mu m$ wide at the nuclear region. It contains thick filaments composed of myosin (diameter $15\,nm$), surrounded by numerous thin filaments composed of actin (diameter $6\,nm$; see Figure 10.1). The actin:myosin ratio is about eight times larger than in striated muscle. The actin filaments insert not into Z lines but into 'dense bands' on the inner surface of the cell, and into 'dense bodies' in the cytoplasm. These structures are composed of α-actinin, the same substance that forms Z-lines in striated muscle. There are no striations in VSM because the contractile units are not aligned in register. A third kind of filament, the intermediate filament, is also abundant in VSM. It is composed of the proteins filamin, actin, α-actinin and desmin, but whether it is involved in the contractile process or merely has a structural role is not clear.

The VSM cell possesses a system of smooth endoplasmic reticulum which forms about 2% of the cell volume and contains a releasable store of calcium ions. The smooth endoplasmic reticulum approaches to within $12–20\,nm$ of the cell membrane in places, and is linked to the surface by periodic dark-staining bands which could be involved in excitation-contraction coupling. The cell surface also has numerous tiny invaginations called caveolae, whose function is uncertain. The membranes of adjacent cells are linked by electrically conduc-

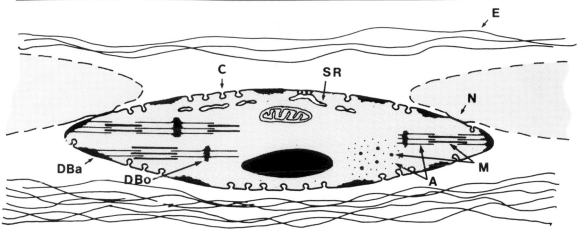

Figure 10.1 Sketch of main elements within a vascular smooth muscle cell. A, actin filaments in longitudinal and transverse section; C, caveola; DBa, DBo, dense band and dense body; E, elastic elements in parallel to cell (collagen and elastin); M, myosin filaments; N, nexus or gap junction, SR, sarcoplasmic reticulum. (After electron micrographs in Gabella, G. (1984) *Physiological Reviews*, **64**, 455–477)

tive 'gap junctions', which allow depolarization to spread from cell to cell, so the cells form a functional syncytium. The spread is decremental, however, and extends only a millimetre or so along the axis of the vessel.

10.2 Mechanism of contraction

As in cardiac muscle, contraction is initiated by a rise in free Ca^{2+} ions in the cytoplasm, as proved by experiments with calcium-sensitive intracellular indicators. The free calcium derives partly from the sarcoplasmic reticulum store and partly from the extracellular fluid via calcium channels in the cell membrane, as proved by electron-probe analysis and electrophysiological methods respectively. The rise in free calcium causes the myosin filaments to form crossbridges with the actin filaments and thereby 'row' themselves into the spaces between the actin filaments, producing shortening and tension. The process in VSM has, however, two major differences from that in myocardium.

1. *Myosin light-chain phosphorylation.* Unlike cardiac or skeletal muscle myosin, VSM myosin can only interact with actin after its light chains have been phosphorylated by ATP. (The light chains are part of the crossbridge head; see Figure 3.3). The phosphorylating enzyme, myosin light-chain kinase, is itself only active in the presence of a complex of calcium and calmodulin, calmodulin being a calcium-binding protein closely related to troponin C. Thus a rise in cytoplasmic calcium causes the formation of the calcium-calmodulin complex, which activates myosin light-chain kinase, leading to myosin phosphorylation and crossbridging.

2. The *latch state.* Striated muscle relies on the continuous, rapid making and breaking of crossbridges to maintain tension (crossbridge cycling), and this is an energy-expensive process. VSM, however, can maintain an active tension for just 1/300th of the energy expenditure of striated muscle. This is achieved by the formation of long-lasting crossbridges called 'latch bridges', which cycle only very slowly (rather like the 'catch' state

of the muscle which closes a mollusc's shell). It is thought that the latch bridges may be crossbridges that become dephosphorylated.

10.3 Ionic channels in the VSM cell membrane

In order to understand excitation-contraction coupling in VSM, we must take account of a minimum of three different ion-conducting channels in the cell membrane, although many more are known.

K^+-conducting channels The resting VSM cell has a negative resting membrane potential of -70 mV to -40 mV, which results chiefly from the membrane being more permeable to potassium ions than to other ions. A Na^+-K^+ exchange pump in the cell membrane maintains a high intracellular K^+ concentration, and an outward flux of K^+ ions through the K^+ channels down the potassium electrochemical gradient sets up a resting potential, rather as in myocytes (see Chapter 3).

Voltage-gated channels for Ca^{2+} (VGCs) These channels are permeable to divalent cations like calcium and barium, and they open when the membrane depolarizes beyond a certain point (see Figure 10.4). They allow calcium to enter the cell, initiating contraction, and since their opening is initiated by depolarization, this is termed *electrochemical coupling*. At least two types of calcium VGC exist in many VSM cells. Some have a low conductance, open at a low threshold (around -50 mV) and then quickly self-inactivate. Others have a larger conductance, open at a more positive threshold (-20 mV) and inactivate more slowly. The channels with thresholds close to the resting potential may be important in regulating basal tone. The VGC calcium channels are important in generating the VSM action potential (see below), as is evident from blockage of the action poten-

tial by dihydropyridines (the calcium channel blockers, e.g. nifedipine).

Receptor-operated channels admitting Ca^{2+} (ROCs) This class of calcium-conducting channel is insensitive to membrane potential but is activated when a chemical transmitter such as noradrenaline binds to its specific receptors on the cell membrane. This provides a second, depolarization-independent way by which noradrenaline can induce contraction, called *pharmacomechanical coupling* (see later). The nature of the link between the pharmacological receptor and the ROC is still uncertain: the two membrane proteins could be physically linked, or the link could involve an intermediary membrane protein called G protein (see Appendix on 'Second messengers'). Substances like vasopressin, angiotensin and histamine are also able to cause contraction by activating ROCs.

10.4 Neuromuscular excitation

Nearly all arteries and arterioles are innervated by sympathetic vasoconstrictor fibres (see Chapter 11), the main exceptions being the aorta and pulmonary arteries in some species, and the fine cerebral vessels. The sympathetic fibres run along the adventitial-medial border, and the inner part of the media is not directly innervated, except in veins. The terminal fibres bear a string of swellings (several hundred per fibre) called 'junctional varicosities'. Each varicosity lies close to a VSM cell membrane; in random sections the gaps appear to range from 50 nm to 10 µm wide, but recent serial reconstructions indicate that most varicosities in fact approach to within 75 nm of a VSM at some point. Each varicosity contains both small and large dense-cored vesicles, which contain a variable mixture of noradrenaline and ATP. On arrival of the nerve action potential there is a rise in intraneural calcium ion concentration which causes some vesicles to discharge their contents into the extracellular space (see Figure 11.9 later). The neurotransmitter then quickly diffuses across to the VSM cell membrane and binds to receptors there.

The effects of neurally-induced receptor activation are best explained by considering the electrical and mechanical responses of the rat tail artery (see Figure 10.2), which is typical of many blood vessels, although not all (see later). When the perivascular sympathetic fibres are excited by a brief small stimulus, they elicit an electrical response with two components. There is an initial

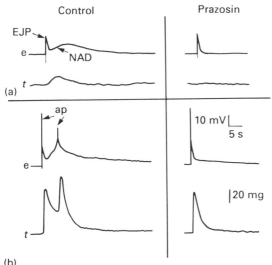

(a)

(b)

Figure 10.2 Simultaneous records of membrane potential recorded by an intracellular microelectrode (e) and tension (t) in a rat tail artery. The perivascular nerves were stimulated by a single external pulse in each frame, in the abscence (left) or presence (right) of the α-adrenoceptor blocker, prazosin. (a) Medium intensity stimulation produced an excitatory junction potential (EJP) and a slower depolarization. Only the latter was blocked by prazosin (neurogenic alpha depolarization, NAD). Note that the slow contraction precedes the NAD. (b) Higher intensity of stimulation evoked a larger EJP and NAD, each of which triggered an action potential (ap) and associated twitch contraction. (After Cheung, D. W. (1984) *Pfluger's Archiv*, **400**, 335–337, by permission)

rapid brief depolarization of the VSM cell (amplitude approximately 10 mV, duration approximately 1 s) which is called an *excitatory junction potential* (EJP); this is not an action potential. The EJP is followed by a smaller, slower depolarization of longer duration (amplitude approximately 2 mV, duration several seconds). The slow depolarization can be blocked by α-adrenoceptor antagonists like prazosin (see Figure 10.2), so it is called a *neurogenic alpha depolarization*, or NAD. The NAD is evidently caused by noradrenaline, but the EJP by contrast is not blocked by adrenoceptor antagonists and is probably initiated by ATP release from the nerve (see Figure 11.10 later for evidence).

The mechanical response, a slow contraction, begins before the NAD and peaks earlier. Although, like the NAD, it is blocked by α-adrenoceptor antagonists, it is clear from the time sequence that the contraction is not caused by the NAD; nor is it caused by the EJP since the latter is unaffected by α-adrenoceptor antagonists. The electrical events are thus **not** the cause of slow contraction: the slow contraction is in fact an example of the pharmacomechanical coupling referred to earlier.

If the stimulus strength is increased, the EJP becomes larger, reaches the threshold of the voltage-gated calcium channels and triggers an action potential: this elicits a short-latency brief contraction, i.e. twitch (see Figure 10.2b). The NAD becomes larger too and triggers another action potential: this produces a second twitch superimposed on the underlying slow contraction. The twitches are examples of electromechanical coupling, while the underlying slow contraction reflects pharmacomechanical coupling. The action potential itself is a simple spike depolarization of rather variable amplitude, lasting 10–100 ms. The depolarization is due mainly to an influx of Ca^{2+} ions into the cell as the voltage-gated calcium channels open. The twitch is due partly to Ca^{2+} ions arriving by this route and partly to Ca^{2+} ions released from the internal stores.

It must be stressed that the response to sympathetic stimulation varies greatly from vessel to vessel, and not all vessels respond as above, though the pattern is a common one. In some vessels, such as the pulmonary artery, NADs can be elicited without a preceding EJP (see Figure 10.3c), while in others under appropriate conditions EJPs are elicited without NADs.

10.5 Pharmacomechanical coupling

Pharmacomechanical coupling, as indicated earlier, is the induction of contraction by a

chemical agent (whether released from a local nerve or circulating) without the necessity of a change in membrane potential or the firing of an action potential. In certain vessels, such as the pulmonary artery, the superfusion of noradrenaline at a low concentration causes a sustained contraction without any detectable membrane depolarization. Superfusion at higher concentrations or sympathetic nerve stimulation causes a NAD without EJPs or action potentials (see Figure 10.3c). Moreover, even action potential-generating vessels such as the portal vein (see Figure 10.3a) can be induced to contract without action potentials if their VGCs are blocked by

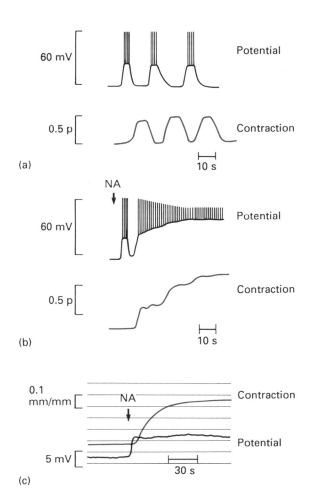

(a)

(b)

(c)

Figure 10.3 Examples of varying characteristics of vascular smooth muscle. (a) Upper trace shows an intracellular record of membrane potential in a spontaneously active vessel *in vitro* (guinea pig portal vein), illustrating automaticity. Regular spontaneous slow depolarizations trigger bursts of action potentials, followed after some delay by contraction (lower trace; p is tension). (b) Response of same preparation to addition of noradrenaline (NA, 10^{-6} g/ml) to bathing medium, illustrating electromechanical coupling. (c) By contrast, response of sheep carotid artery to superfused noradrenaline demonstrates pharamacomechanical coupling. There is only a small depolarization and no action potentials, yet a sustained contraction occurs. ((a) and (b) from Golenhofen, Hermstein and Lammel (1973) *Microvascular Research*, **5**, 73–80. (c) from Keatinge, W. R. and Harman, C. M. (1980) *Local Mechanisms Controlling Blood Vessels*. Academic Press, London, by permission)

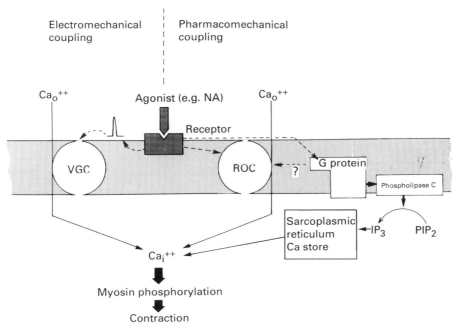

Figure 10.4 Schematic diagram of three membrane processes activated by noradrenaline (NA) leading to contraction: voltage-gated channels (VGC), receptor-operated channels (ROC) and the G protein-second messenger system. Membrane GTP-binding protein activates an enzyme, phospholipase C which breaks down phosphatidyl inositol bisphosphate (PIP_2) to yield the intracellular second messenger inositol triphosphate (IP_3)

verapamil. Pharmacomechanical coupling is initiated by activation of the specific receptors in the VSM membrane such as α-receptors. This sets two mechanisms in operation (see Figure 10.4). One mechanism is the opening of the receptor-operated channels, which allows extracellular calcium ions to flow into the cell and initiate contraction. The second mechanism involves a chain of biochemical reactions, in which a membrane protein called a 'G protein' activates the enzyme phospholipase C. The latter catalyses the production of an intracellular 'second messenger', inositol triphosphate (IP_3), which acts on the sarcoplasmic reticulum to induce the release of stored Ca^{2+} (see Appendix, 'Second messengers').

10.6 Automaticity

In most arteries and veins the VSM cell have a stable resting potential but in some vessels (portal vein, terminal pial arteries and arterioles) the resting potential is unstable and depolarization occurs spontaneously, triggering action potentials and spontaneous contractions (see Figure 10.3a). The process bears some resemblance to that in SA node cells, though it is less regular. The excitation then spreads from cell to cell via the gap junctions. If most of the VSM cells contract synchronously the vessel undergoes rhythmic variations in calibre producing the 'vasomotion' often seen in small arterioles (Section 8.1). The frequency of the spontaneous discharges increases up to threefold when an unstable VSM cell is stretched and this probably contributes to the myogenic response described in the next chapter.

Further reading

Bolton, T. B. and Large, W. A. (1986) Are junction potentials essential? Dual mechanism of smooth muscle activation by transmitter release from autonomic nerves. *Quarterly Journal of Experimental Physiology*, **71**, 1–28

Hirst, G. D. S. and Edwards, F. R. (1989) Sympathetic neuroeffector transmission in arteries and arterioles. *Physiological Reviews*, **69**, 546–604

Murphy, R. A. (ed.) (1989) Contraction in smooth muscle cells. *Annual Review in Physiology*, **51**, 275–331

Somlyo, A. P. (1985) Excitation-contraction coupling and the ultrastructure of smooth muscle. *Circulation Research*, **57**, 497–507

Chapter 11
Control of blood vessels

11.1 Overview of vascular control

The tunica media of blood vessels contains smooth muscle cells arranged in a predominantly circumferential pattern and the active tension of these cells controls the radius of the vessel, as explained previously (see Figure 7.12). It is worth reiterating here that vasodilatation is a passive, not active process, being powered by the blood pressure and recoil of elastic elements in the wall as vascular smooth muscle relaxes.

What does vessel radius influence?

The *arterioles* are the chief resistance vessels of the systemic circulation and even quite small changes in arteriolar radius cause large changes in vascular resistance, owing to the fourth-power term in Poiseuille's law (Section 7.5). This has the following effects. (1) Local arteriolar resistance regulates blood flow to the tissue downstream of the arteriole. The range of flows that can be produced is enormous in some organs, as shown in Figure 11.1, and as a general rule the flow is varied to match the metabolic activity of the tissue. (2) The total arteriolar resistance, acting in concert with the cardiac output, regulates the systemic arterial pressure. (3) Capillary recruitment and capillary filtration pressure are both regulated by local arteriolar tone, as shown in Figures 8.15 and 9.4 respectively. Thus, arteriolar radius exerts both local effects (control of nutritive supply and organ fluid balance) and central effects (homeostasis of blood pressure and plasma volume).

The *veins and venules* normally contain around 60% of the blood volume, and a decrease in the average radius of the peripheral veins and venules can displace a considerable volume of blood into the central veins. Venous smooth muscle can thus influences the cardiac filling pressure, and hence stroke volume.

Basal tone and its regulation

The total active tension of the vascular smooth muscle (VSM) in a segment of wall is called the vessel tone. Owing to the automaticity of their VSM cells, the arterioles and some larger vessels retain a degree of tone (i.e. remain partially contracted) even when their sympathetic innervation is interrupted, and this is called the basal tone. A high basal tone is vital if a vessel is to be capable of substantial dilatation because dilatation is simply a reduction in

tone: the greater the basal tone the greater the potential for vasodilatation. Tissues which are capable of producing large increases in blood flow, such as skeletal muscle and the salivary gland (see Figure 11.1), have a high basal tone. By contrast, basal tone is slight in most veins.

The tone of a vessel *in vivo* is influenced by numerous factors which fall into two broad categories: local or intrinsic control mechanisms and extrinsic control mechanisms (see Figure 11.2). *Intrinsic control mechanisms* are located entirely within the organ and include physical factors (temperature, pressure), the myogenic response, tissue metabolites and autocoids. The *extrinsic control mechanisms* are the autonomic nerves and circulating endocrine secretions.

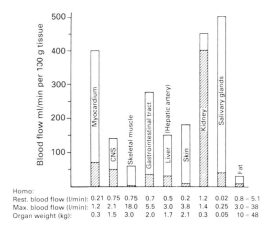

Figure 11.1 Range of flows between rest (▨) and maximal function (▢) in the various organs of a 70 kg man. Maximal flow through all organs at the same time is impossible because the total flow would be 38 litres/min, which exceeds the output capactiy of the heart. (From Mellander, S. and Johansson, B. (1968) *Pharmacological Reviews*, **20**, 117–196, by permission)

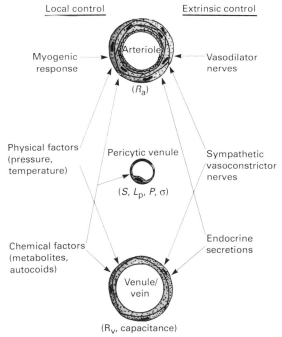

Figure 11.2 Schema for control of the peripheral vessels. Ra, R_v, pre- and postcapillary resistances respectively; S, perfused area of capillary bed, influenced by metabolites; L_p, P and σ are the microvascular permeability parameters (see Chapter 8) which are altered by the mediators of inflammation

11.2 Local control mechanisms
Local temperature

This is particularly important in the skin. High ambient temperatures cause the cutaneous arterioles and veins to dilate producing the familiar reddening of a limb immersed in hot water. Conversely, skin temperatures down to 10–15°C cause vasoconstriction which conserves heat and safeguards core temperature. The vasoconstrictor response to cold is attributed partly to slowing of the Na^+-K^+ pump in the VSM membrane, which leads to depolarization, and partly to an increase in the affinity of the cutaneous VSM adrenoceptors for noradrenaline as temperature falls. Cooling below approximately 12°C, by contrast, leads to paradoxical cold vasodilatation (see Chapter 12), which is caused by the impairment of neurotransmitter release and by the release of vasodilator substances like prostaglandins (see later) from the underperfused tissue.

Transmural pressure and the myogenic response

A high *external pressure* compresses the vessels and impairs blood flow, as in skeletal muscle during its contraction phase (see Figure 11.6 later). Similarly, flow through the skin is impaired by compression during sitting, kneeling, lying etc., and should a patient be bedridden by age or paralysis, the prolonged impairment of skin nutrition can result in large ulcerating bed sores over the buttocks and heels.

The immediate mechanical effect of raising *internal pressure* is to distend the vessel and reduce its resistance but most systemic arterioles and some arteries (e.g. cerebral arteries) then react to the distension by contracting (see Figure 11.3). This is called the myogenic response, and was first described by Sir William Bayliss, brother-in-law of Ernest Starling, in 1902. The myogenic response safeguards blood flow and capillary filtration pressure in the face of

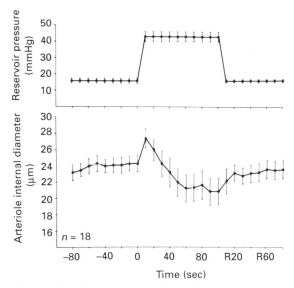

Figure 11.3 Change in arteriolar diameter upon raising the luminal pressure in cat mesentery. The initial diameter increase is due to passive distension. The active myogenic response at about 10 s then reduces the diameter to less than its original value. (From Johnson, P. C. and Intalietta, M. (1976) *American Journal of Physiology*, **231**, 1686–1698, by permission)

changes in blood pressure (see 'Autoregulation', Section 11.3), and also contributes to the basal tone of the vasculature.

The immediate cause of the myogenic response in spike-generating vessels, like the arterioles and portal-mesenteric vein, is that stretch increases the frequency of the spontaneous action potentials resulting in increased active tension. The nature of the stretch sensor is, however, a more difficult issue since the myogenic response is sometimes so strong that vessel radius is actually reduced, as shown in Figure 11.3, so the VSM cell is no longer stretched. This has led to the suggestion that a region of VSM membrane in series with the contractile proteins (possibly at the dense bands) might be sensitive to local stress; stress, unlike vessel radius, is increased during the myogenic response (see Figure 7.12). Such a region might initiate spike activity. It is relevant to note that stress-activated ionic channels have been identified in the membranes of a wide variety of cells. Another possibility is raised by the recent discovery that arterial endothelium can release vasoconstrictor substances in response to mechanical stress (see later); indeed, endothelial stripping abolishes the myogenic response of large arteries like the basilar and cerebral arteries. On the other hand, the myogenic response of the spike-generating portal vein is not abolished by endothelial stripping, so conceivably there might be more than one mechanism underlying the myogenic response.

Local metabolites

Many of the chemical by-products of metabolism cause vascular relaxation, so any increase in tissue metabolic rate causes arteriolar dilatation and automatically increases the local tissue perfusion. This process is called *metabolic vasodilatation.* The increase in blood flow is almost linearly proportion to metabolic rate in tissues such as exercising muscle, myocardium and brain, provided that the arterial pressure is constant. The vasodilator influences include hypoxia;* acidosis due to the release of carbon dioxide and lactic acid; breakdown products of ATP, namely adenosine diphosphate (ADP), adenosine monophosphate (AMP), adenosine and inorganic phosphate; potassium ions released by contracting muscle and (in the brain) by active neurones; and a rise in interstitial osmolarity due to the release of lactate, K^+ and other small solutes.

The relative importance of each factor varies from organ to organ: cerebral vessels for example are particularly sensitive to H^+ and K^+ ions, while coronary vessels are influenced chiefly by hypoxia and/or adenosine (see Chapter 12).

---

* Hypoxia causes large arteries to constrict however. This is due to the arterial endothelium secreting a prostanoid vasoconstrictor agent.

The mechanisms of action of the above factors are varied and can only be touched on here. Hypoxia and ADP stimulate the endothelium to produce endothelium-derived relaxing factor (EDRF, see later). The dilator effect of adenosine, by contrast, is endothelium-independent; adenosine inhibits neurotransmitter release by a prejunctional effect (see Figure 11.9), and also reduces Ca^{2+} influx across the VSM membrane. Acidosis (H^+ ions) may act by inactivating membrane calcium channels. The action of K^+ ions is complicated. A moderate increase in extracellular K^+ concentration (up to 10 mM) causes the membrane potassium conductance to increase; this brings the potential closer to the potassium equilibrium potential, so the membrane hyperpolarizes and relaxation follows. In addition, a rise in interstitial K^+ stimulates the electrogenic Na^+-K^+ exchange pump and may also stimulate EDRF production. Higher extracellular potassium concentrations (e.g. 20 mM) reduce the potassium gradient across the VSM membrane to such an extent that the membrane depolarizes (see 'Nernst equation', Equation 3.1) and vasoconstriction ensues.

Autocoids

Autocoids can be defined as vasoactive chemicals that are produced locally, released locally and act locally (unlike endocrine secretions). They include histamine, bradykinin, 5-hydroxytryptamine and the prostaglandin-thromboxane-leukotriene group. They are involved chiefly in special local responses such as inflammation and haemostasis.

Histamine This is produced from the amino acid histidine by decarboxylation, and is found in granules within mast cells and basophils. Histamine is one of the chemical mediators of inflammation, being released in response to trauma and certain allergic reactions (urticaria, anaphylaxis). Histamine dilates arterioles, constricts veins and increases venular permeability.

Bradykinin Exocrine glands like the salivary gland and sweat glands secrete the enzyme kallikrein into their ducts when stimulated by cholinergic nerves. If kallikrein back-diffuses into the interstitium (e.g. following duct obstruction) it acts on an interstitial plasma globulin, kininogen, to produce a decapeptide, kallidin. This is converted by an aminopeptidase into the 9 amino acid peptide, bradykinin. Bradykinin is strong vasodilator agent and also increases venular permeability.

5-hydroxytryptamine (serotonin, 5-HT) This derivative of the amino acid tryptophan is found in platelets, the intestinal wall and the central nervous system. 5-HT is released from platelets during clotting and its vasoconstrictor effect on large vessels contributes to haemostasis. In the intestinal tract, 5-HT occurs in argentaffin cells and may be involved in the regulation of local blood flow, and gastrointestinal smooth muscle. The argentaffin cells occasionally form a tumour (carcinoid tumour) and the release of large quantities of 5-HT into the circulation by the tumour causes attacks of hypertension and diarrhoea. In the brain, 5-HT occurs in neurones close to cerebral vessels; it markedly potentiates the effect of noradrenaline and may be involved in the vasospasm associated with migraine and subarachnoid haemorrhage. It is also a central neurotransmitter, and the hallucinogenic drug lysergic acid diethylamide (LSD) is an antagonist of 5-HT.

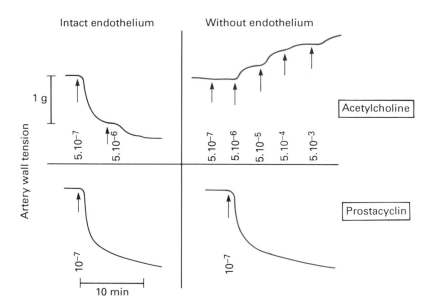

Figure 11.4 Response of canine intrapulmonary arteries *in vitro* to addition of acetycholine and prostacyclin to bathing medium. Artery strips precontracted with 5-hydroxytryptamine, and isometric force recorded. Both agonists produce a potent relaxation when the vessel endothelium is intact. When the endothelium is removed by rubbing, the acetylcholine-induced response changes to contraction. The prostacyclin-induced relaxation, however, is not endothelium-dependent. (After Altura, B. (1988) *Microcirculation Endothelium and Lymphatics*, **4**, 91–110, by permission)

Prostaglandins These are vasoactive agents synthesized from a fatty acid precursor, arachidonic acid, via the enzyme cyclo-oxygenase. Prostaglandins are produced by macrophages, leucocytes, fibroblasts and endothelium. The different prostaglandins have different actions; the F series (PGF) are mainly vasoconstrictor agents, while the E series (PGE) and prostacyclin (PGI_2) are vasodilator substances (see Figure 11.4). The vasodilator prostaglandins contribute to inflammatory vasodilatation and reactive hyperaemia (see later), and their synthesis is inhibited by aspirin and indomethacin. The related arachidonic-acid derivative, thromboxane A_2, is a powerful vasoconstrictor substance found in platelets; it is involved in platelet aggregation and haemostasis. A general term for all these arachidonic acid derivatives is 'eicosanoid'.

Leukotrienes This group of vasoactive substances is produced from arachidonic acid by a different pathway, involving the enzyme lipoxygenase. They are synthesized by leucocytes and are important mediators of the inflammatory response. They cause vasoconstriction, leucocyte margination and emigration, and the formation of gaps in the walls of venules. Their gap-inducing action develops at 1000th the concentration at which histamine acts.

Contrariness of pulmonary vessels The response of pulmonary resistance vessels to chemical factors is often different from the response of systemic vessels. Alveolar hypoxia, for example, causes pulmonary vasoconstriction, not dilatation. This is physiologically important, for it reduces the perfusion of underventilated regions and thereby helps to maintain a normal ventilation-perfusion ratio. If hypoxia is generalized, as at high-altitude, pulmonary hypertension results from this response. The pulmonary vascular response to many other agents is reversed also: histamine and bradykinin, for example, causes pulmonary vasoconstriction.

Endothelium-dependent relaxation and contraction

In 1980, Furchgott and Zawadski discovered that whereas a normal artery ring relaxes in response to carbachol (a stable analogue of acetylcholine) the response changes to contraction when the endothelial lining is rubbed away. The same is true for human hand veins if their endothelial lining is destroyed by local perfusion with distilled water. The explanation is that arterial and venous endothelium can synthesize a dilator substance, *endothelium-derived relaxing factor* (EDRF), in response to stimulation by acetylcholine (see Figure 11.4). The EDRF diffuses directly from the endothelial lining into the underlying smooth muscle, activating the enzyme guanylyl cyclase and initiating a rise in the intracellular 'second messenger', cyclic guanosine monophosphate (cGMP); the latter induces VSM relaxation by as yet unknown mechanisms. EDRF production is also stimulated by a number of other substances, the clearest examples being bradykinin, ADP, substance P and, in certain tissues/species, histamine. EDRF is produced too in response to endothelial shear stress, and this accounts for the long-recognized phenomenon of flow-induced vasodilatation in arteries: when flow through a large artery is increased (due to a fall in downstream arteriolar resistance), the artery dilates, facilitating blood flow to the arterioles. It should be noted, however, that the vasodilator effect of many substances is not EDRF-dependent (e.g. adenosine, AMP, isoprenaline, papaverine and nitrodilators), while for other substances EDRF involvement varies between tissues and species. It may also be doubted whether cholinergic parasympathetic nerves act via EDRF release (see legend to Figure 11.13).

The half-life of EDRF is extremely short (approximately 6 s) and it is now clear that EDRF is in fact nitric oxide (NO), which is cleaved-off the amino acid arginine by an endothelial enzyme. Blockage of this enzyme by the stable analogue L-methyl arginine causes a rise in arterial pressure in the intact animal, from which it is deduced that EDRF production occurs continuously and exerts a tonic vasodilator influence on the resistance vessels. Thus vessel tone represents a balance between the opposing effects of EDRF and vasoconstrictor influences like the myogenic response and sympathetic nerve activity. Organic nitrates like glyceryl trinitrate have long been used as vasodilator drugs in the treatment of angina, and it is now perceived that they act by mimicking EDRF and releasing nitric oxide into the tissue.

Evidence is now appearing for the existence of a second kind of EDRF besides nitric oxide. Nitric oxide acts without altering the VSM membrane potential, whereas cases of endothelium-dependent vasodilatation are being documented that involve membrane hyperpolarization. The existence of an endothelium-derived hyperpolarizing factor (EDHF) is therefore postulated.

Endothelium can produce not only dilator substances but also vasoconstrictor compounds. One vasoconstrictor substance appears to be a prostanoid, i.e. a product of cyclo-oxygenase; it mediates the contractile response of larger arteries to hypoxia. Another agent, discovered very recently is a peptide called endothelin which causes a strong vasoconstriction lasting 2–3 h. Its physiological role is still under investigation. It should be noted that most of the work on endothelium-derived substances has been carried out on large vessel endothelium, and we know much less about endothelial-dependent responses in arterioles, which are the vessels controlling blood flow *in vivo*.

11.3 Circulatory adjustments due to local mechanisms

Several major circulatory adjustments can be initiated by the above local mechanisms rather than by extrinsic neural control. The most important examples are autoregulation, metabolic hyperaemia and reactive hyperaemia.

Autoregulation

The relation between perfusion pressure and blood flow through skeletal muscle, myocardium, intestine, kidney or brain is rather remarkable because, over a certain

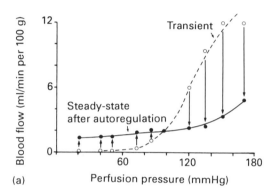

(a)

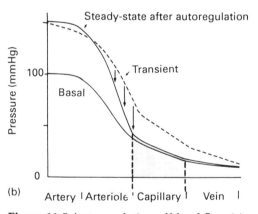

(b)

Figure 11.5 Autoregulation of blood flow (a) and of capillary pressure (b) in isolated, perfused skeletal muscle. The dashed line in each plot shows the transient flow or pressure immediately after changing perfusion pressure from its control level of 100 mmHg. Arteriolar contraction/dilatation (arrows) then adjusts the flow and downstream pressure to a steady-state value (solid curve) that is only slightly different from the control value. ((a) from Jones, R. D. and Berne, R. M. (1964) *Circulation Research*, **14**, 126, by permission)

range, changes in the perfusion pressure have relatively little effect on blood flow in the steady state, seemingly in defiance of Poiseuille's law (see Figure 11.5a). This relative constancy of tissue perfusion in the face of pressure changes is called autoregulation. Autoregulation occurs independently of the nervous system; the resistance vessels respond directly to the changes in arterial pressure, a rise in pressure evoking vasoconstriction and a rise in vascular resistance while a fall in pressure evokes vasodilation and fall in resistance. This accounts for the near constancy of the flow. Cerebral blood flow, for example, is well maintained during spinal anaesthesia, despite the concomitant systemic hypotension induced by this procedure. Although autoregulation holds flow almost constant in the steady state, the underlying change in arteriolar radius takes 30–60 s to develop fully, so there is initially a brief rise in flow with pressure which affords a glimpse of the unregulated pressure-flow relation (see Figure 11.5a, dashed line).

As well as stabilizing the tissue perfusion, autoregulation also stabilizes capillary filtration pressure. Capillary pressure depends on the pre- to postcapillary resistance ratio (Section 9.2), and the autoregulatory changes in precapillary resistance protect the capillaries from changes in arterial pressure. In one study (see Figure 11.5b), in which skeletal muscle was perfused artificially by a pump at various pressures, the pressure in the muscle capillaries changed by only 2 mmHg as the pump pressure was varied between 30 mmHg and 170 mmHg. This autoregulation of capillary pressure assists the homeostasis of plasma and interstitial volumes, and protects the tissue against oedema if arterial pressure rises. Autoregulation of capillary pressure is particularly important in the kidney where it ensures a virtually constant glomerular filtration rate. Autoregulation operates only over a limited range of pressures, however, and cerebral blood flow, renal function etc. fall off during severe hypotension.

Two mechanisms mediate autoregulation in most organs, namely the myogenic response and vasodilator washout. The Bayliss myogenic response was described in Section 11.2. 'Vasodilator washout' refers to the effect of blood flow on locally-produced vasodilator substances: if blood flow is increased transiently by a rise in arterial pressure, vasodilator products of tissue metabolism are washed out and their interstitial concentration declines, allowing vessel tone to increase. The relative importance of the myogenic response and vasodilator washout has been assessed by the response to venous congestion. Venous back-pressure stretches the arterioles and this would be expected to elicit a myogenic constriction and an increase in vascular resistance, but venous congestion also reduces the pressure gradient driving flow, which would be expected to temporarily reduce the flow, reducing the washout of vasodilators and producing vasodilatation. In practice, venous congestion elicits vasoconstriction in the brain, intestine, colon, liver and spleen, indicating a myogenic predominance, while in skin and skeletal muscle the effect is variable, indicating an approximate equipotence of the two mechanisms.

It must be emphasized that while autoregulation is an intrinsic property of most vascular beds (except the lungs), this does not mean that blood flow and capillary pressure are necessarily constant in the intact animal. On the contrary, they are frequently altered by changes in sympathetic drive and changes in local metabolic rate, which reset autoregulation to operate at a new level. The steady-state line in Figure 11.5a, for example, is raised to a higher level (higher flow) during moderate exercise, but the line remains relatively flat, i.e. autoregulation continues at a re-set higher flow.

Metabolic hyperaemia

Blood flow increases almost linearly with metabolic rate in exercising skeletal muscle, myocardium and secreting exocrine glands (see Figure 12.3). This is called metabolic or active hyperaemia, and since perfusion pressure changes little (or not at all) the hyperaemia is evidently caused by a fall in vascular resistance. The fall is induced by vasodilator substances that are released by the tissue, e.g. adenosine, adenosine nucleotides, K^+ and H^+; their varied modes of action on VSM cells were described in Section 11.2. In addition, most of these agents inhibit the release of noradrenaline from sympathetic fibres ('neuromodulation', see Figure 11.9). Dilatation of the main conduit artery to the tissue often accompanies the dilatation of the resistance vessels due to flow-induced production of EDRF by the arterial endothelium.

In *rhythmically* exercising skeletal muscle, mean blood flow is increased but the flow oscillates violently, decreasing during each contraction phase as the vessels become compressed by the muscle (see Figure 11.6). Most of the hyperaemia occurs during the

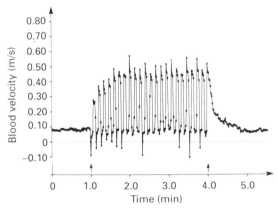

Figure 11.6 Blood flow in human femoral artery during 3-min period of quadriceps muscle exercise. Each contraction and relaxation phase lasted 4 s. Flow was measured continuously by the Doppler ultrasound method, and each cross marks the average flow during one cardiac cycle. Note the slow build-up to maximum response, creating a nutritional debt; the post-exercise period of hyperaemia repays the nutritional debt. (From Walloe, L. and Wesche, J. (1988) *Journal of Physiology*, **405**, 257–273, by permission)

resting phases: myoglobin within the muscle fibres provides a small oxygen reserve for the poorly-perfused contraction phase. Figure 11.6 also shows that the hyperaemic response takes a minute or so to develop fully, creating a blood-flow deficit or 'debt' over the first minute; the muscle's store of high-energy creatine phosphate is drawn on during this period, until oxygen supply catches up with demand. When the exercise stops, the hyperaemia takes 2–3 min to die away and this 'post-exercise hyperaemia' repays the metabolic debt. During *static* exercise, hyperaemia during the active period is less pronounced because the sustained muscle pressure upon the vessels limits their dilatation. Consequently an oxygen debt builds up more rapidly, leading to lactic acidosis and rapid muscle fatigue.

It is stressed that the sustained vasodilatation seen in active muscle and myocardium is caused by the locally-produced vasodilator substances and not by vasomotor nerves. It should also be noted that while the increase in blood flow is a local or 'automatic' process, it is not an example of autoregulation, for autoregulation is by definition a relative *constancy* of flow in the face of changes in arterial pressure.

Reactive hyperaemia

If the blood flow to a tissue is stopped completely for a period by compressing the supplying artery, or is slowed to the point where it is inadequate for tissue nutrition (a state called 'ischaemia'), the blood flow immediately after releasing the compression is much higher than normal, and then decays exponentially (see Figure 11.7). This phenomenon is called reactive hyperaemia or postischaemic hyperaemia: it is particularly obvious in the skin, which flushes a bright pink after a period of compression.

The myogenic response probably contributes significantly to the hyperaemia that follows brief arterial occlusions (<30 s), the arterioles dilating in response to the fall in transmural pressure. With longer periods of

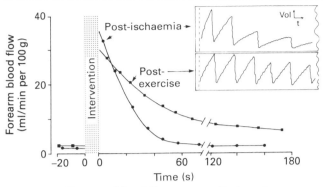

Figure 11.7 Forearm blood flow in a medical student measured by venous occlusion plethysmography (see Section 7.3) after 120 s of ischaemia (brachial artery occlusion) or after 30 s of strenuous forearm exercise. Note that blood flow remains elevated for a longer period after exercise. *Insets* show typical forearm volume traces during plethysmographic measurements. (Data from a first-year practical class)

occlusion, however, vasodilator substances accumulate and after 3 min of occlusion the vasodilatation is near-maximal in human limbs. Prostaglandins too play a part for the hyperaemia is reduced (but not abolished) by indomethacin, an inhibitor of cyclo-oxygenase. The greater the duration of the ischaemic period the greater is the accumulation of vasodilator substances and the greater the total cumulative blood flow afterwards; indeed, a hyperaemic plateau precedes the exponential decay after long occlusions. The functional importance of reactive hyperaemia lies in resupplying oxygen and nutrients to the ischaemic tissue as rapidly as possible.

Oxygen-derived free radicals and ischaemia-reperfusion injury

Recent work has indicated that reperfusion after long periods of ischaemia (over 30 min) can initiate tissue damage even though reperfusion is of course essential for the tissue's long-term survival. During prolonged anoxic periods the enzyme xanthine dehydrogenase is converted to xanthine oxidase, an enzyme capable of oxidizing hypoxanthine to xanthine when oxygen becomes available. Hypoxanthine itself is formed during the anoxic period as a breakdown product of ATP. Upon reperfusion, oxygen becomes available and the xanthine oxidase catalyses the oxidation of the hypoxanthine, but this oxidation reaction also results in the conversion of ordinary molecular oxygen into superoxide radicals ($O_2^-\cdot$) and hydroxyl radicals ($OH\cdot$). Radicals are highly reactive particles owing to a lone electron in the outer shell, and the hydroxyl radical in particular readily attacks cell membrane lipids, proteins and glycosaminoglycans. This damages the tissue and the capillary wall, causing leucocytes to adhere to the wall and obstruct flow ('no-reflow' phenomenon). Such damage can be attenuated by pretreatment with allopurinol (an inhibitor of xanthine oxidase) or superoxide dismutase (a superoxide scavenger) or dimethyl sulphoxide (a hydroxyl radical scavenger). Reperfusion injury is thought to contribute to myocardial, intestinal and brain damage after thrombotic episodes, and possibly to joint damage in rheumatoid arthritis.

11.4 Nervous control: sympathetic vasoconstrictor nerves

Local mechanisms serve only local needs. To serve the more general needs of the whole organism, such as the homeostasis of arterial pressure and core temperature, the central nervous system superimposes a sentient control system over the circulation. The efferent limb of this extrinsic system comprises autonomic vasomotor nerves and endocrine secretions, and the afferent limb involves sensory inputs which are described in Chapter 13. The autonomic vasomotor nerves fall into three classes: sympathetic vasoconstrictor fibres, sympathetic vasodilator fibres and parasympathetic vasodilator fibres; this terminology refers to the effect of stimulating (cf. inhibiting) the nerve. Of these the sympathetic vasoconstrictor fibres are the most widespread and important. Students accustomed to associating the sympathetic system with alarm and dilatation should note that the *vast majority of sympathetic vasomotor fibres are in fact vasoconstrictor fibres.*

Anatomy of the sympathetic vasoconstrictor system

The pathway controlling the sympathetic noradrenergic vasoconstrictor fibres begins in the brainstem. From here, descending excitatory and inhibitory fibres called bulbospinal fibres pass down the spinal cord and synapse with *sympathetic preganglionic neurones* in the intermediolateral columns of the grey matter (thoracicolumbar segments T1 to L3, see Figure 11.8). The output of these spinal neurones depends on the interplay of excitatory and inhibitory inputs by the bulbospinal fibres, plus local spinal inputs. The sympathetic preganglionic axons travel via the ventral roots of the spinal nerves and white rami communicantes into the sympathetic chains and may travel up or down the chain for several segments before synapsing with postganglionic neurones

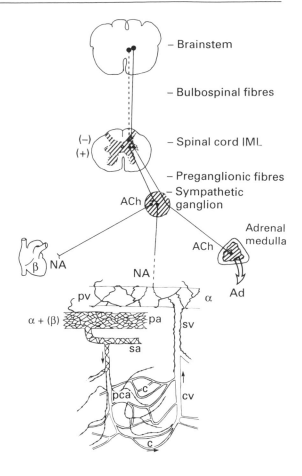

Figure 11.8 Schematic diagram of sympathetic noradrenergic innervation of the cardiovascular system. α and β refer to predominant adrenoceptors on the end-organs. NA, noradrenaline; Ad, adrenaline; ACh, acetylcholine: postganglionic cotransmitters not shown; IML, intermediolateral horn of grey matter in segments T1-L3. One descending tract (bulbospinal tract) excites the sympathetic IML cells by noradrenaline release; other bulbospinal fibres inhibit the cell, but the transmitters involved are complex and are omitted. pa, pv, primary artery and vein to an organ; sa, sv, small artery and vein; pca, precapillary arteriole; c, capillary; cv, collecting venule. (Drawing of gradient of innervation from Furness, J. B. and Marshall, J. M. (1974) *Journal of Physiology*, **239**, 75–88, by permission)

located in the sympathetic ganglia. Some fibres do not synapse until they reach more distant ganglia (coeliac and hypogastric ganglia) or the adrenal medulla. The preganglionic fibres are cholinergic and the receptors on the cell bodies of the postganglionic neurones are of the nicotinic variety, being blocked by hexamethonium.

The *postganglionic cells* of the sympathetic chain send non-myelinated axons through the grey rami communicantes for distribution in the mixed peripheral nerves; some fibres also course directly over the major vessels. The terminal fibres run along the outer border of the tunica media but, except in veins, they do not penetrate the inner part owing to the higher pressure there. Most small arteries and large arterioles are richly innervated whereas terminal arterioles are poorly innervated and are probably controlled chiefly by local tissue metabolites. This pattern is repeated in the splanchnic venous system, although venous vessels are in general less densely innervated, and the venous system of skeletal muscle receives almost no innervation.

Pharmacology of sympathetic vasoconstrictor nerves

Noradrenergic transmission The classical, long-recognized transmitter released by postganglionic sympathetic fibres is noradrenaline (norepinephrine in the American literature). This quickly diffuses across the junctional gap and binds to α-adrenoreceptors on the VSM cell membrane. There are two kinds of α receptor (see Table 11.1): α_1-receptors are widely distributed over the VSM membrane, while α_2-receptors occur on the nerve fibre as 'pre-junctional receptors' (see below) and on certain blood vessels (e.g. human skin arterioles) along with the α_1-receptors. Activation of the postjunctional receptors elicits vasoconstriction by pharmacomechanical and electro-

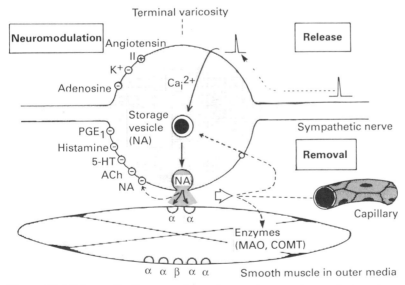

Figure 11.9 Schematic diagram of noradrenergic neurotransmission at the junctional varicosity of a sympathetic vasoconstrictor nerve. Not to scale (the varicosity is actually approximately 2 μm long × 1 μm wide, the axon 0.1–0.5 μm wide and the smooth muscle cell 4 μm wide at the centre). NA, noradrenaline; for other abbreviations see text. ATP and neuropeptide Y, which act as co-transmitters with NA in some fibres are omitted for clarity.

Table 11.1 Adrenergic receptors in the cardiovascular system

Receptor	Sub-type	Principal location and effect	Agonist	Antagonist	Medical use of antagonist
α		Vascular smooth muscle: vasoconstriction	Noradrenaline (NA) Adrenaline (Ad)	Ergotamine Phentolamine Phenoxybenzamine	Migraine Raynaud's vasospasm Acute hypertension (phaeochromocytoma)
	α_1	Postjunctional receptor of vascular smooth muscle: vasoconstriction	Specific agonist = phenylephrine Also NA and Ad	Prazosin	Anti-hypertension drug
	α_2	Prejunctional receptors of nerve varicosity: inhibition of NA release. Also vascular smooth muscle in skin	NA (and Ad) Clonidine	Yohimbine Rauwolscine	–
β		SA node, myocardium and arterioles of coronary and skeletal muscle and liver. Increased heart rate, contractility and vasodilatation	Specific agonist = isoprenaline Also NA and Ad	Propranolol Oxprenolol	Relief of angina by reducing cardiac work Hypertension
	β_1	Subtype found in pacemaker and myocardium	NA Ad	Practolol (toxic) Atenolol Metoprolol	Angina relief Hypertension Arrhythmia control
	β_2	Arterioles of skeletal muscle heart and liver: also bronchiole smooth muscle	Ad		

mechanical coupling as described in Chapter 10. About 80% of the noradrenaline is then taken back into the nerve by an active membrane process which terminates its action and restocks the terminal. To a lesser extent transmitter action is also terminated by diffusion into the nearby capillaries and by postjunctional degrading enzymes, namely catechol-O-methyltransferase and monoamine oxidase (Figure 11.9).

Local modulation of noradrenaline release

The amount of neurotransmitter released at the junction depends not just on impulse frequency but also on the chemical environment of the nerve ('neuromodulation'). As mentioned earlier, agents like H^+, K^+ and adenosine act on the nerve membrane to depress the release of transmitter (see Figure 11.9), and this contributes to metabolic hyperaemia. Most autocoids have a similar effect. Noradrenaline too binds to prejunctional α_2-receptors ('autoreceptors') and this inhibits further release of noradrenaline. By contrast, angiotensin II facilitates transmitter release and thereby amplifies vasoconstriction (see later).

Non-adrenergic transmission: role of ATP and neuropeptide Y

The vasoconstrictor reponse of the pulmonary artery and most veins to sympathetic stimulation is completely abolished by drugs that block α-adrenoreceptors (e.g. phentolamine, phenoxybenzamine), but the vasoconstrictor response of many systemic arteries and arterioles to sympathetic nerve activity is only partially prevented by α-blockers. This led to the discovery of additional neurotransmitters (*co-transmitters*) in the sympathetic varicosities, namely the purine ATP and the peptide 'neuropeptide Y' (see Table 11.2). Their relative abundance and importance vary from tissue to tissue. ATP is synthesized in the nerve terminal and is released along with the noradrenaline in some large arteries and small mesenteric arteries. It stimulates postjunctional purinergic receptors (P_2 receptors) and

Table 11.2 Main transmitter agents co-existing in perivascular nerves*

Sympathetic vasoconstrictor fibre	Noradrenaline (NA) Adenosine triphosphate (ATP) Neuropeptide Y (NPY)
Parasympathetic dilator fibre	Acetylcholine (ACh) Vasoactive intestinal polypeptide (VIP)
Sensory-dilator axons (C fibre)	Substance P (SP) Calcitonin-gene related peptide (CGRP) ATP

* From Burnstock, G. (1988) *Acta Physiologica Scandavica*, **133**, Suppl. 571, 53–57. The ratio of transmitter substances within the fibre varies from tissue to tissue

evokes a fast, brief depolarization (excitatory junction potential; Figure 11.10). Neuropeptide Y by contrast is synthesized in the postganglionic cell body and is transported slowly along the axon to the sympathetic terminals where it is probably stored along with noradrenaline in the large dense-cored vesicles. Neuropeptide Y has been identified in the vasomotor nerves of skeletal muscle, kidney, salivary gland, spleen and nasal mucosa. It is released chiefly in response to high frequency stimulation, which occurs naturally only under stress conditions, and it produces a much slower, more prolonged depolarization than ATP. Neuropeptide Y also acts as a neuromodulator, exerting a prejunctional inhibitory effect on noradrenalin release.

Tonic sympathetic activity and the effects of altered impulse frequency

Sympathetic vasoconstrictor nerves discharge continually at about 1 impulse per second or less in resting subjects: the maximum frequency is only 8–10 per s *in vivo*. The tonic activity of the system at rest, though low, contributes substantially to vessel tone and if it is interrupted by nerve sectioning or pharmacological blockade vasodilatation ensues. In resting skeletal

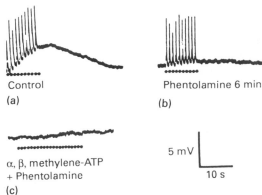

Control
(a)

Phentolamine 6 min
(b)

α, β, methylene-ATP
+ Phentolamine
(c)

5 mV

10 s

Figure 11.10 Evidence for co-transmission by
noradrenaline and ATP in sympathetic
vasoconstrictor nerves to the rat tail artery.
Intracellular potential recorded in smooth muscle
while the sympathetic nerve was stimulated at
each dot. Each stimulus produces a brief
excitatory junction potential (spike) plus a slower
depolarization (baseline under the spikes). The
spikes are not action potentials (see voltage
scale). Phentolamine, an α-adrenoceptor blocker,
abolishes the slow response, which is therefore a
'neurogenic alpha depolarization' (see Chapter
10). α, β-methylene ATP, a desensitizer of
purinergic receptors, abolishes the fast EJPs.
(After Sneddon, P. and Burnstock, G. (1984)
European Journal of Pharmacology, **106**, 149–152)

muscle, for example, interruption of the
tonic sympathetic drive increases the blood
flow from 2–5 ml min^{-1} 100 g^{-1} to
6–9 ml min^{-1} 100 g^{-1}; the latter flow,
however, is far from maximal owing to the
persistence of basal tone. Vasodilatation
induced by a *fall* in sympathetic noradrenergic nerve activity is physiologically very
important, being part of the baroreceptor
reflex which prevents excessive rises in
blood pressure (see Chapter 13). It is also
important in producing cutaneous vasodilatation during the regulation of body temperature. Neurogenic tone is in fact especially
well developed in the skin of the extremities, and it was the flushing of the rabbit ear
upon cutting the cervical sympathetic nerve
which led Claude Bernard in 1851 to
discover the sympathetic vasomotor nerves.

The effects of *increased* sympathetic vasomotor activity are illustrated in Figure 11.11.

1. Local blood flow is reduced. This can be
   sustained for hours in some tissues (e.g.
   skin) but in the intestine the arterioles
   quickly 'escape' from the vasoconstriction; the veins however do not.

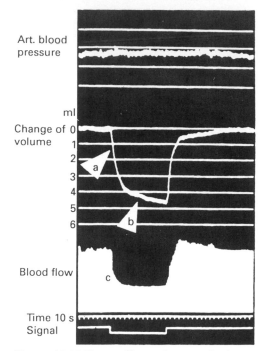

Art. blood
pressure

ml
Change of 0
volume 1
2
3
4
5
6

a

b

Blood flow c

Time 10 s
Signal

Figure 11.11 Three effects of sympathetic nerve
stimulation on the vasculature of cat
hindquarters. The volume of the hindquarters
was measured by a displacement method
(plethysmography), and blood flow was
measured by collecting the venous outflow.
Lumbar sympathetic nerves were stimulated at
just 2 impulses/s (signal). Arrow (a) indicates the
decrease in volume due to reduction in
capacitance vessel size. This is mostly secondary
to a fall in venous pressure induced by arteriolar
contriction; skeletal muscle veins have little
direct innervation. Arrow (b) indicates the slow
fall in volume due to capillary absorption of
interstitial fluid, secondary to fall in capillary
pressure induced by arteriolar contraction.
Arrow (c) indicates the fall in blood flow due to
contraction of resistance vessels. (From
Mellander, S. (1960) *Acta Physiologica
Scandinavica*, **50** (Suppl.), 176, by permission)

2. The volume of blood in an organ is reduced by active venoconstriction, which can displace 28 ml blood per kg in the intestinal tract and liver, and 15 ml per kg in skin. In skeletal muscle, the venous system lacks an effective innervation but nevertheless up to 7.5 ml/kg can be displaced passively because venous pressure falls secondarily to arteriolar contraction.

3. Capillary pressure is reduced by the arteriolar constriction causing a transient absorption of interstitial fluid into the plasma compartment.

4. If the increase in sympathetic outflow is widespread (which is not inevitably the case), the total peripheral resistance and cardiac output rise, altering the arterial blood pressure. The regulation of blood pressure is perhaps the single most important function of the sympathetic vasomotor system.

The above four effects (reduced peripheral flow, reduced peripheral blood volume, fluid translocation and blood pressure maintenance) together form part of a life-preserving response to haemorrhage and shock (see Chapter 15). It should be appreciated, however, that while the changes in sympathetic activity are sometimes widespread, as during a haemorrhage, altered discharge can also be confined to a single tissue (e.g. skin during temperature changes) and confined even to a particular kind of vessel (e.g. arteriovenous anastomoses in skin, see Chapter 12): the sympathetic activity is far from an 'all-or-none' affair.

Sympathetic activity fluctuates in phase with respiration. In conjunction with sinus arrhythmia, this produces small oscillations in blood pressure in phase with respiration, called Traube-Hering waves, but these have no known functional significance.

11.5 Vasodilator nerves

In a limited number of tissues, the arterioles are innervated by vasodilator fibres as well as by the ubiquitous sympathetic vasoconstrictor fibres. Vasodilator fibres occur within the sympathetic, parasympathetic and sensory systems and, unlike the vasoconstrictor fibres, they are not tonically active.

Sympathetic vasodilator nerves

Sympathetic cholinergic fibres to carnivore muscle vessels: the alerting response

In carnivores such as the dog and cat the arterioles of skeletal muscle are innervated not only by sympathetic vasoconstrictor nerves but also by sympathetic vasodilator nerves whose neurotransmitter is acetylcholine. Selective excitation of the sympathetic cholinergic nerves causes vascular relaxation and increased muscle blood flow. The effect is mediated by muscarinic receptors, being blocked by atropine, but whether the acetylcholine is acting directly on the VSM or is acting indirectly by stimulating the endothelial cell to produce EDRF is less clear. A direct effect on VSM in the outer media seems more likely, in that (1) the dilator fibres terminate at the adventitia-media border, not the endothelial layer, (2) acetylcholine is rapidly degraded *in vivo* by cholinesterase, so is unlikely to survive the journey to the endothelium, and (3) the relaxant effect of acetylcholine released by dilator fibres to lingual artery VSM persists after endothelial destruction (see Figure 11.13). Evidently, not all VSM responds identically to acetylcholine, for aortic and pulmonary artery VSM contracts in response to acetylcholine after endothelial destruction (see Figure 11.4). This puzzle awaits resolution.

The sympathetic cholinergic system differs from the vasoconstrictor system in more ways than one (see Table 11.3). The cholinergic system is controlled by the forebrain and is activated solely as part of the 'alerting response' to fear and danger; the central fibres do not synapse in the brainstem vasomotor regions; the distribution is confined to the skeletal muscle vasculature of some species; the response is

Table 11.3 Comparison of sympathetic vasoconstrictor and vasodilator nerves

Feature	Sympathetic constrictor nerve	Sympathetic dilator nerve
Main neurotransmitter	Noradrenaline	Acetylcholine (and VIP)†
Distribution	Most organs and tissue	Restricted; skeletal muscle and sweat glands
Tonically active	Yes	No
Central control	Brainstem	Forebrain
Involvement in baroreceptor reflex	Major factor governing activity	Negligible
Role in blood pressure homeostasis	Very important	–
Duration of effect	Mostly well sustained	Transient

† Vasoactive intestinal polypeptide occurs in vasomotor nerves to skin that contains sweat glands

only transient, and the fibres take no part in the baroreflex control of blood pressure.

The alerting or defence response, which is the natural stimulus for sympathetic vasodilator activity in carnivores, is induced by fear-flight-fight situations. It is tempting to assume that the increased blood flow 'improves' muscle nutrition in readiness for action, but measurements of microvascular permeability-surface area products (see Chapter 8) show that cholinergic vasodilatation does not increase the permeability-surface area product (unlike metabolic vasodilatation), so the nutritional benefit is very limited. Metabolic hyperaemia, by contrast, causes capillary recruitment which facilitates the transfer of all solutes. It cannot be emphasized too strongly that local metabolic factors, not sympathetic vasodilator nerves, cause the hyperaemia associated with normal, non-emotional exercise. Perhaps a greater advantage of the rapid cholinergic vasodilatation is that it prevents an excessive rise in blood pressure and afterload when the heart rate suddenly increases, as it does in the alerting response.

Stress-induced vasodilatation in the human forearm

Acute mental stress (e.g. mental arithmetic for most of us) causes a marked vasodilatation in human forearm muscle (see Figure 11.12), though not in calf muscle. The explanation of the vasodilatation is unclear for the net discharge frequency recorded directly in the sympathetic nerves to human forearm muscle is unchanged by mental stress. This could conceivably be the result of a simultaneous reduction in sympathetic vasoconstrictor fibre discharge and increase in sympathetic vasodilator fibre discharge, if the latter exist in man. The view that cholinergic vasodilator fibres innervate human (cf. carnivore) muscle is based on the partial inhibition of stress-induced vasodilatation by the local infusion of atropine, and by unilateral sympathectomy. Another factor contributing to human forearm dilatation in severe stress is the secretion of adrenaline by the adrenal medulla (see later).

Sympathetic NANC vasodilatation

A few tissues, such as the cat paw pad, are innervated by sympathetic vasodilator nerves whose neurotransmitter is neither acetylcholine nor noradrenaline. The non-cholinergic non-adrenergic (NANC) transmitter seems to be a neuropeptide called vasoactive intestinal polypeptidce (VIP). Immunocytochemistry shows that VIP is also present in vasomotor fibres close to human sweat glands, and VIP probably contributes to the intense atropine-resistant vasodilatation that accompanies sweating.

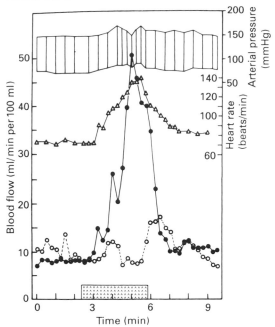

Figure 11.12 Increase in human forearm blood flow in response to stress. Closed circles, forearm blood flow (dominated by skeletal muscle blood flow); Open circles, hand blood flow (greater contribution from skin); triangles, heart rate. During the time represented by the rectangle the experimenters alarmed the subject by hinting at a sudden leak and severe blood loss. The increase in forearm flow is comparable with that in severe exercise, and greatly exceeds the maximum flow that withdrawal of sympathetic vasoconstrictor tone can produce. The response is attributed partly to the secretion of adrenaline and perhaps also to the existence of a sympathetic cholinergic (vasodilator) innervation, although the latter is more controversial. (From Blair, D. A., Glover, W. E., Greenfield, A. D. M. and Roddie, L. C. (1959) *Journal of Physiology*, **148**, 633–647, by permission)

(In the past, this hyperaemia was attributed to the kallikrein-bradykinin system, as was the analogous hyperaemia in salivary glands and pancreas – see later).

Parasympathetic vasodilator nerves

Parasympathetic preganglionic fibres are much longer than their sympathetic counterparts and leave the central nervous system in two outflows: the cranial nerve outflow (e.g. vagus) and the sacral spinal outflow. Their distribution is less universal than that of sympathetic vasoconstrictor fibres. They innervate the salivary glands and exocrine pancreas, the gastric and colonic mucosa, the genital erectile tissue, and cerebral and coronary arteries. The long preganglionic fibres synapse with postganglionic neurones within the end-organ, and these send short postganglionic fibres to the arterioles. The fibres are not tonically active, firing only when organ function demands a rise in blood flow.

Cholinergic and NANC-induced vasodilatation

The postganglionic fibres release the classical neurotransmitter acetylcholine, which hyperpolarizes some VSM cells to cause vascular relaxation (see Figure 11.13), resulting in an increased blood flow. The issue of whether neurally-released acetylcholine acts directly, or indirectly via EDRF is confused because in some vessels the vasodilator response appears to be endothelium-independent.

It has become clear in recent years that most parasympathetic postganglionic fibres can release not only acetylcholine but also non-cholinergic, non-adrenergic (NANC) transmitters with a vasodilator action. For example, nerve-induced vasodilatation in the rabbit lingual artery, which supplies the tongue and submandibular salivary gland, is only partially prevented by atropine (see Figure 11.13). The predominant NANC vasodilator transmitter is a neuropeptide, *vasoactive intestinal polypeptide* (VIP; see Table 11.2), which also occurs in sympathetic cholinergic fibres.

Dilatation in salivary glands and pancreas

Blood flow through the cat submandibular gland can be increased tenfold by stimulating the parasympathetic supply (chorda tympani nerve). The increased blood flow supplies the water for saliva formation,

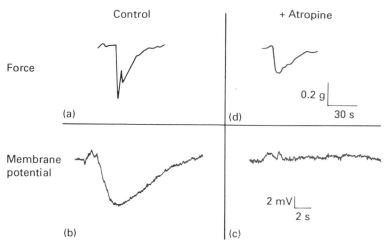

Figure 11.13 Mechanisms underlying neurogenic vasodilatation in the rabbit lingual artery. The noradrenergic fibres were blocked by guanethidine so that perivascular stimulation excited only parasympathetic responses. (a) Mechanical response, showing dilatation. (b) Membrane potential showing hyperpolarization (baseline −51 mV). (c) Total abolition of electrical response by atropine, establishing that the electrical response is purely cholinergic. (d) Dilatation, however, is only partially blocked by atropine, revealing the existence of non-cholinergic dilator transmitter, whose action does not involve hyperpolarization. This might be vasoactive intestinal polypeptide, which acts by stimulating membrane adenyl cyclase to produce an intracellular 'second messenger', cyclic adenosine monophosphate (cAMP). Removal of the arterial endothelium by rubbing did not abolish these responses. (From Brayden, J. E. and Large, W. A. (1986) *British Journal of Pharmacology*, **89**, 163–171, by permission)

which requires an enormous rate of fluid filtration across the gland's fenestrated capillaries; in fact the gland can secrete up to its own weight of fluid in just 1 min. The vasodilatation is due partly to acetylcholine (being partially blocked by atropine), and, at high stimulation frequencies, to the release of VIP. In addition, a vasodilator neuropeptide called substance P is released from parasympathetic fibres in the rat salivary gland. In the pancreas, VIP seems to be the main parasympathetic transmitter, rather than acetylcholine, i.e. these are true 'peptidergic' vasodilator nerves.

Role in erectile tissue
Sacral parasympathetic fibres innervate the vasculature of genital erectile tissue and the colonic mucosa. Stimulation of the pelvic nerve in a dog (parasympathetic fibres) causes a profound vasodilatation of the arterioles feeding the corpus cavernosum of the penis, reversing the usual balance of resistances: inflow resistance becomes less than outflow resistance. The sinuses of the corpus therefore fill with blood at a pressure close to the arterial level, creating distension and erection. The vasodilatation is mostly atropine-resistant, implying that transmission is mainly NANC. Immunocytochemistry reveals the presence of VIP in the parasympathetic nerves innervating the penile artery, and assays of the venous effluent show that VIP is released on pelvic nerve stimulation. Withdrawal of sympathetic vasoconstrictor tone may be a

supplementary factor in penile erection. Whatever the exact mechanism, the sacral parasympathetic nerves are truly essential to the existence of our species!

Vasodilatation induced by sensory nerves

The curious phenomenon of a sensory nerve having a motor function is illustrated by Lewis's triple response, which is the response of human skin to a mild trauma such as a scratch. This response was noted and investigated by Sir Thomas Lewis in 1927. The three components are: (1) a local redness along the line of the scratch caused probably by the release of K^+ ions and the vasodilator mediators of inflammation from activated cells, (2) a spreading flare, which is an area of redness that gradually extends laterally from the scratch line, for 2–3 cm, and (3) a local swelling or wheal along the scratch line, caused by inflammatory oedema. The flare is evidently mediated by sensory nerves because it is abolished by local anaesthetics, like lignocaine, and by sensory denervation. The ability of sensory nerves to cause cutaneous vasodilatation was demonstrated long ago by Bayliss, who stimulated the spinal nerve dorsal root antidromically, sending action potentials in the 'wrong' direction down the sensory nerves; this elicits a cutaneous vasodilatation. Antidromic activity probably explains how infection of a dorsal root by herpes zoster virus causes the segmental cutaneous hyperaemia characteristic of 'shingles'.

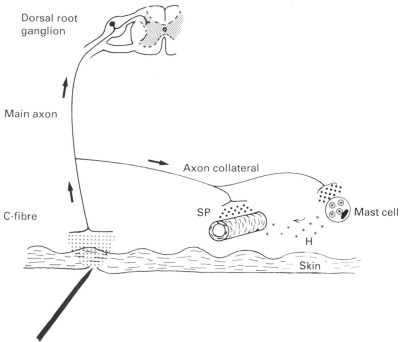

Figure 11.14 Vasodilatation induced by the release of a neuropeptide, substance P (SP, black dots) from a sensory terminal (nociceptive C fibre). Some fibres end close to mast cells, which are found around blood vessels; SP from such fibres stimulates the mast cell to release the vasodilator substance histamine (H shown as asterisks). An 'axon reflex' is believed to cause the rapid lateral spread of the response (the flare) when skin is traumatized with a needle. Since the axon spreads over only 0.5–1 cm (the receptive field), a flare of 2–3 cm is attributed to histamine triggering adjacent fibres. (After Foreman, J. C. (1987) *Allergy*, **42**, 1–11)

The sensory nerves mediating the vasodilatation are the nociceptive C fibres, whose terminals contain two vasodilator neuropeptides: *substance P* and *calcitonin-gene related peptide*. Substance P seems to be the main agent released on nerve activation, whether the fibre is excited antidromically or orthodromically (e.g. by capsaicin, the hot ingredient of chillis). To explain the rapid lateral spread of the flare in human skin (which does not occur in all species, and which is too fast to be explained by the diffusion of vasodilator), an 'axon reflex' was proposed by Lewis. It is supposed that the C fibre action potential, elicited by trauma in the skin, propagates centrally but also passes antidromically down a side branch of the axon to influence a blood vessel up to a 0.5–1 cm away (see Figure 11.14).

Histamine too is involved in mediating the flare. Many of the substance P-containing fibres terminate on mast cells, which possess substance P receptors. Activation of these receptors causes the mast cell to release its granules of histamine, a vasodilator agent. Since histamine also stimulates local C fibres, this could explain the lateral spread of the flare over 2–3 cm.

Substance P and histamine not only causes vasodilatation but also causes a pathological rise in microvascular permeability, leading to plasma exudation and high-protein oedema. This provides an explanation for the *neurogenic inflammation* which is elicited in skin and joints by antidromic stimulation of sensory C fibres.

11.6 Hormonal control of the circulation

Several endocrine secretions have acute effects on the heart and circulation, but these need to be viewed in perspective; in normal healthy animals hormones are of less importance for short-term cardiovascular regulation than is neural control. If however neural control is impaired, as in transplanted hearts or if a pathological event such as haemorrhage arises, then endocrine secretions become very important. Hormones such as aldosterone are also of major importance in the long-term regulation of plasma volume.

Adrenaline

The adrenal gland is situated at the upper pole of the kidney. Its medulla (core) secretes adrenaline (epinephrine) and noradrenaline (norepinephrine), which are known collectively as the catecholamines; adrenaline is a methylated form of noradrenaline. The medulla develops, embryologically, from postganglionic sympathetic neurones and it retains an innervation by preganglionic sympathetic fibres which run in the splanchnic nerve and control the gland. Although both adrenaline and noradrenaline are secreted, adrenaline forms over three-quarters of the secretion in man. (In diving mammals, by contrast, the secretion is mainly noradrenaline: this induces muscle vasoconstriction during dives and thereby conserves oxygen.) The plasma levels of adrenaline and noradrenaline are 0.1–0.5 nM and 0.5–3.0 nM respectively at rest, the higher level of noradrenaline being due to 'spillage' from the tonically active sympathetic terminals rather than glandular secretion.

The catecholamines are secreted in response to exercise, fear-flight-fight situations, hypotension and hypoglycaemia. During exercise the plasma adrenaline level can ready 5 nM and the noradrenaline level 10 nM, the latter again due chiefly to spillover from the increasingly active sympathetic terminals. Adrenaline affects both the heart and vasculature but these effects are quite small at physiological concentrations compared with the effects of the autonomic nerves and local factors. (The metabolic effects of adrenaline are at least as important as its cardiovascular effects,

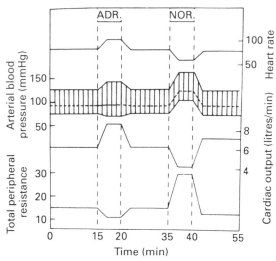

Figure 11.15 Comparison of effects of intravenous adrenaline and noradrenaline on the human circulation in the steady state. CO, cardiac output; HR, heart rate. For explanation, see text. An initial transient drop in blood pressure that occurs during adrenaline infusion is not shown here. (From the classic monograph of Barcroft, H. and Swan, H. J. C. (1953) *Sympathetic Control of Human Blood Vessels*, Edward Arnold, London)

namely the stimulation of liver glycogenolysis and fat lipolysis, which releases glucose into the blood stream.)

Adrenaline and noradrenaline show some similarities and some differences in their effect on the circulation (see Figure 11.15). Both hormones stimulate the cardiac β-adrenoceptors, so their direct action is to increase heart rate and contractility. Both hormones at physiological concentrations cause the arterioles and veins to contract in many tissues and at high concentrations both cause vasoconstriction in all tissues. This is due to activation of the VSM α-adrenoceptors. (Many students have an ingrained belief that adrenaline necessarily causes vasodilatation but this is simply not true.) As an exception to the rule 'catecholamines cause vasoconstriction', adrenaline at physiological concentrations causes vasodilatation in three tissues, namely skeletal

muscle, myocardium and liver. This is due to the abundance of β-adrenoceptors in these tissues, coupled with the high affinity of adrenaline for β-receptors. After β-blockade by propranolol, adrenaline causes vasoconstriction even in skeletal muscle because it activates α-receptors too. Noradrenaline normally causes vasoconstriction because it has a higher affinity for α-receptors than β-receptors.

Adrenaline and noradrenaline thus have opposite effect on skeletal muscle, the single most abundant tissue in the body (approximately 40% body weight), and as a result their overall effects on the systemic circulation differ considerably in an intact animal. Intravenous *noradrenaline* causes a generalized vasoconstriction which *raises* the peripheral resistance and blood pressure markedly (see Figure 11.15). This elicits a baroreceptor reflex which reduces the sympathetic drive to the heart and increases the parasympathetic drive. These reflexes slow the heart and reduce its output, offsetting the direct stimulatory effect of noradrenaline on the myocardium. Intravenous *adrenaline* by contrast *reduces* the total peripheral resistance slightly, because muscle vasodilatation outweighs vasoconstriction in other tissues. Mean blood pressure therefore changes little, and the direct stimulation of the heart by circulating adrenaline proceeds without significant opposition by the baroreflex. Since adrenaline is the predominant catecholamine secreted by the human medulla, the overall effect of adrenal gland stimulation is to increase cardiac output.

A rare tumour of the adrenal medulla, the phaeochromocytoma, secretes a mixture of catecholamines, causing hypertension. The latter can be treated with α-antagonists like phentolamine (see Table 11.1).

Vasopressin (antidiuretic hormone)

The remaining important hormones all have roles in the regulation of renal fluid excretion as well as vascular tone. Vasopressin is a peptide produced by the magnocellular neurones in the supraoptic and paraventricular nuclei of the hypothalamus. From the cell bodies the vasopressin is transported along the axons, through the pituitary stalk and into the posterior lobe of the pituitary gland. There the vasopressin is released into the blood stream.

The secretion of vasopressin is regulated partly by hypothalamic cells sensitive to tissue fluid osmolarity (osmoreceptors) and partly by cardiovascular pressure receptors. The main action of vasopressin at normal plasma levels is to promote water retention by the kidney. The cardiovascular effects of vasopressin are seen at higher concentrations, such as occur during haemorrhagic hypotension in response to reduced pressure receptor traffic (see Chapter 13). High concentrations of vasopressin cause a strong vasoconstriction in most tissues which helps to support arterial pressure and contributes to the pallor of the hypovolaemic patient. The cerebral and coronary vessels, by contrast, respond to vasopressin with an EDRF-mediated dilatation; vasopressin thus produces a redistribution of the cardiac output in favour of the brain and heart, as is appropriate in hypovolaemia. In dogs with diabetes insipidus and in Brattleboro rats, both of which lack vasopressin, blood pressure is abnormally depressed during dehydration or haemorrhage.

Renin-angiotensin-aldosterone system

Angiotensin II is a circulating octapeptide with a powerful vasoconstrictor action. Its production is initiated by an enzyme, renin, which is secreted into the bloodstream by the juxtaglomerular cells of the kidney in response to hypotension and renal sympathetic nerve activity. Renin enzymatically cleaves an α_2-globulin in plasma (angiotensinogen) to produce a peptide, angiotensin I. Angiotensin I is then modified by an enzyme on the surface of endothelial cells (converting enzyme) to form angiotensin II, this process taking place mainly in the lungs. At normal plasma concentrations, the main role of angiotensin II is to stimulate the secretion of aldosterone, an adrenal cortical hormone that causes the kidneys to retain salt and water. At higher concentrations, however, angiotensin II elicits vasoconstriction and does so in a rather interesting fashion. Not only does it directly stimulate the VSM cell, but it also enhances noradrenaline release by the sympathetic fibres (neuromodulatory action, see Figure 11.9), and also penetrates the blood-brain barrier at the area postrema to stimulate sympathetic activity (central action). Renin and angiotensin levels are particularly high after a haemorrhage and help to support the blood pressure in this situation.

Atrial natriuretic peptide (ANP)

This peptide is a relatively recent discovery, secreted by specialized myocytes in the atria in response to high cardiac filling pressures. In contrast to vasopressin and the renin-angiotensin-aldosterone system, ANP enhances the renal excretion of salt and water and has a modest relaxing effect on resistance vessels. It also reduces plasma volume to a greater extent that can be accounted for by diuresis alone, which may be due to a rise in capillary pressure and an increase in the hydraulic conductance of the capillary wall.

11.7 Special features of venous control

The venous system has been called 'the Cinderella of the circulation' because it is often neglected. The control of the peripheral capacitance vessels is however very important, because these vessels govern the distribution of blood volume between the periphery and thorax; they therefore regulate cardiac filling pressure and thereby influence stroke volume.

Differentiation of the venous system

With regard to control and function, the venous system may be divided into four parts (see Figure 11.16): the passive thoracic vessels and three peripheral systems.

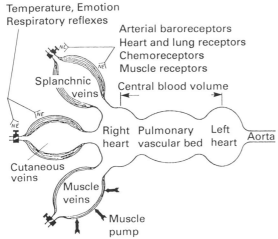

Figure 11.16 Differentiation of the venous system. Active changes in central blood volume and cardiac filling pressure are brought about by contraction of peripheral veins, especially the splanchnic (intra-abdominal) veins. (From Shepherd, J. T. and Vanhoutte, P. M. (1979) *The Human Cardiovascular System, Facts and Concepts,* Raven Press, New York, by permssion)

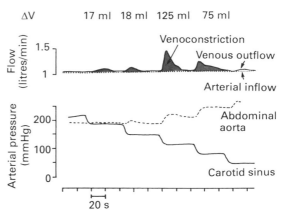

Figure 11.17 Active venoconstriction in the splanchnic circulation in response to reflex activation of the sympathetic vasoconstrictor nerves in the anaesthetized dog. Step reductions in carotid sinus pressure elicited a reflex increase in sympathetic outflow (see Chapter 13). Arterial inflow and the outflow from the inferior vena cava were measured. Each transient excess of outflow over inflow marks an episode of venoconstriction. ΔV is the total volume of blood displaced by each venoconstriction. Aortic perfusion pressure increases due to a sympathetically-mediated arteriolar constriction. (After Hainsworth, R. and Karim, F. (1976) *Journal of Physiology,* **262**, 659–677, by permission)

Splanchnic veins The veins of the gastrointestinal tract, liver and spleen contain about 20% of the total blood volume at rest. They are well innervated by sympathetic constrictor nerves and possess α-adrenoreceptors. They contract actively during exercise and hypotension due to increased sympathetic activity and circulating catecholamines (see Figure 11.17). This helps to maintain CVP at times of circulatory stress.

Skeletal muscle veins The intramuscular veins are very poorly innervated. Their volume is influenced chiefly by body posture (i.e. gravity) and by the muscle pump

(Section 7.7). Although direct sympathetic control of these veins is almost non-existent, their volume is nevertheless affected indirectly by sympathetic activity because arteriolar constriction reduces the downstream pressure, allowing the venous system to recoil elastically and displace blood centrally (see Figure 11.11).

Cutaneous veins These are richly innervated by sympathetic noradrenergic fibres. Their tone is greatly influenced by temperature: high core temperatures elicit a reduction in cutaneous sympathetic nerve activity producing venodilatation. Conversely, raised sympathetic activity causes the skin veins to contract strongly, as happens in hypotensive patients as part of the baroreflex. A sympathetic-mediated constriction of the skin veins is also elicited by emotional stress and by deep inspiration.

Comparison between venous control and arteriolar control

Veins response in the same way as arterioles to many stimuli – in the skin for example both are constricted during hypotension – but there are also differences between venous and arteriolar behaviour. Most veins and venules have little basal tone in the absence of sympathetic activity, unlike arterioles, and they show little myogenic response to stretch (except for the portal vein). The responses to certain hormones, autocoids and drugs also differ. Angiotensin II for example has little direct effect on veins but a powerful effect on arterioles. It does however act on the sympathetic terminals to potentiate the effect of sympathetic nerves on human veins (neuromodulatory action), and this action contributes to the intense venoconstriction seen in patients with cardiac failure, who commonly have high angiotensin levels. Histamine causes veins to constrict but arterioles to dilate. Glyceryl trinitrate has a greater dilator effect on veins than on arterioles, and its efficacy in relieving angina is now recognized as being due partly to the reduction of cardiac filling pressure (and therefore cardiac work) following venodilatation.

11.8 Summary

Although very complex in detail, the control of blood vessels can be conceptualized fairly simply as a hierarchy of three control systems, each able to override the lower one. The lowest level of control is the Bayliss myogenic response, which tends to maintain a constant blood flow and capillary pressure in the face of arterial pressure fluctuations (autoregulation). The second level of control is that exerted by local metabolites, which reset blood flow and autoregulation to a level appropriate to the tissue's metabolic activity. This process is predominant in the heart and brain, and in exercising skeletal muscle. The third level of control is the neuroendocrine system, which manipulates the vasculature of tissues like the skin and the splanchnic region for the benefit of the organism as a whole, or rather, for the benefit of the brain, which has ultimate control over the vasculature.

Further reading

Burnstock, G. (1986) The changing face of autonomic neurotransmission. *Acta Physiologica Scandinavica*, **126**, 67–91

Furchgott, R. F. and Vanhoutte, P. M. (1989) Endothelium-derived relaxing and contracting factors. *FASEB Journal*, **3**, 2007–2018

Johnson, P. C. (1986) Autoregulation of blood flow. *Circulation Research*, **59**, 483–495

Lisney, S. J. W. and Bharali, L. A. M. (1989) The axon reflex: an outdated idea or a valid hypothesis? *NIPS*, 4, 45–48

Lundberg, J. M., Pernow, J. and Lacroix, J. S. (1989) Neuropeptide Y: sympathetic cotransmitter and modulator? *NIPS*, **4**, 13–17

Lundgren, O. (1984) Microcirculation of the gastrointestinal tract and pancreas. In *Handbook of Physiology, Cardiovascular System, Vol. 4, Part 2, The Microcirculation* (eds E. M. Renkin and C. C. Michel), American Physiological Society, Bethesda, pp. 799–864

McCord, J. M. (1985) Oxygen radicals in ischaemic injury. *New England Journal of Medicine*, **312**, 159–163

Moncada, S., Palmer, R. M. J. and Higgs, E. A. (1988) The discovery of nitric oxide as the endogenous vasodilator. *Hypertension*, **12**, 365–372

Renkin, E. M. (1984) The control of the microcirculation. In *Handbook of Physiology, Cardiovascular System, Vol. 4, Part 2, The Microcirculation* (eds E. M. Renkin and C. C. Michel), American Physiological Society, Bethesda, pp. 627–688

Rothe, C. F. (1983) Venous system; physiology of the capacitance vessels. In *Handbook of Physiology, Cardiovascular System, Vol. 3, Peripheral Circulation*, Part 1 (eds J. T. Shepherd and F. M. Abboud), American Physiology Society, Bethesda, pp. 397–452

Share, L. (1988) Role of vasopressin in cardiovascular regulation. *Physiological Reviews*, **68**, 1248–1284

Sparks, H. V. (1980) Effect of local metabolic factors on vascular smooth muscle. In *Handbook of Physiology, Cardiovascular System, Vol. 2, Vascular Smooth Muscle* (eds D. F. Bohr, A. P. Somlyo and H. V. Sparks), American Physiological Society, Bethesda, pp. 443–474

Wallin, B. G. and Fagius, J. (1988) Peripheral sympathetic neural activity in conscious humans. *Annual Reviews in Physiology*, **50**, 565–576

Chapter 12
Specialization in individual circulations

A catalogue of all the factors controlling the circulation would be long and daunting, but fortunately rather fewer factors predominate in the day-to-day regulation of individual circulations, and one function of this chapter is to highlight which factors predominate in which tissue. Each tissue has its own special function, and this often calls for specialized vascular control. The skin for example regulates body heat, and its blood flow is controlled largely by temperature; whereas in the heart, metabolic rate predominates in vascular control. Five circulations are covered here (heart, skeletal muscle, skin, brain and lung), these being selected partly for their physiological importance and partly for their contrasting characteristics. Space prevents the inclusion of other major circulations (renal, hepatic, splenic and gastrointestinal). To unify the topic, the same approach is adopted for each system. First, the special tasks imposed by the tissue on its circulation are outlined. How these tasks are accomplished is considered under the headings 'structural adaptation' and 'functional adaptation'. Special vascular problems presented by the organ are assessed, and finally the measurement of blood flow in man is addressed.

12.1 Coronary circulation

Flow during basal cardiac output:
70–80 ml min^{-1} 100 g^{-1}
Flow during maximal cardiac work:
300–400 ml min^{-1} 100 g^{-1}

The right and left coronary arteries arise from the aorta immediately above the cusps of the aortic valve, the left coronary artery supplying mainly the left ventricle and septum and the right artery mainly the right ventricle – though this distribution is somewhat variable in man. Most of the venous blood drains via the coronary sinus directly into the right atrium (95%) and the rest

203

drains into the cardiac chambers via the anterior coronary and Thebesian veins. The coronary circulation is the shortest in the body, the mean transit time of the coronary blood being only 6–8 s at rest.

Special tasks

The coronary circulation must deliver oxygen at a high rate to keep pace with cardiac demand. Even in a resting subject the myocardial oxygen consumption is very high, namely approximately 8 ml of oxygen per min per 100 g: this is twenty times greater than in resting skeletal muscle. In exercise, where cardiac work rate can increase over fivefold, the coronary circulation must increase its delivery of oxygen correspondingly.

Structural adaptation

Myocardial capillary density is very high, there being 3000–5000 capillaries per mm^2 cross-section, or roughly one capillary per

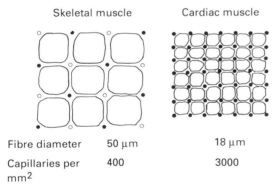

	Skeletal muscle	Cardiac muscle
Fibre diameter	50 μm	18 μm
Capillaries per mm^2	400	3000

Figure 12.1 Relative density of capillaries in skeletal and cardiac muscle on same scale. Each tissue has approximately 1 capillary per fibre but since the myocardial fibres are smaller, the capillary density is greater in myocardium and the diffusion distances are shorter. Open circles in skeletal muscle represent capillaries not perfused with blood at any given moment in resting muscle. (After Renkin, E. M. (1967) In *International Symposium on Coronary Circulation* (eds. Marchetti, G. and Taccardi, B.), Karger, Basel, pp. 18–30, by permission)

myocyte (see Figure 12.1). This facilitates the efficient delivery of oxygen and nutrients to the cell, partly by creating a very large endothelial area for exchange and partly by reducing the maximum diffusion distance to only 9 μm (the myocyte being approximately 18 μm wide).

Functional adaptation

High basal flow and high oxygen extraction In a resting subject, the blood flow per unit weight of myocardium is roughly ten times the average value for the whole body. Even so, the myocardium extracts 65–75% of the oxygen from the coronary blood, in contrast to the whole body average of 25% at rest (see Figure 12.2). The high extraction reduces the oxygen content from 195 ml/litre in arterial blood to only 50–70 ml/litre in coronary sinus blood. The corresponding oxygen pressure (P_{O_2}) in coronary venous blood is only 20 mmHg, and in the myocardial fibre itself the P_{O_2} is only about 6 mmHg, falling further during exercise. In heavy exercise, coronary oxygen extraction can rise to 90% leaving just 20 ml oxygen in each litre of venous blood at a P_{O_2} of only 10 mmHg. The extraction of fatty acid from coronary blood is also high (40–70%), but glucose extraction is usually low (2–3%), reflecting the substrate preference of myocardium.

Metabolic hyperaemia, the dominant control process The extra oxygen required at high work rates is supplied chiefly by an increase in blood flow rather than extraction; the latter can increase only modestly. Coronary blood flow increases in almost linear proportion to myocardial oxygen consumption at light to moderate work rates (see Figure 12.3), whilst at high work rates the flow increase lags a little and oxygen extraction rises. Myocardial metabolism evidently generates vasodilator messages in a quantitative manner – in other words this is a fine example of metabolic hyperaemia (Section 11.2). The nature of the vasodilator substance(s) however 'remains a

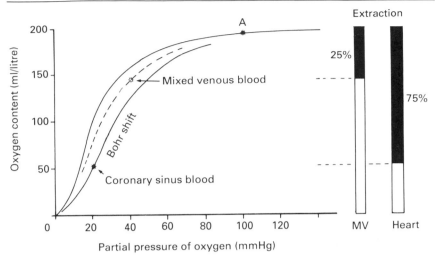

Figure 12.2 Oxygen carriage curves for arterial blood (P_{CO_2} 40 mmHg), mixed venous blood (P_{CO_2} 46 mmHg) and coronary sinus blood (P_{CO_2} 58 mmHg). Carbon dioxide displaces the oxyhaemoglobin dissociation curve to the right (the Bohr shift): this markedly enhances oxygen unloading in myocardium. A, arterial point; MV, mixed venous blood. Oxygen extraction from mixed venous blood and from coronary blood at rest are compared on right

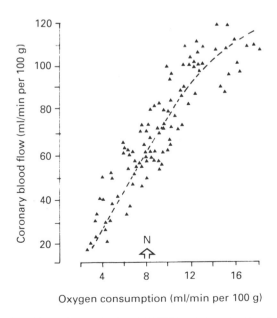

Figure 12.3 Effect of cardiac metabolic rate, as measured by oxygen consumption, on coronary blood flow in the dog. Arrow (N) marks normal values at resting cardiac output. Cardiac work was either increased above this point by intravenous infusions of adrenaline or reduced below normal by haemorrhage. The dashed line was fitted by eye to highlight the curvilinear tendency at high oxygen consumptions. (After Berne, R. M. and Rubio, R. (1979), see Further reading list, by permission)

well-sought but carefully guarded secret of nature'. The major contenders for the role are interstitial hypoxia and adenosine released by myocytes following ATP degradation. Both candidates dilate coronary arterioles, and the adenosine *versus* hypoxia controversy remains unresolved.

Autoregulation is well developed in the coronary circulation and is reset by metabolic vasodilatation to operate at a higher flow. Autoregulation protects myocardium against underperfusion during periods of low blood pressure, though only down to about 50 mmHg.

The coronary vasomotor nerves Myocardial arteries and arterioles are well innervated by sympathetic vasoconstrictor fibres, whose tonic discharge contributes to the arteriolar tone. This effect is overcome, in a graded fashion, during metabolic vasodilatation. If all the sympathetic fibres to the heart are excited, including those to the pacemaker and myocytes, the ensuing increase in heart rate and contractility raises the cardiac work and the concomitant metabolic vasodilatation outweighs the increased vasoconstrictor nerve activity: blood flow increases. This metabolic 'over-riding' of vasoconstriction ensures that any generalized activation of the sympathetic system does not adversely affect the blood flow to this vital organ. Adrenaline, secreted at times of stress, reinforces the coronary hyperaemia by preferentially activating β_2-adrenoreceptors on the coronary VSM, causing dilatation. There is some evidence that parasympathetic cholinergic fibres too can dilate coronary arteries but this is not thought to be important in exercise.

Special problems

Mechanical obstruction during systole
The branches of the coronary arteries within the myocardium are compressed during each systole. This effect is at its worst in the left ventricle during isovolumetric contraction when pressure within the ventricular wall reaches approximately 240 mmHg and

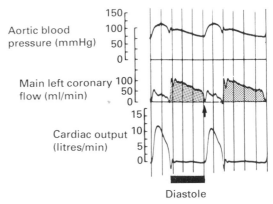

Figure 12.4 A record of flow in left coronary artery monitored by an electromagnetic flow meter in a conscious dog. Note the sharp curtailment of flow at the onset of systole (arrow); most coronary flow occurs during diastole (shaded area). Time lines 0.1 s. (After Khouris, E. M., Gregg, D. E. and Rayford, C. R. (1965) *Circulation Research*, **17**, 427–437)

coronary blood pressure is simultaneously at its nadir (approximately 80 mmHg). The reversal of the transmural pressure gradient across the coronary vessels transiently closes them so that coronary blood flow ceases briefly and even reverses in early systole (see Figure 12.4). A modest coronary flow is restored during the ejection phase as stress in the ventricle wall eases and arterial pressure rises, but flow is only restored fully during diastole. Roughly 80% of coronary flow occurs in diastole at basal heart rates.

Functional end-arteries, myocardial infarction and angina

Sudden obstruction Although cross-connections exist between the branches of the coronary arteries they are few in number and small in diameter in man (35 μm–500 μm), and can transmit only a low flow. Consequently, when an atheromatous artery is suddenly blocked by thrombosis, the residual blood flow to the tissue downstream is less than 10% of normal, which is insufficient to support normal contraction and metabolism (ischaemia). Human coronary arteries have therefore been called 'functional' end-arteries. The ischaemic myocardium becomes acidotic, causing severe cardiac pain and malfunction. The residual flow may be so low that myocytes begin to die after a few hours (necrosis see Figure 1.2). This sequence of events is called myocardial infarction or a 'heart attack', and is the single commonest cause of death in the West. Myocardial infarcts are most frequent and largest in the subendocardium (inner tissue) because the wall stress during systole is greatest here and selectively curtails endocardial blood flow at low perfusion pressures.

Slow obstruction If an atheromatous narrowing develops gradually, without abrupt thrombosis, the small anastomoses have time to enlarge and maintain a precarious nutritional flow. These vessels cannot, however, supply the additional flow needed during exercise or emotional stress, and such events trigger local ischaemia leading to acidosis and chest pain (angina pectoris). Angina can be relieved by nitrodilator drugs like glyceryl trinitrate which causes peripheral venodilatation and vasodilatation; this lowers the filling pressure and arterial pressure, which in turn reduces cardiac work and oxygen demand. β-adrenergic blockers like propranolol reduce oxygen demand by reducing heart rate and contractility.

Assessment of the human coronary circulation

The site of an atheromatous obstruction in a major coronary artery can be located by arteriography (coronary angiography), usually as a preliminary to vascular surgery. Blood flow can be measured more quantitatively by the coronary sinus thermodilution method, in which the sinus is cannulated, a bolus of cold saline injected and temperature recorded downstream (see Section 5.2 for thermodilution principle). To assess the

evenness of myocardial perfusion, a γ-emitting isotope such as thalium is injected at the root of the aorta and the isotope's appearance in myocardium is followed by an external γ-camera. Images of slices across the isotope-laden myocardium (tomograms) can be generated with this technique. Positron emission tomography (PET) is another method for imaging myocardial perfusion but it is available in very few centres.

Table 12.1 Summary of characteristics of the coronary circulation

Special tasks
   Maintain a high basal oxygen supply.
   Oxygen supply must keep pace with cardiac work.

Structural adaptation
   High capillary density, short diffusion path.

Functional adaptation
   High oxygen extraction (>60%).
   Metabolic vasodilatation dominant controlling factor.

Special problems
   Functional end-arteries, so risk of infarction and angina.
   Mechanical interference during systole.

Measurement in man
   Coronary angiography for sites of stenosis.
   Coronary sinus thermal dilution method for absolute flow.
   Isotope imaging for distribution of perfusion

12.2 Circulation through skeletal muscle

Flow in resting tonic (postural) muscle: about 15 ml min^{-1} 100 g^{-1}
Flow in resting phasic muscle: 3–5 ml min^{-1} 100 g^{-1}
Maximal flow during phasic exercise: >60 ml min^{-1} 100 g^{-1}

The circulations of skeletal muscle and cardiac muscle have much in common, for example the predominance of metabolic control, but there are also some marked differences, such as the greater role of nervous reflexes in the regulation of resting skeletal muscle flow.

Special tasks

1. During exercise, the circulation must deliver oxygen and glucose to the muscle fibres at an increased rate and remove waste products and heat at an increased rate.
2. The regulation of arterial pressure is a less obvious but very important duty of the skeletal muscle vasculature. Skeletal muscle constitutes about 40% of the adult body mass, and the resistance of this large vascular bed has a substantial effect on blood pressure.

Structural adaptation

Most human muscle fibres are phasically-active white fibres (twitch fibres), as in the forearm and gastrocnemius muscles. About 15% of human fibres, however, are tonically-active red fibres (slow fibres), and these predominate in postural muscles like the soleus. They are continuously active during the maintenance of posture and have a higher blood flow and capillary density than phasic muscle.

Functional adaptation

Importance of sympathetic vasoconstrictor innervation Skeletal muscle arterioles are richly innervated by sympathetic vasoconstrictor fibres whose tonic discharge enhances arteriolar tone in resting muscle: a high arteriolar tone is of course a prerequisite if dilatation (loss of tone) is to be possible. The vasoconstrictor nerve activity is controlled reflexly by blood pressure receptors in the thorax and neck (see Chapter 13), and when vasoconstrictor

activity is increased by these reflexes (as in orthostasis and hypovolaemia) the resistance of the muscle circulation increases. Aided by similar changes in the splanchnic and renal circulations this plays an important part in the regulation of arterial pressure, which depends on peripheral resistance and cardiac output. After a severe haemorrhage, the vasoconstrictor discharge to skeletal muscle reaches its maximum rate, namely 6–10 per s, and flow is reduced to around one-fifth the normal level – an exceedingly low perfusion indeed.

Dominance of metabolic vasodilatation during exercise During strenuous exercise the mean flow through phasic muscle can increase more than 20-fold; indeed, it is estimated that if all the muscle groups were maximally vasodilated at the same time the output capacity of the heart would be greatly exceeded. During strenuous exercise the muscle blood flow in fact accounts for up to 80–90% of the cardiac output (cf. 18% at rest). The muscle hyperaemia is due almost entirely to a fall in vascular resistance occasioned by metabolic vasodilatation, rather than to the relatively modest changes in arterial pressure. As in myocardium, the flow rises almost linearly with local metabolic rate, and the nature of the vasodilator agents is again controversial. During the first few minutes of exercise, potassium ions released locally by the contracting muscle raise the interstitial potassium concentration around the vessels, causing dilatation (see Section 11.2 for mechanisms). The local venous concentration of potassium is up to 0.1 mM higher than the arterial value, and this causes a gradual rise in mixed plasma potassium from approximately 4 mM (rest) to 5 mM over a 10-min period of submaximal exercise. A rise in interstitial osmolarity is important in the early stages of exercise too; the venous effluent osmolarity can increase by 20–30 mOsm. The vasodilator effect of both these factors is potentiated by local hypoxia. The osmolarity level soon decreases, and there are conflicting reports on the potassium levels over long periods, so it is far from clear what maintains the vasodilatation during prolonged exercise.

Only about a third of the capillaries in resting skeletal muscle is well perfused at any one instant (see Figure 11.1). Metabolic vasodilatation increases the well perfused fraction by dilating the terminal arterioles. This 'capillary recruitment' shortens the extravascular diffusion distances and thus speeds up the exchange process (see Figure 8.16 for the Krogh cylinder model of this process).

Variable oxygen extraction and 'oxygen debt' Resting skeletal muscle extracts only 25–30% of the oxygen from blood, whereas in severe exercise the extraction can reach 80–90%. Intracellular oxygen tension, which is only a few mmHg at the edge of the Krogh cylinder even at rest, falls so low in severe exercise that anaerobic glycolysis begins to predominate and lactic acid production increases. The quantity of lactate formed is an index of the deficit in oxygen supply and this 'oxygen debt' can reach several litres. The local lactic acidosis stimulates nociceptive C fibres, causing pain and the termination of violent exercise. The lactate also stimulates muscle 'work receptors' involved in the reflex control of the circulation (see Chapter 13). At the end of the exercise, a period of post exercise hyperaemia re-supplies the muscle with oxygen (which takes only seconds) and more gradually washes out the accumulated lactate and other vasodilator substances (see Figure 11.7). Only a little of the lactate is oxidized locally; most diffuses into the bloodstream and is either taken up by the liver for resynthesis into glycogen or by the heart as a primary substrate.

The skeletal muscle pump Although exercise hyperaemia is due chiefly to a fall in vascular resistance, the massaging effect of rhythmic muscle contraction on the deep veins assists limb perfusion, particularly in the calf. In a standing, stationary adult the arterial and venous pressures in the calf are each elevated by approximately 70 mmHg (effect of gravity), to approximately 165 mmHg and 80 mmHg respectively, giving a calf perfusion pressure (pressure difference) of 85 mmHg. During walking etc. the muscle pump lowers the venous pressure to approximately 35 mmHg (see Figure 7.21), which increases the pressure drop driving flow to 130 mmHg (i.e. 165–35), an increase of over 50%.

Two other special adaptations of the skeletal muscle circulation are the sympathetic vasodilator nerves found in predators like cats and dogs (see Section 11.5), and the vasodilator response of skeletal muscle arterioles to adrenaline (Section 11.6).

Special problems

Mechanical interference When skeletal muscle contracts with 30–70% of its maximum voluntary force, it compresses the intramuscular vessels sufficiently to impair the blood flow. In rhythmic exercise such as walking, this causes blood flow to oscillate (see Figure 11.6). In a sustained strong contraction, however, the impairment of flow is maintained: and since the store of oxygen in the muscle's myoglobin only suffices for 5–10 s of ischaemia, the fibres quickly become hypoxic and lactate accumulates, leading to pain and fatigue. The rapid loss of strength during a strong sustained contraction will be familiar to anyone who has struggled along with a heavy suitcase.

Fluid translocation across capillaries The problem of fluid translocation across capillaries in exercising muscle, leading to a 10–15% fall in plasma volume, was described in Section 9.9.

Measurement of human muscle blood flow

The soft tissue of a human limb is mainly skeletal muscle, and the limb blood flow can be measured by venous occlusion plethysmography (see Figures 7.4 and 11.7) or by a Doppler velocity meter over the principal artery (as in Figure 11.6). Local capillary perfusion rate can be estimated by the Kety tissue clearance method (see Figure 7.5) following an intramuscular injection of xenon-133.

Table 12.2 Summary of characteristics of the circulation through skeletal muscle

Special tasks
   Delivery of oxygen and nutrients in proportion to exercise intensity.
   Contributes to homeostasis of arterial pressure.

Structural adaptation
   High capillary density in tonic (postural) muscle.

Functional adaptation
   Sympathetic vasoconstrictor nerves dominate resting flow.
   Participates in vascular reflexes.
   Metabolic vasodilatation dominant during exercise.
   Sympathetic vasodilator nerves activated in fear-flight-fight response.
   Vasodilator response to adrenaline.
   Skeletal muscle pump.
   Variable oxygen extraction.

Special problems
   Mechanical interference during contraction.
   Increased capillary filtration in exercise.

Measurement in man
   Venous occlusion plethysmography.
   Doppler velocity meter.
   Kety's isotope clearance method.

12.3 Cutaneous circulation

Flow in a thermoneutral environment (27°C): 10–20 $ml\,min^{-1}\,100\,g^{-1}$
Minimal flow: 1 $ml\,min^{-1}\,100\,g^{-1}$
Maximal flow: 150–200 $ml\,min^{-1}\,100\,g^{-1}$

The skin of a 70 kg adult weighs 2–3 kg; its area is approximately 1.8 m^2 and its thickness 1–2 mm (epidermis and dermis combined). In contrast to the coronary circulation where metabolic rate dominates blood flow, the skin's metabolic requirements are modest, and blood flow is primarily controlled by the sympathetic vasomotor fibres whose activity is linked to temperature regulation.

Special tasks

1. Regulation of internal temperature. The temperature of the human 'core' (the brain, thoracic and abdominal organs) is normally kept within a degree or so of 37°C by balancing the internal heat production and the heat loss from the surface. The surface in question depends on the species: the skin in man, the tongue in dogs and the ears in rabbits. Heat is lost by four processes: radiation, conduction, convection and evaporation (see Figure 12.5). With radiation, the rate of heat loss is linearly proportional to the difference between ambient temperature and skin temperature. Skin temperature is influenced by the rate at which blood delivers heat from the core, i.e. by blood flow. In conduction-convection, warm skin heats up the adjacent air by conduction, and the warmed air is removed by convection (air currents). Again the rate of heat loss increases as a function of skin temperature. In the evaporation of sweat, 2.4 kJ of heat energy are consumed per gram of water evaporated (latent heat of evaporation), and both the water and heat are delivered to the skin by the blood. Cutaneous blood flow is thus a key factor in each case. The skin itself is a poikilothermic, not homeothermic tissue, and over short periods it can tolerate temperatures as extreme as 0°C and 45°C without damage.

2. Protection against the environment is the other major role of skin, and the circulation plays a part here too.

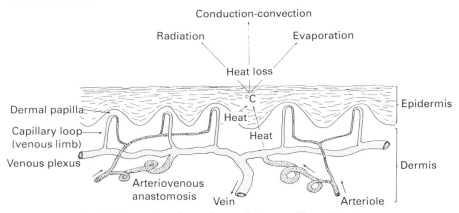

Figure 12.5 Sketch illustrating dermal vasculature and heat flux in an extremity. Arterial system stippled. See text for details

Structural adaptation: the arteriovenous anastomosis

Certain specific areas of skin possess abundant direct connections between dermal arterioles and venules, called arteriovenous anastomoses (AVAs, see Figure 12.5). These occur at exposed sites with a high surface area/volume ratio, namely the fingers and toes, palm and sole, lips, nose, ears and, in panting animals, the tongue. AVAs, which were first discovered in the rabbit ear by R. T. Grant in 1930, are coiled, muscular-walled vessels of average diameter 35 μm. They have little basal tone and are controlled almost exclusively by sympathetic vasoconstrictor fibres, whose activity is controlled by a temperature-regulating centre in the hypothalamus. When core temperature is high, vasomotor drive is reduced and the AVAs dilate. Being much wider than the terminal arterioles or capillaries, the dilated AVAs offer a low-resistance shunt pathway, which increases the cutaneous blood flow and delivers more heat to the skin. Heat readily crosses the walls of the dermal venous plexus that is fed by the AVAs, so skin temperature rises and heat loss increases. Conversely, AVAs are constricted to conserve heat under cold conditions. It is worth reiterating that the *dilatation* of AVAs *increases* heat loss, since students sometimes believe the opposite. The following 'aide

memoire' might well have been sung by the mud-loving hippopotamus in Flanders and Swan's famous song:

A – V – A's, let 'em flood,
Nothing quite like them
For cooling the blood.
So dilate them widely,
Lets lose heat right blithely –
But close them up tightly
When chill is the mud.

Functional adaptation

The response of skin blood flow to temperature is mediated by several different mechanisms: (1) ambient temperature influences skin temperature, which directly influences cutaneous vessel tone, (2) ambient temperature alters the activity of skin temperature receptors and this initiates a reflex change in sympathetic vasoconstrictor activity to the skin, and (3) core temperature is sensed by receptors in the brain, which modulate the sympathetic discharge to the skin.

Direct responsiveness to ambient temperature In a thermoneutral environment, namely 27–28°C for a naked man, skin temperature is around 33°C and the sympathetic vasoconstrictor nerves to the hands and feet (but not the limbs) are tonically active; this is proved by the hyperaemia of the extremities following surgical sympathectomy or a local nerve block. Local heating of the skin causes dilatation of the cutaneous arterioles, venules and small veins, and the skin reddens due to an increase on the volume of well-oxygenated blood in the dermal venular plexus. Conversely, local cooling to 10–15°C causes vasoconstriction and venoconstriction, which conserves heat (Figure 12.6a). The

mechanisms underlying this effect were described in Section 11.2.

The constriction of the large vein on the back of the hand in cold weather is easily seen, and this superficial venoconstriction diverts returning blood away from the cold surface into the deep veins. In the flippers of whales and the feet of wading birds, this is developed into an elaborate countercurrent heat-conserving mechanism: the cool cutaneous blood from the extremity drains into deep veins that ramify around the limb artery so that heat can pass directly from the warm arterial blood into the cool venous blood. The extremity is thus fed pre-cooled arterial blood, and heat-loss is reduced. The same short-circuiting of heat occurs in human limbs too, though to a lesser degree.

When the hand is placed in water at 10°C or less there is an initial cold-induced vasoconstriction followed by an increasingly painful sensation. After 5–10 min a dilatation sets in, with reddening of the skin and relief of pain – the 'paradox' of cold-induced vasodilatation (see Figure 12.6a). Paradoxical cold vasodilatation occurs in regions rich in AVAs and accounts

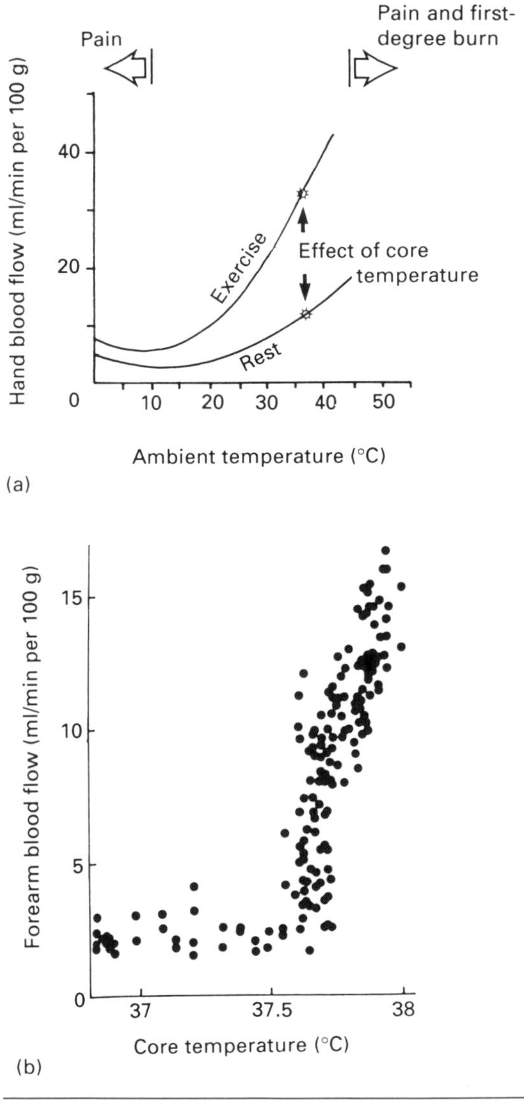

(a)

(b)

Figure 12.6 Effect of external and internal temperature on skin blood flow (a) Response of the hand to immersion in water at various temperatures, when the internal heat load is either low (rest) or high (exercise). Note the paradoxical vasodilatation at very low ambient temperatures. (b) Response of forearm blood flow, measured by plethysmography, to changes in internal temperature induced by leg exercise. The increase in blood flow arises in the skin; xenon clearance studies showed that forearm muscle flow actually decreased slightly. ((a) After Greenfield, A. D. M. (1963) *Handbook of Physiology, Cardiovascular System, Vol III, Part II, Peripheral Circulation*, (eds W. F. Hamilton and P. Dow), American Physiological Society, Bethesda, pp. 1325–1352; and (b) From Johnson, J. M. and Rowell, L. B. (1975) *Journal of Applied Physiology*, **39**, 920–924, by permission)

for the cold, red noses and hands seen in frosty weather. The cause of the vasodilatation is thought to be a paralysis of noradrenergic transmission. Its value lies in preventing skin damage during prolonged exposure to cold, and the response is particularly well developed in manual workers in cold climates such as Arctic Indians and Norwegian fishermen. If the exposure to cold persists, vasoconstriction recurs after a while and cutaneous perfusion oscillates with a 15–20 min cycle time (the 'hunting reaction').

Vasomotor reflexes initiated by skin temperature receptors If skin blood flow is measured in one hand while the opposite hand is immersed in cold water, a modest vasoconstriction is observed in the unimmersed hand. This is sympathetically-mediated bilateral reflex initiated by temperature receptors in the immersed hand.

Regulation of cutaneous perfusion by the hypothalamic temperature-regulating centre The activity of cutaneous vasomotor nerves is to a large extent governed by core temperature. Core temperature is sensed by warmth receptors in the anterior hypothalamus, and these hypothalamic neurones influence the brainstem neurones that control sympathetic vasomotor discharge to the skin. They also control the neurones governing sweating, i.e. the sympathetic cholinergic sudomotor nerves. A rise in core temperature thus leads to cutaneous vasodilatation (see Figure 12.6b) and sweating. In regions containing AVAs, the vasodilatation is due mainly to inhibition of sympathetic vasoconstrictor discharge to the AVAs. In regions lacking AVAs, namely the limbs, trunk and scalp, dilatation is associated with sweating and is due mainly to an active sympathetic vasodilator discharge, for it is abolished by local nerve block. This 'active' vasodilatation is impaired but not abolished by atropine, and is probably mediated partly by VIP (Section 11.5). Overall, the cutaneous response is finely graded: small central heat-loads lead to hyperaemia chiefly in the hands, feet and facial extremities while higher heat loads (core temperatures above 37.5°C) recruit the skin of the limbs and trunk.

Maximum cutaneous vasodilatation produces blood flows in excess of 5 litres/min in a 70 kg man. This necessitates a substantial rise in cardiac output and also a compensatory vasoconstriction in the splanchnic, renal and skeletal muscle circulations (to maintain blood pressure). If strenuous exercise is undertaken in a hot environment, the massive hyperaemia in both skin and muscle lowers the peripheral resistance greatly, and at the same time plasma volume declines due to sweating and fluid filtration into exercising muscle. As a result, the capacity of the heart to maintain blood pressure can be exceeded leading to hypotension and collapse (heat exhaustion).

In a cold environment, by contrast, the cutaneous blood falls as low as approximately $1\,\mathrm{ml\,min^{-1}}\ 100\,\mathrm{g^{-1}}$, which is only 20 ml/min for the entire body surface. This allows the full insulating action of the subcutaneous fat to operate, and protects the core temperature. If the vasoconstrictor response is prevented by inflammatory vasodilatation, as can happen with severe eczema and psoriasis, temperature regulation becomes very unstable and can even necessitate admission to hospital.

Role in regulation of arterial pressure and CVP The cutaneous circulation participates in many cardiovascular reflexes. Hypotension, caused by hypovolaemia or acute cardiac failure (shock) reflexly elicits a constriction of the cutaneous veins and arterioles producing the pale, cold skin characteristic of shock. The rise in cutaneous vascular resistance helps to support arterial pressure while the venoconstriction displaces blood centrally and helps to support central venous pressure. The life-preserving value of these responses became clear on the battlefields of France during World War I, when it was noticed that wounded men who were rescued quickly and warmed in blankets (producing cutaneous dilatation) survived severe haemorrhages less successfully than the men who could not be reached for some time and therefore inadvertently retained their natural cutaneous vasoconstriction.

Exercise initially evokes a sympathetically mediated vasoconstriction in skin, but this later gives way to dilatation if core temperature rises. Stimuli that elicit a defence (alerting) response, such as mental arithmetic, cause a transient cutaneous vasoconstriction: so too does a deep inspiration.

Dependency below heart level causes a strong vasoconstriction in the dependent skin (see Figures 7.5 and 9.4); cutaneous blood flow in the dependent foot falls to under a third of the supine value. The vasoconstriction seems to be caused by local rather than central mechanisms, namely the myogenic response and perhaps also a 'sympathetic axon reflex'. During orthostasis there is little sustained vasoconstriction in the non-dependent skin (e.g. forearm skin kept at heart level during an upright tilt), and this agrees with the finding that the impulse frequency in human cutaneous sympathetic nerves is unaffected by direct stimulation of the baroreceptor nerve in the neck.

Skin, mirror of the soul Poets and playwrights have long emphasized the responsiveness of the skin to emotion – we blush with embarrassment and blanch in response to stress or fear (defence response, above)*. Blushing is ill-understood because it is difficult to produce on demand. Folkow and Neil recall, in their classic text 'The Circulation', how they were unable to make a habitual blusher blush in a laboratory setting, either by insults or rude jokes, but when they disconnected their equipment and thanked the subject, she blushed violently. Blushing is often associated with emotional sweating, so it might be mediated by a similar mechanism, namely activation of sympathetic vasodilator fibres. A hyperaemic response to emotional stimuli is not confined to skin, having been observed in the gastric and colonic mucosa too.

Response to injury The Lewis triple response (Section 11.5) illustrates the cutaneous vascular response to trauma. The ensuing hyperaemia and increased capillary permeability enhance the delivery of the defensive elements (white cells and immunoglobulins) to the injured tissue.

Special problems

Mechanical interference Skin gets sat on, stood on and leaned on for long periods, and such compression impairs its blood flow. Ischaemic damage is normally prevented by the high tolerance of skin to hypoxia, combined with reactive hyperaemia on removal of the external stress and the onset of restlessness as metabolites accumulate and stimulate the skin nociceptors this makes any long-held position uncomfortable. The last mechanism fails to operate in certain categories of patient, namely paraplegics, the elderly, the enfeebled and the comatose. If such patients are not turned regularly they can develop severe ischaemic necrosis of the skin in the weight-bearing area (bed sores).

Problems during hot weather Dilatation of cutaneous veins in a hot environment can lower the central venous pressure and thereby predispose the subject to postural fainting, the classic example being the

*Juliet was evidently a blusher; her nurse, announcing Romeo's desire to marry her, remarked:

There stays a husband to make you a wife;
Now comes the wanton blood up in your cheeks,
They'll be in scarlet straight at any news!
(Romeo and Juliet, Act 2, Sc. 6)

Emotional trauma has the opposite effect; when Salisbury tells King Richard of his army's desertion, the King pales, crying:

But now the blood of twenty thousand men
Did triumph in my face, and they are fled,
And till so much blood hither comes again,
Have I not reason to look pale, and dead?
(Richard II, Act 3, Sc. 2)

guardsman who faints while standing at attention in hot weather. Cutaneous vasodilatation also increases the local capillary filtration pressure leading to interstitial swelling: this is why a ring often feels tighter on the finger during hot weather.

Measurement of human cutaneous blood flow

Measurements of skin surface temperature (thermography) have been used as an index of flow but this is unreliable because skin temperature depends on ambient temperature as well as blood flow. Venous occlusion plethysmography of a digit is often used as a measure of skin blood flow. Kety's isotope

Table 12.3 Summary of characteristics of the cutaneous circulation

Special tasks
  Temperature regulation.
  Response to trauma.

Structural adaptation
  Arteriovenous anastomoses in extremities.

Functional adaptation
  Sympathetic control dominant and regulated
    by core temperature receptors.
  Vessel tone also directly sensitive to local
    temperature.
  Dependent vasoconstriction by local
    mechanisms.
  Reflex vaso- and venoconstriction in response
    to hypotensive shock.
  Triple response to cutaneous trauma.

Special problems
  Compression when weight-bearing; bedsores.
  Hot weather causes local swelling and
    venodilatation, aggravating postural
    hypotension.

Measurement in man
  Digital plethysmography.
  Kety's isotope clearance method.
  Laser-Doppler flow probe.

clearance technique can be used to measure nutritive flow, and the new laser-Doppler method is gaining popularity as a rapid semi-quantitative measurement of superficial flow.

12.4 Cerebral circulation

Average flow (whole brain): $55\,\mathrm{ml\,min^{-1}}$ $100\,\mathrm{g^{-1}}$
Basal flow to grey matter: $100\,\mathrm{ml\,min^{-1}}$ $100\,\mathrm{g^{-1}}$

The adult human brain forms about 2% of the body mass and consists of 40% grey matter (mostly neurones) and 60% white matter (mostly myelinated tracts). The brain receives 14% of the resting cardiac output and most of this goes to the grey matter. One peculiarity of the cerebral circulation is that the arterioles are rather short and the arteries account for an unusually high proportion of the vascular resistance, namely 40–50%. Moreover, the arteries receive a richer autonomic innervation than the arterioles.

Special tasks

1. Need for a totally secure oxygen supply. Grey matter has a very high rate of oxidative metabolism, and its oxygen consumption (approximately $7\,\mathrm{ml\,min^{-1}}$ $100\,\mathrm{g^{-1}}$) accounts for nearly 20% of human oxygen consumption at rest. Grey matter is exquisitely sensitive to hypoxia and in man consciousness is lost after just a few seconds of cerebral ischaemia, with irreversible cell damage following within minutes. The primary task of the cerebral circulation, and indeed of the entire cardiovascular system, is therefore to secure an uninterrupted delivery of oxygen to the brain.
2. Adjustment of local supply to local demand. Many mental functions are localized in well-defined regions; for

example visual interpretation is located in the occipital visual cortex. External monitoring of the uptake of radiolabelled glucose and oxygen in the human brain has proved that local neuronal activity increases the local metabolic rate. Illumination of the retina, for example, increases the metabolic rate of the occipital visual cortex. The cerebral circulation must therefore be capable of regional adjustment to meet the varying metabolic rate of each region.

Structural adaptation

1. The circle of Willis. The arteries that enter the cranial cavity (the basilar and internal carotid arteries) anastomose around the optic chiasma to form a complete arterial circle, called the circle of Willis. (The young assistant employed by Thomas Willis to illustrate the circle later designed St. Paul's cathedral in London, being none other than Christopher Wren.) The anterior, middle and posterior cerebral arteries all arise from the circle of Willis, and this arrangement ought in principle to preserve cerebral perfusion even when one carotid artery becomes obstructed. This is so in young subjects but in elderly subjects the anastomoses are less effective. The main cerebral arteries divide into pial arteries running over the surface of the brain from where finer arteries penetrate into the parenchyma and give rise to the short arterioles.
2. High capillary density. Grey matter contains on average 3000–4000 capillaries per mm^2 cross-section (similar to myocardium). This provides a large exchange area and minimizes the extravascular diffusion distance.

Functional adaptation

High basal blood flow The grey matter receives about 100 ml blood $min^{-1} 100 g^{-1}$,

more than ten times the average for the whole body, and it extracts about 35% of the delivered oxygen.

Protection of cerebral blood flow by reflex control of other circulations Cerebral blood flow, like that to other tissues, depends on vascular conductance and arterial pressure. Unlike any other organ, however, the brain is able to safeguard its own blood supply by controlling the cardiac output and the vascular resistance of other organs via its autonomic outflow. Perfusion of peripheral organs (except the heart) is sacrificed to preserve cerebral perfusion, when necessary.

Cerebral autoregulation Autoregulation is very well developed in the brain; a fall in blood pressure causes the resistance vessels to dilate and thereby maintain flow (see Figure 12.7). Below approximately 50 mmHg, however, autoregulation fails and cerebral blood flow declines steeply

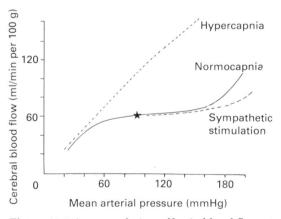

Figure 12.7 Autoregulation of brain blood flow at normal arterial P_{CO_2} (solid line); star represents normal operating point. Flow in the autoregulated range changes by only approximately 6% per 10 mmHg. High carbon dioxide tension causes vasodilatation (upper dashed line). Sympathetic nerve stimulation only affects flow significantly when arterial pressure is abnormally high (lower dashed line). (After Heistad, D. D. and Kontos, H. A. (1983, see Further reading list)

leading to mental confusion and syncope. The upper limit of autoregulation is probably around 175 mmHg. Cerebral autoregulation seems to involve both myogenic and metabolic mechanisms (Section 11.3).

Sensitivity to interstitial H^+ and arterial carbon dioxide Cerebral vessels have a well-developed response to arterial carbon dioxide: hypercapnia causes vasodilatation and hypocapnia vasoconstriction (see Figure 12.7). These effects are mediated by changes in the carbonic acid concentration, and therefore hydrogen ion concentration, in the interstitial fluid around the cerebral vessels. Other intravascular acids like lactic acid are ineffective because unlike carbon dioxide they cannot penetrate the blood-brain barrier. If arterial P_{CO_2} is reduced to 15 mmHg by hyperventilation (normal value 40 mmHg), cerebral blood flow halves and vasoconstriction can be observed directly in the retina, the retina being embryologically an extension of the brain. Owing to this effect, hysterical hyperventilation in adults or hyperventilation 'for fun' by schoolchildren can lead to disturbed vision, dizziness and even fainting. The traditional remedy is said to be a paper bag over the subject's head to cause rebreathing of expired carbon dioxide.

Cerebral vessels dilate in response to local hypoxia, but if the arterial blood is hypoxic it stimulates ventilation too, and the ensuing hypocapnia causes vasoconstriction.

Regional active hyperaemia in the brain It has long been known that shining a light onto one retina causes a rise in temperature in the corresponding occipital cortex. This was an early clue that neuronal activation evokes a local increase in blood flow. Deeper parts of the visual pathway, such as the lateral geniculate body, also display active hyperaemia. More recently, computer-assisted isotope imaging techniques have greatly extended our knowledge of active hyperaemia in the human cortex. Radioactive xenon-133 solution is injected into the internal carotid artery and its arrival and washout in the brain are monitored externally by an array of over 200 gamma-counters, each covering 1 cm² of skull. The rate of arrival and washout of the xenon is proportional to the local cortical blood flow, and the local counts can be converted by computer into a colour-coded map of regional flow. This has not only enabled flow to be studied but has also led to the localization of sites of mental function in the conscious human brain (see Figure 12.8). The cause of cortical metabolic hyperaemia is in part an increase in interstitial K^+ concentration, which can rise from its normal level of 3 mM to as much as 10 mM owing to outward currents from the active neurones. The mechanism of action of K^+ was described in Section 11.2. Other causative factors include a rise in interstitial H^+ concentration and adenosine concentration, secondary to increased neuronal metabolism.

Nervous innervation of intracerebral and extracerebral vessels The intracerebral arterioles are innervated rather poorly whereas the cerebral arteries outside the substance of the brain are well innervated by sympathetic vasoconstrictor nerves. Participation of brain vessels in the baroreceptor reflex is, however, negligible, which is clearly a 'good thing' teleologically. The maximal cerebral response to cervical sympathetic stimulation in anaesthetized humans is a mere 37% rise in vascular resistance (cf. 500–600% in skeletal muscle). The cerebral vessels have few α-adrenoreceptors and the vasoconstriction is probably mediated chiefly by neuropeptide Y, which is abundant in cerebral sympathetic fibres. The role of the vasoconstrictor innervation may be to protect the blood-brain barrier against disruption should arterial pressure rise suddenly.

Perivascular nerve fibres also contain 5-hydroxytryptamine, possibly co-stored with noradrenaline; 5-HT has a powerful vasoconstrictor effect on cerebral arteries

(a) Control:rest

(b) Hand movement

(c) Reasoning

Figure 12.8 Active hyperaemia in human cortex revealed by the xenon-133 imaging method (see text). The filled circles on the computer-calculated images show flows 20% above mean. In the resting, pensive subject (a) there is frontal lobe hyperaemia. On moving the contralateral hand voluntarily (b) there is hyperaemia of the hand area of the upper motor, premotor and sensory cortex. The reasoning test (c) evokes hyperaemia in the precentral and postcentral areas. (After Ingvar, D. H. (1976) *Brain Research*, **107**, 181–197 and Lassen, N. A. *et al.* (1978) see Further reading list, by permission)

and, along with a high K$^+$ level, is thought to contribute to the vasospasm that follows a subarachnoid haemorrhage.

Cerebral arteries are innervated by dilator fibres, probably of parasympathetic origin. They contain acetylcholine and VIP but their role is obscure. There are also abundant perivascular sensory fibres, which are thought to be the nociceptor fibres responsible for the vascular headache pain that accompanies strokes and migraines. The sensory fibres contain the vasodilator neuropeptides substance P and calcitonin-gene related peptide (CGRP), which in skin are implicated in antidromic vasodilatation (Section 11.5).

Blood-brain barrier Lipid-soluble molecules like oxygen, carbon dioxide and xenon diffuse freely between the plasma and brain interstitium but ionic solutes like the dye Evans blue fail to penetrate from plasma into most regions of the brain, demonstrating the existence of a barrier to lipid-insoluble solutes (see Section 8.9). Because of the blood-brain barrier, the neuronal environment is the most tightly-controlled cellular environment in the body, and the neurones are protected from the fluctuating levels of ions and catecholamines in the bloodstream. As J. Barcroft eloquently wrote in his book, *Architecture of Physiological Function*, 'To look for high intellectual development in a milieu whose properties have not become stabilized is to seek music amongst the crashings of a rudimentary wireless, or ripple patterns on the surface of the stormy Atlantic'. The blood-brain barrier can be disrupted experimentally by hyperosmotic infusions, and is disrupted clinically by acute hypertension or cerebral ischaemia. There are a few regions where the barrier is normally absent, these being regions where plasma solutes have access to receptors (e.g. to the osmoreceptors in the circumventricular region).

Special problems

Effect of gravity Gravity has no direct effect on cerebral flow during orthostasis because the cerebral circulation behaves like an inverted U-tube and the drag of gravity on the arterial limb is offset by the drag on the venous limb and on the cerebrospinal fluid (see Figure 7.19 and related text). Gravity does, however, influence cerebral blood flow indirectly in that central venous pressure and stroke volume are reduced in the upright posture. The ensuing postural hypotension can reduce cerebral flow to the point of dizziness or fainting (postural syncope) in the absence of brisk autonomic reflexes (see Chapter 14).

Effect of encasement in a rigid cranium
Except in the neonate, the brain is enclosed in a rigid bony box. Any space-occupying lesion (a cerebral tumour or haemorrhage) raises the intracranial pressure and forces the brainstem down into the foramen magnum. As the brainstem becomes compressed, altered activity in the neurones controlling the sympathetic system causes a rise in sympathetic vasomotor drive, and thus a rise in arterial blood pressure (Cushing's reflex). The rise in blood pressure maintains the cerebral blood flow for a while despite the raised intracranial pressure. The elevated blood pressure also evokes a bradycardia via the baroreceptor reflex, and the combination of bradycardia and acute hypertension is recognized by neurologists as the hallmark of a large, space-occupying lesion.

Measurement of human cerebral blood flow

The qualitative technique of carotid angiography (arteriography) is widely used to assess the patency and course of major cerebral vessels. A trans-cranial Doppler velocity meter is on trial for more quantitative measurement of flow in major cerebral arteries. SPECT imaging (single photon emission compound tomography) is a recent development of the xenon-imaging method for assessing regional perfusion. A lipophilic, high energy γ-emitting isotope is injected into the arterial supply and distributes within the brain in proportion to local blood flow. Owing to a chemical reaction within the brain, the radiolabelled solute is then trapped in the brain tissue for 4–6 h, thus 'freezing' a representation of the flow distribution. The pattern of distribution can then be mapped by using a large number of short-focus γ-cameras. This enables images of 'slices' (tomograms) through the brain to be generated.

Table 12.4 Summary of characteristics of the cerebral circulation

Special tasks
    Maintain oxygen supply to hypoxia-intolerant grey matter.
    Adapt local perfusion to local activity.

Structural adaptation
    Circle of Willis.
    High capillary density.
    Tight endothelial junctions (blood-brain barrier)

Functional adaptation
    High basal blood flow.
    Brain controls heart and peripheral resistance to maintain its perfusion pressure.
    Cerebral vessels 'excused' from baroreflex vasoconstriction.
    Good autoregulation in face of pressure changes; sensitive to P_{CO_2}.
    Local metabolic hyperaemia in response to local cortical activity.
    Blood-brain barrier provides a highly stable neuronal environment.

Special problems
    Postural syncope if baroreflex impaired.
    Space-occupying lesions lead to bulbar ischaemia.

Measurement in man
    Carotid angiography.
    Transcranial Dopplerimetry.
    Xenon-133 uptake method.
    SPECT imaging.

12.5 Pulmonary circulation

The lung circulation differs very substantially from the various systemic circulations. The entire output of the right ventricle flows through the alveoli, so their perfusion vastly exceeds their nutritional needs and metabolic factors exert no influence on flow. The metabolic needs of the bronchi are met by an independent systemic, bronchial circulation. There is only a low basal tone in the pulmonary circulation and no autoregulation. Sympathetic vasomotor nerves exist but have no well-defined physiological role. The features of importance are as follows.

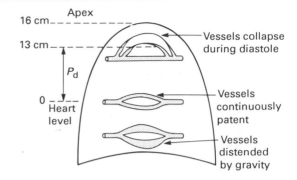

Figure 12.9 Sketch illustrating vertical gradient of perfusion in an adult lung in the upright position. Scale on left shows vertical distance above heart level. The diastolic arterial pressure (P_d, 13 cmH₂O at heart level) falls to atmospheric pressure at 13 cm above heart level (arrows). Vessels higher than this are therefore only perfused during systole. The vertical gradient has been demonstrated by injecting a radioisotope (xenon-133 or labelled microspheres) into the right heart and monitoring its arrival or deposition at various vertical positions by external gamma-counting

Low vascular resistance and pressures

Pulmonary vascular resistance is about an eighth that of the systemic circulation so the arterial pressure is low, typically 20–25 mmHg in systole and 6–12 mmHg in diastole. Pulmonary arteries and arterioles are shorter and have thinner, less muscular walls than systemic vessels and whereas arterioles dominate vascular resistance in most systemic organs, resistance in the lungs is shared between the arterial vessels (30%), microvasculature (arterioles to venules, 50%) and veins (20%). Consequently, the capillary pressure (8–11 mmHg) is roughly midway between mean arterial pressure (12–15 mmHg) and left atrial pressure (5–8 mmHg). The effect of the low capillary pressure on fluid balance is described in Section 9.6. Elevation of the left atrial pressure to 20–25 mmHg raises capillary pressure sufficiently to create pulmonary oedema, but smaller rises are within the safety margin against oedema (Section 9.10).

Vertical distribution of blood flow

Blood flow is distributed unevenly in the lung of an upright subject. The mean arterial pressure at heart level is around 15 mmHg, but owing to the effect of gravity the mean arterial pressure falls to about 3 mmHg at the apex of the lung and rises to 21 mmHg at the base of the lung. The high basal pressure distends the thin-walled vessels lowering their resistance and thus increasing flow through the base. Conversely, vessels at the apex actually collapse during diastole because diastolic pressure at heart level is only approximately 9 mmHg (13 cmH₂O), which is insufficient to distend the apical vessels, the apex being typically 16 cm above heart level. Apical perfusion therefore occurs only during systole and the mean apical flow is about a tenth the basal flow at rest. This is an important point, because the efficiency of oxygen transfer in the alveoli depends on the *ventilation/ perfusion ratio*, which should ideally be 0.8–1.0. Although ventilation too is greater at the base than at the apex, the vertical variation in ventilation does not fully offset the vertical variation in flow. A standing subject, therefore, has a higher ventilation/perfusion ratio at the apex than the base, and this ventilation/perfusion mismatch slightly impairs the efficiency of blood oxygenation in the upright, resting subject. In addition the capillary blood flow is pulsatile, especially at the apex, so oxygen uptake is pulsatile. When a subject is supine, the apex-to-base gradients are abolished and the ventilation/perfusion ratio becomes more even throughout the lung, increasing the effectiveness of blood oxygenation.

Vasoconstrictor response to airway hypoxia

The lungs possess a local mechanism that tends to optimize the local ventilation/perfusion ratio. If the ventilation to a local region is reduced or if local blood flow is increased, the alveolar oxygen content falls and its carbon dioxide content increases. The small pulmonary arteries which pass close to the surface of small airways respond to airway hypoxia by vasoconstriction, while the bronchiolar smooth muscle responds to airway hypercapnia by relaxation; together these changes maintain an optimal ventilation/perfusion ratio. The vasoconstriction is thought to be mediated by some unidentified vasoconstrictor intermediary rather than by hypoxia itself because pulmonary arterioles isolated from the surrounding lung tissue are unresponsive to hypoxia. At high altitude, hypoxia-induced pulmonary vasoconstriction causes high-altitude pulmonary hypertension, even in native residents (see Figure 12.10b).

Pressure-flow curve and the effect of exercise

Pulmonary vessels are essentially passive conduits and the downstream vessels in particular are very distensible. When the pressure driving fluid through an isolated, perfused lung is raised, with pulmonary venous pressure set below airway pressure, the pressure-flow is concave (see Figure

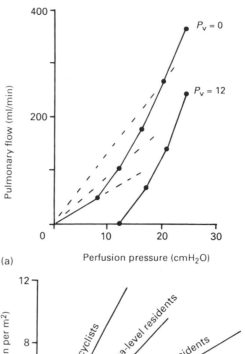

(a)

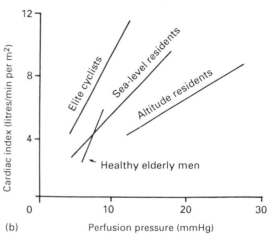

(b)

Figure 12.10 Pressure-flow relations for lung. (a) Isolated cat lung perfused with a plasma-dextran solution at varying arterial perfusion pressures. Venous pressure (P_v) was zero or 12 cmH$_2$O, and airway pressure was 4 cmH$_2$O. Dashed lines are lines of constant conductance; the lung curve cuts across lines of increasing conductance. (b) Conscious human subjects during 10-min periods of supine exercise. Perfusion pressure here means pulmonary artery pressure minus left atrial pressure. Venous pressures were greater than zero and airway pressure close to zero (atmospheric). For explanation, see text. ((a) From Banister, R. J. and Torrance, R. W. (1960) *Quarterly Journal of Experimental Physiology*, **45**, 352–357 and (b) from Grover, R. F. *et al.* (1983), see Further reading list, by permission)

12.10a), indicating that vascular conductance increases with perfusion pressure. This is probably due to vascular distension and to the opening up ('recruitment') of some venous vessels that were initially closed by the airway pressure. If venous pressure is set above airway pressure, as is normal during quiet respiration, the relation is less curved because all the downstream vessels are open; but there is still a modest rise in vascular conductance with perfusion pressure due presumably to vessel distension.

In human subjects, pressure and flow have been investigated during supine exercise, which causes both the flow (cardiac output) and pulmonary artery pressure to vary. A supine posture avoids the complication of changes in apical perfusion. In fit but elderly subjects, the curve again reveals an increase in vascular conductance as pressure rises (see Figure 12.10b). This is attributed to the increase in pulmonary venous pressure that these subjects develop during the exercise, which produces distension of the pulmonary vessels. The increased vascular conductance somewhat buffers the rise in pulmonary pressure during such exercise. In younger subjects, by contrast, left atrial pressure remains nearly constant during moderate supine exercise, and the pulmonary pressure-flow relation is virtually a straight line that projects almost through the origin. This indicates that the change in pulmonary vascular resistance with cardiac output is not very great in supine exercise, contrary to what is often believed. Physical training by endurance athletes (top cyclists) produces a chronic increase in pulmonary vascular conductance.

During upright exercise, a modest rise in pulmonary artery pressure is beneficial for it improves the perfusion of the apices and thereby increases the area of capillaries available for gas exchange. Exercise can increase the oxygen transfer capacity of the lungs by around 70% partly due to a rise in the perfused capillary area (by up to 40%; maximum area 90 m^2 in the adult lung), and

partly due to an increase in capillary blood volume (from approximately 85 ml to approximately 180 ml).

Capacitance function of pulmonary vessels

Pulmonary blood volume is about 600 ml in a recumbent man. Because the pulmonary vessels are thin-walled, their compliance is high. If intrathoracic airway pressure is raised by a forced expiration against a closed glottis (the Valsalva manoeuvre), the external pressure on the vessels can expel up to half of the blood content. Conversely, forced inspiration, which lowers intrathoracic pressure, can increase the human pulmonary blood volume to about 1 litre. The high capacitance of the pulmonary circulation allows it to act as a variable blood reservoir; for example, the pulmonary capacitance vessels act as a transient source of blood for the left ventricle when output begins to increase at the start of exercise. The capacitance may be influenced by sympathetic vasoconstrictor nerve activity.

Further reading

Berne, R. M. and Rubio, R. (1979) Coronary circulation. In *Handbook of Physiology, Cardiovascular System, Vol. 1, The Heart* (eds R. M. Berne and N. Sperelakis), American Physiological Society, Bethesda, pp. 873–952

Edvinsson, L. (1985) Functional role of perivascular peptides in the control of cerebral circulation. *Trends in Neurosciences*, **8**, 126–131

Grover, R. F., Wagner, W. W., McMurty, I. F. and Reeves, J. T. (1983) Pulmonary circulation. In *Handbook of Physiology, Cardiovascular System, Vol. 3, Peripheral Circulation* (eds J. T. Shepherd and F. M. Abboud), American Physiological Society, Bethesda, pp. 103–136

Heistad, D. D. and Kontos, H. A. (1983) Cerebral circulation. In *Handbook of Physiological, Cardiovascular System, Vol. 3, Peripheral Circulation, Part 1* (eds J. T. Shepherd and F. M. Abboud), American Physiological Society, Bethesda, pp. 137–181

Johnson, J. M., Brengelmann, G. L., Hales, J. R. S., Vanhoutte, P. M. and Wenger, C. B. (1986) Regulation of the cutaneous circulation. *Federal Proceedings*, **45**, 2841–2850

Laird, J. D. (1983) Cardiac metabolism and control of coronary blood flow. In *Cardiac Metabolism* (eds A. J. Drake-Holland and M. I. M. Noble), John Wiley, Chichester, pp. 257–278

Lassen, N. A., Ingvar, D. H. and Skinhoj, E. (1978) Brain function and blood flow. *Scientific American,* **239**(4), 50–59

Neuwelt, E. A. (ed.) (1989) *Breaking Through the Blood-Brain Barrier. Vol. 1 Basic Science Aspects.* Plenum Press, New York

Roddie, I. C. (1983) Circulation to skin and adipose tissue. In *Handbook of Physiology, Cardiovascular System, Vol. 3, Peripheral Circulation* (eds J. T. Shepherd and F. M. Abboud), American Physiological Society, Bethesda, pp. 285–317

Shepherd, J. T. (1983) Circulation to skeletal muscle. In *Handbook of Physiology, Cardiovascular System, Vol. 3, Peripheral Circulation, Part 1* (eds J. T. Shepherd and F. M. Abboud), American Physiological Society, Bethesda, pp. 319–370

Schelbert, H. R. and Buxton, D. (1988) Insights into coronary artery disease gained from metabolic imaging. *Circulation Research,* **78**, 496–505

Sobel, B. E. (1988) Coronary artery and ischemic heart disease. In *Cardiovascular Pathophysiology* (ed. G. G. Ahumada), Oxford University Press, Oxford

Chapter 13
Cardiovascular receptors, reflexes and central control

Overview

The heart and blood vessels are controlled by sympathetic and parasympathetic nerves whose activity is regulated and coordinated by the brain. The brain itself is guided by sensory information from peripheral receptors located both within the circulation and outside it. The three elements, namely afferent (sensory) fibres, central relays and efferent fibres, form reflex arcs, as illustrated in Figure 13.1. The two most important groups of sensors are the high-pressure receptors located around the walls of systemic arteries and the low-pressure receptors located in the heart. Together, their afferent fibres transmit information about arterial pressure and cardiac filling pressure to the brainstem, where it is integrated with information from other sensors such as chemoreceptors and muscle receptors. The process of integration and the computation of an appropriate cardiovascular response involves considerable up-and-down traffic between the brainstem, hypothalamus, cerebellum and cortex. Sympathetic and parasympathetic outflows then initiate appropriate responses in the heart and blood vessels and these responses are often (but not inevitably) directed at stabilizing blood pressure. The reflex response to a rise in blood pressure, for example, is bradycardia and peripheral vasodilatation which tends to lower the blood pressure back to its control level.

13.1 Arterial baroreceptors and the baroreflex

The prefix 'baro-' means pressure. Baroreceptors are sprays of non-encapsulated nerve endings found in the adventitial layer of arteries at two main locations: the aortic arch and the carotid sinus (see Figure 13.2). The *carotid sinus* is a thin-walled dilatation at the bottom of the internal carotid artery.

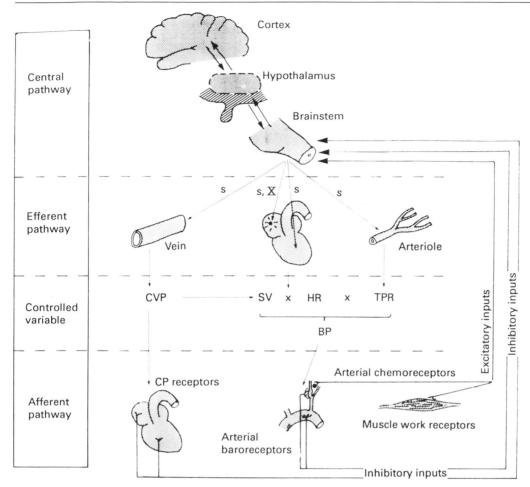

Figure 13.1 Schematic overview of the reflex and central control of the circulation. CP, cardiopulmonary receptor group (heterogeneous); CVP, central venous pressure; SV, stroke volume; HR, heart rate; TPR, total peripheral resistance; BP, arterial blood pressure; s, sympathetic fibres (noradrenergic); X, vagal cardiac fibres (cholinergic). Neuroendocrine reflexes and cerebellar relay are not shown. Terms 'inhibitory input' and excitatory input' refer to net effect of receptor activation on cardiac output and blood pressure

The afferent fibres form the fine carotid sinus nerve and then ascend in the glossopharyngeal nerve (IXth cranial nerve) to the petrous ganglion, where the parent cells are located. Like all afferent neurones, the petrous ganglion cells are bipolar and their centrally-directed axons continue with the glossopharyngeal nerve to enter the brainstem, where they terminate in the nucleus tractus solitarius (see later). The *aortic baroreceptors* are located mainly around the transverse arch of the aorta and their fibres form the aortic or 'depressor' nerve (in some species) before ascending in the vagus (Xth cranial nerve). The cell bodies lie in the nodose ganglion and the central axons again terminate in the nucleus tractus solitarius.

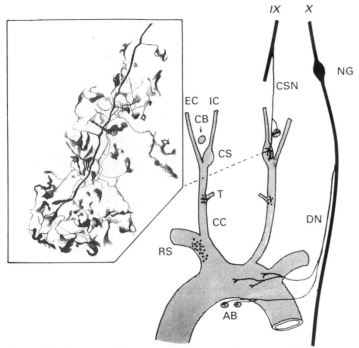

Figure 13.2 Sketch of the arterial reflexogenic areas. CS, carotid sinus; CB, carotid body; AB, aortic bodies. Fibres are shown innervating the two major receptor areas (left side only shown). Minor receptor areas are represented by dots. Nerves are shown only for the left side: CSN, carotid sinus nerve (nerve of Hering); DN, aortic nerve (depressor nerve); IX, glossopharyngeal; X, vagus; NG, nodose ganglion. Arteries: RS, right subclavian; CC, common carotid; IC, internal carotid; EC, external carotid. *Inset* shows extensive ramification of a single baroreceptor fibre-ending in the human carotid sinus, ×300. (From Abraham, A. (1969) *Microscopic Innervation of the Heart and Blood Vessels in Vertebrates including Man* (Pergamon Press, Oxford)

Receptor properties: dynamic sensitivity, threshold and range

The baroreceptors are actually mechanoreceptors; they respond to stretch rather than pressure, and if stretch is prevented by applying a plaster cast around the artery the baroreceptors fail to respond to pressure. Normally, however, a rise in arterial pressure causes arterial distension which in turn excites the mechanoreceptors. In the carotid sinus the transduction of the pressure signal

into stretch is enhanced by the thinness of the tunica media which makes the wall stretchy and allows a 15% change in diameter for a normal pulse pressure.

Static and dynamic responses When the carotid sinus is distended by a controlled, sustained rise in blood pressure (see Figure 13.3, fibre 1), an individual baroreceptor fibre fires an initial burst of action potentials as the pressure is changing (the dynamic response), and this signals the rate of

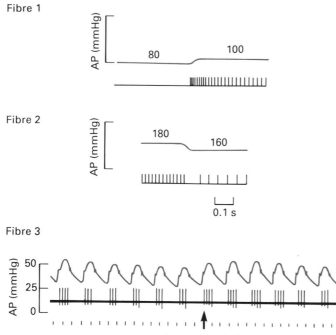

Figure 13.3 Characteristics of three single baroreceptor afferent fibres. Fibres 1 and 2. Action potentials recorded from myelinated afferents of the cat carotid sinus. Fibre 1 illustrates response to a sustained elevation of arterial pressure (AP); numbers refer to sinus pressure in mmHg. Fibre 2 illustrates response to a fall in pressure. A dynamic response is followed by adaptation to a quasi steady state in each case. Fibre 3. Single baroreceptor fibre from the rabbit aorta subjected to a normal pulsatile pressure. There are 4 impulses per pulse when blood pressure is high (inspiration, arrow) and only 3 when it is lower (expiration). Time intervals 0.1 s. (Fibres 1 and 2 after Landgren, S. (1952) *Acta Physiologica Scandinavica*, **26**, 1–34 and fibre 3 from Downing, S. E. (1960) *Journal of Physiology*, **150**, 210–213, by permission)

change of pressure. The activity then falls back somewhat and settles down to a sustained rate (the adapted response) that signals the new pressure level: the adaptation is probably due to mechanical 'creep' of the receptor within its visco-elastic environment. Baroreceptors are also dynamically sensitive to a fall in pressure; they transiently fall silent when pressure is reduced, then resume activity at a new slower rate. As a result of this dynamic sensitivity many baroreceptors *in vivo* fall silent during

diastole and their normal activity pattern consists of bursts of potentials during systole (see Figure 13.3, fibre 3).

Threshold 'Threshold' refers here to the pressure at which the afferent fibre first begins to discharge action potentials. Baroreceptors fall into two main classes in this respect. The large-diameter, myelinated fibres (*A fibres*) have variable but low thresholds (often approximately 50–60 mmHg) and a maximum saturated response at 80–100 mmHg; such fibres are best suited for signalling moderate hypotension. There are also more numerous, thinner non-myelinated fibres (*C fibres*) that have higher activation thresholds (mostly >80 mmHg), and are well suited for signalling acute hypertension. In general, the carotid sinus receptors have lower thresholds and greater sensitivities than the aortic receptors.

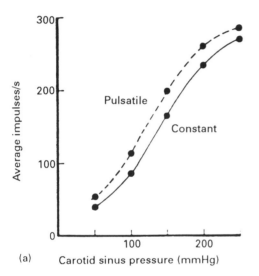

(a)

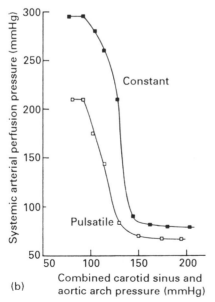

(b)

Range Each individual afferent fibre only responds over a limited pressure range, usually 30 mmHg or so. The baroreceptor nerve trunk, however, has a much wider operating range because it contains many A fibres and even more C fibres: when pressure rises, fibres of progressively higher threshold are 'recruited' and thereby extend the range of pressures signalled in the nerve trunk (see Figure 13.4a). The effective range of the entire baroreceptor reflex (cf. receptor range) can be defined as the steep part of the stimulus-response curve, and this range usually extends over 60–100 mmHg (see Figures 13.4b and also Figure 13.6).

Sensitivity to pulse pressure Carotid baroreceptors are sensitive to the pulse pressure as well as to mean pressure: the greater the oscillation in pressure about a mean the greater the aggregate activity in the nerve trunk at that pressure (see Figure 13.4a), and the greater the ensuing reflex (Figure 13.4b). This is partly due to the dynamic sensitivity of the individual receptors and partly to recruitment of high-threshold fibres during each systole. The signalling of pulse pressure is probably very important during orthostasis and during small haemorrhages, when there is often a fall in pulse pressure (due to a reduced stroke volume) without any fall in mean arterial pressure.

Figure 13.4 Effect of perfusion pressure on baroreceptor traffic and the evoked reflex in anaesthetized dogs. (a) Average rate of discharge in a multifibre preparation of the carotid sinus nerve when sinus pressure is varied. Pulsatile pressure evokes a higher net activity than a steady pressure. (b) Reflex fall in systemic pressure (due to bradycardia and vasodilatation) when pressure in a vascularly-isolated perfused baroreceptor region is increased. Pulsatility strengthens the reflex. ((a) From Korner, P. I. (1971) *Physiological Reviews*, **51**, 312–367 and (b) from Angell-James, J. E. and De Burgh Daly, M. (1970) *Journal of Physiology*, **209**, 257–293, by permission)

The baroreceptor reflex and short-term homeostasis of blood pressure

Reflex response to baroreceptor stimulation
The key features of the reflex evoked by baroreceptor activation (the baroreflex) were discovered by Ludwig and Cyon in 1866 who found that stimulating the aortic nerve caused a reflex depression of heart rate and blood pressure. Hering later observed the same reflex changes on stimulating the carotid sinus nerve (see Figure 13.5). In the intact animal any acute rise in arterial pressure increases the baroreceptor discharge rate and this, via a polysynaptic central pathway, enhances the vagal output to the heart and inhibits the sympathetic outflow to the heart and vasculature. Multifibre recordings from sympathetic nerves to human muscle show that the regular burst of activity accompanying each pulse disappears completely if blood pressure is rapidly raised above 150/90 mmHg. The inhibition is better sustained in laboratory animals than in man. The reduced activity in sympathetic vasoconstrictor nerves leads to vasodilatation and a fall in peripheral resistance. At the same time, the reduced cardiac sympathetic nerve activity and increased vagal activity cause a bradycardia and reduced contractility. Since blood pressure is the product of cardiac output and peripheral resistance these changes together tend to return arterial pressure to normal (see Figure 13.1). The baroreflex thus functions as a 'buffer', stabilizing the blood pressure against acute change.

Reflex response to baroreceptor unloading
From a clinical point of view, the baroreflex to a fall in pressure is perhaps more important than the response to a rise, since acute hypotension is a common medical emergency. As an experimental manoeuvre in man, external compression of the common carotid arteries is often used to reduce pressure in the carotid sinuses (which lie higher in the neck, close to the angle of the jaw). This manoeuvre reduces the baroreceptor input to the brainstem, as does hypotension in clinical situations, with the following reflex effects:

1. Tachycardia and increased myocardial contractility. Heart rate increases due to reduced vagal inhibition of the pacemaker and increased sympathetic activity. Increased contractility is mediated sympathetically.
2. Splanchnic venoconstriction. Reflex venoconstriction actively displaces blood from the gut and liver into the central veins (see Figure 11.17), which enhances stroke volume via the Frank-Starling mechanism. In skeletal muscle, the volume of blood in the veins (which are poorly innervated) declines secondarily to the fall in venous pressure caused by arteriolar contraction (see Figure 11.11). Both skin and muscle veins can also be constricted by reflexly secreted adrenaline, vasopressin and angiotensin (see later).
3. Arteriolar constriction. In most laboratory animals, the baroreflex elicits a

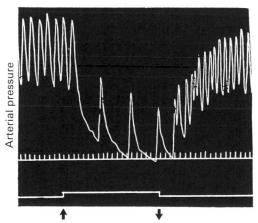

Figure 13.5 Effect of carotid sinus nerve activity on blood pressure and heart rate in the dog. Electrical stimulation of the carotid sinus nerve between the two arrows elicited a reflex hypotension and bradycardia. This classic smoked-drum recording is taken from the work of Hering, who discovered the carotid baroreflex in 1923. Time intervals approximately 0.2 s

sympathetically mediated vasoconstriction involving the skeletal muscle, skin, kidneys and the splanchnic vasculature: this raises peripheral resistance. In man, compression of the common carotid artery causes a strong reflex vasoconstriction in the splanchnic and renal circulations but relatively little change in the forearm and none in the calf. Direct stimulation of the human carotid sinus nerve, on the other hand, does inhibit directly-recorded activity in sympathetic nerves to muscle although sympathetic activity to the skin is unchanged. It seems, therefore, that carotid sinus baroreceptors do exert some reflex control over human muscle vessels (but not over the skin), but the sympathetic output to human muscle is probably influenced to a greater extent by the cardiopulmonary receptors (see later).

4. Adrenaline is secreted by the adrenal medulla in response to increased splanchnic nerve activity. Adrenaline stimulates the heart and enhances glucogenesis (Section 11.6).

5. Effects on extracellular fluid volume. The baroreflex affects fluid volume and distribution by several mechanisms.

    (a) Arteriolar vasoconstriction lowers capillary pressure producing an absorption of interstitial fluid and expansion of the plasma volume (Section 9.12).

    (b) Increased renal sympathetic nerve activity stimulates renin secretion, activating the angiotensin-aldosterone system (Section 11.6), which cause generalized vascular contraction (angiotensin II) and renal retention of salt and water (aldosterone).

    (c) In primates, a fall in baroreceptor traffic evokes the release of vasopressin (ADH) from the posterior pituitary gland, producing an antidiuresis and peripheral vasoconstriction.

The net result of the baroreflex to acute hypotension is thus to stimulate cardiac output, raise TPR, promote fluid retention and enhance plasma volume, thereby countering the hypotension.

Sensitivity and 'setting' of the baroreflex

In animal experiments, the sensitivity of the baroreflex can be assessed by isolating the carotid sinus and/or aortic arch and measuring the reflex change in heart rate or systemic pressure in response to controlled distension of the baroreceptors. In human subjects, baroreceptor activity can be altered by injecting the vasoconstrictor drug, phenylephrine, which raises arterial pressure, or by applying suction to a rigid cuff around the neck, which distends the carotid sinus directly (but not the aortic region). From such measurements, stimulus-response curves can be constructed as in Figures 13.4b and 13.6. The responses prove to be sigmoidal functions of the applied pressure change and the maximum slope of the response curve reveals the optimal sensitivity or 'gain' of the reflex. In man, the sensitivity declines with age and in chronic hypertension, due to a fall in artery wall distensibility. The pressure which the reflex strives to maintain is called its setting or set point, and this can be altered either by neural interactions within the central nervous system ('central resetting') or by physical changes in the receptor region ('peripheral resetting'). The advantages and disadvantages of resetting are as follows.

Central resetting Exercise offers an example of central resetting. The moderate rise in arterial pressure during exercise does not reduce the heart rate even though the baroreflex continues to function (see Figure 13.6), because the reflex is reset centrally to operate around a higher pressure. Thus the cardiac output is not suppressed, yet blood pressure is still buffered around the new level. Central modulation also occurs during the defence response (see later), which involves a sharp rise in blood pressure and suppression of the baroreflex. Sinus

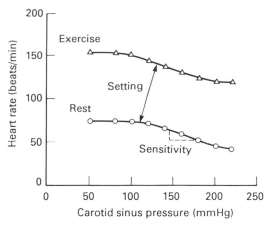

Figure 13.6 The baroreflex in a dog during mild exercise (5 km/h, 7% gradient). The carotid sinus was vascularly-isolated from the rest of the circulation and perfused at controlled, steady pressures to explore the response curve at rest and during exercise. The level of the baroreflex is reset in exercise, whereas its sensitivity (slope) changes little. Systemic arterial pressure increased by approximately 10 mmHg during the exercise. (After Melcher, A. and Donald, D. E. (1981) *Amercian Journal of Physiology*, **241**, H838–849, by permission)

arrhythmia (see Figure 4.9) is an example of regular central modulation of the baroreflex: the brainstem neurones that drive inspiration also inhibit the cardiac vagal motor neurones, rendering them temporarily unresponsive to the baroreceptor input (see Figure 13.13). The resulting fall in vagal activity largely explains the tachycardia associated with each inspiration.

Peripheral resetting If a rise in arterial pressure is imposed on the circulation the baroreceptor threshold rises to a new, higher pressure within 15 min or so. This shifts the whole stimulus-response curve to the right, and in this way the receptors regain a position on the steep part of the stimulus-response curve where they can operate most effectively. Although this extends the pressure-range over which the reflex can effectively buffer sudden pressure fluctuations, it also means that the

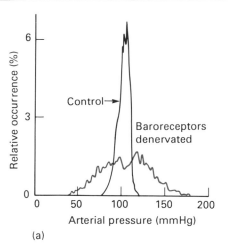

(a)

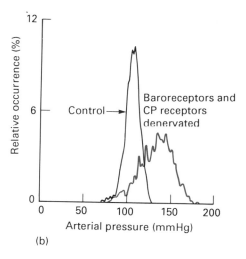

(b)

Figure 13.7 Frequency distribution for arterial pressure. (a) Effect of chronic arterial baroreceptor denervation in dogs. After some days, when these data were obtained, mean pressure has not changed significantly but the fluctuations about the mean increase, i.e. pressure is less stable. (b) Both the cardiopulmonary and arterial baroreceptors were denervated. There is now an increase in mean pressure, as well as pressure instability. ((a) From Cowley, A. W., Liard, J. F. and Guyton, A. C.; (b) in Persson, P. B., Ehmke, H., and Kirchheim, H. R. (1989) *NIPS*, **4**, 56–59, by permission)

baroreceptors cannot, over long periods, provide the brain with reliable information about the absolute blood pressure: the new pressure may produce the same signal as the old pressure, after resetting has occurred. The long-term ambiguity of the signal is also exacerbated by sympathetic motor nerves which innervate the carotid sinus and can enhance the baroreceptor activity. Because the baroreceptors do not transmit unambiguous information about the absolute blood pressure to the brain they cannot control absolute pressure in the long term.

Short-term role of the baroreflex

The baroreflex is important chiefly in buffering acute changes in arterial pressure in the short-term, but not in regulating the absolute pressure in the long term. If the baroreceptor nerves of a dog are cut under anaesthesia and the dog allowed to recover, it is hypertensive for a few days but the average blood pressure then settles down to a normal level, indicating that arterial baroreceptors are not necessary for the long-term setting of blood pressure. But while the pressure averaged over a period of time returns to normal, the pressure is nevertheless very unstable from minute to minute in the denervated animal, fluctuating over a much wider range than normal (see Figure 13.7b). Walking up a 21-degree incline, for example, raises a normal dog's arterial pressure by approximately 10 mmHg, whereas the baroreceptor-deprived dog experiences a 50 mmHg rise in pressure. The major role of the baroreflex is thus to buffer short-term fluctuations in arterial pressure.

13.2 Cardiopulmonary receptors

The heart and pulmonary artery are richly innervated by afferent fibres (see Figure 13.8). Like the arterial baroreceptors these fibres have, *overall*, a tonic inhibitory effect on heart rate and peripheral vascular tone: this is known from the effects of total cardiac deafferentation, which causes a tachycardia, peripheral vasoconstriction and rise in blood pressure. The cardiopulmonary afferents can be divided into three main functional classes, namely (1) mechanoreceptors around the right and left veno-atrial junctions, connected to myelinated vagal fibres, (2) mechanoreceptors that are scattered diffusely throughout the atria, ventricles and pulmonary artery and are served by non-myelinated fibres travelling in the vagus and cardiac sympathetic nerves, and (3) chemosensitive fibres travelling in the vagus and cardiac sympathetic nerves. (Both the vagus and cardiac sympathetic nerves are 'mixed nerves', i.e. carry both motor and sensory fibres.) Of these, the veno-atrial receptors are normally the most active.

1. The veno-atrial stretch receptors

These are branched, non-encapsulated nerve endings around the junctions of the great veins and atria, served by large myelinated vagal afferents. As Figure 13.8 shows, the receptor discharge coincides with either atrial systole (type A pattern) or with the V wave of atrial filling (type B pattern), or with both (intermediate pattern, not illustrated). The receptors therefore signal cardiac filling. If they are stimulated by inflating small balloons at the veno-atrial junction in experimental animals, they elicit a reflex tachycardia and a modest increase in urine flow. The tachycardia is induced, rather unusually, by a selective increase in sympathetic outflow specifically to the pacemaker, leaving myocardial contractility unaltered; vagal efferent activity is unaltered too. This reflex is probably the basis of the 'Bainbridge reflex' (discovered in 1915), which is a tachycardia induced by a large, rapid infusion of saline into a dog's venous system. The reflex diuresis and natriuresis (salt excretion) are still something of a puzzle. Reduced renal sympathetic nerve activity plays a role, for it causes renal

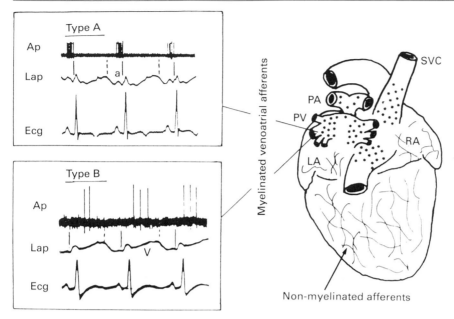

Figure 13.8 Posterior view of heart to illustrate the distribution of cardiopulmonary receptors. Position of venoatrial receptors is indicated by small asterisks. LA, RA, left and right atrium; PA, PV, pulmonary artery and vein; SVC, superior vena cava. The relation between the venoatrial fibre action potentials (Ap) and left atrial pressure (Lap) is shown on the insets for type A and type B receptors. (Canine recordings from Kappagoda, C. T., Linden, R. J. and Sivananthan, N. (1979) *Journal of Physiology*, **291**, 393–412, by permission)

vasodilatation, but in addition a hormone must be involved, since blood from a dog subjected to atrial distension can induce diuresis in a denervated kidney or even in an isolated insect Malpighian tubule. Some atrial muscle cells secrete a natriuretic hormone (atrial natriuretic peptide, ANP; Section 11.6) as a direct response to stretch: but this is not a reflex, and plasma ANP concentration does not correlate well with the reflex natriuresis. It is possible therefore that a further diuretic hormone awaits discovery. Vasopressin (antidiuretic hormone) too may be involved in the reflex diuresis, for in some species its release can be inhibited by an atrial receptor reflex, as well as by the baroreflex (see earlier). Changes in plasma vasopressin do not, however, seem to explain the reflex diuresis in dogs.

2. Unmyelinated mechanoreceptor fibres

Around 80% of the cardiac vagal afferents are small-diameter, unmyelinated fibres, and many of these subserve mechanoreception: similar fibres also travel in the cardiac sympathetic nerves. Such mechanoreceptors form a meshwork of fine fibres throughout the heart. Those in the atria respond to distension, and fire during the 'V' wave of atrial diastole, though only during inspiration, when atrial filling is greatest. The ventricular fibres, by contrast, fire mainly during ventricular systole, monitoring the speed and force of contraction. The net effect of these atrial and ventricular mechanoreceptors is to induce a reflex bradycardia and peripheral vasodilatation, the dilatation being particularly marked in the kidney. The bradycardia contrasts with the reflex tachycardia evoked by the myelinated veno-atrial group.

3. Chemosensitive fibres

Some unmyelinated vagal afferents are chemosensitive, and discharge in response to capsaicin, bradykinin and prostaglandins, the last of which is known to be released by ischaemic myocardium. Some unmyelinated sympathetic afferents are chemosensitive too, responding to bradykinin and to other substances released directly by hypoxic myocardium, such as lactic acid and K^+ ions. The sympathetic chemosensitive afferents are thought to mediate the pain of angina and myocardial infarction because surgical interruption of the cardiac sympathetic pathway relieves chronic ischaemic pain from the heart. The sympathetic afferents ascent the spinal cord in the spinothalamic tract, in which there is considerable convergence with somatic afferent fibres, and this convergence could explain why cardiac pain is usually experienced as emanating from the chest wall and arms ('referred pain').

Interplay of cardiopulmonary receptors and arterial baroreceptors in the regulation of arterial pressure

As indicated earlier, the *overall* effect of cardiopulmonary receptor activity is tonic inhibition of heart rate and vasoconstrictor tone, depressing blood pressure. This depressor effect seems to be involved in the long-term regulation of blood pressure, for if dogs are deprived of their cardiopulmonary receptor input as well as their arterial baroreceptor input they experience a sustained hypertension (see Figure 13.7b), as well as the excessive fluctuations in pressure that characterize baroreceptor denervation alone. Nevertheless, patients with transplanted, denervated hearts, have no problem in long-term blood pressure regulation.

Cardiopulmonary reflexes in man

It is difficult to establish the relative importance of the various intrathoracic receptors in human subjects, for obvious reasons, but it is clear that the cardiopulmonary group as a whole acts as a sensor of the volume of blood within the low-pressure, central compartment of the cardiovascular system. The cardiopulmonary receptors are, therefore, often referred to as 'central volume receptors' or 'low-pressure receptors' (cf. the high-pressure arterial baroreceptors). The central volume receptors have several important roles in man, as follows.

Reflex control of skeletal muscle vasculature during orthostasis and hypovolaemia The cardiopulmonary receptors can detect changes in blood volume which are too small to affect arterial pressure significantly. This is thought to explain how a rise in intrathoracic blood volume, without significant accompanying change in arterial pressure, causes a reflex vasodilatation in the human forearm (see Figure 13.9). Direct recordings from sympathetic vasomotor fibres to human forearm muscle have confirmed that the central receptor group can reflexly inhibit sympathetic vasomotor activity. Conversely, a fall in central blood volume (e.g. in response to orthostasis or a haemorrhage) leads to a reflex peripheral vasoconstriction, which complements the concomitant baroreceptor reflex. Human cardiopulmonary receptors are thought to be as important as the baroreflex in controlling the muscle circulation and also the splanchnic veins.

Reflex control of human extracellular fluid volume The role of central volume receptors in controlling extracellular fluid volume has been investigated by immersing human subjects in water in a feet-down, head-out position: this shifts about 700 ml of blood into the thorax owing to the pressure of the water on the lower limbs (see Figure 7.18), and thus simulates an expansion of body

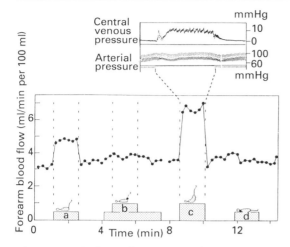

Figure 13.9 Reflex influence of intrathoracic volume on resistance vessels of human forearm muscle. (a) Legs alone raised, raising intrathoracic blood volume: vasodilatation follows. (b) Legs raised but pneumatic cuff around thigh at 180 mmHg prevents blood translocation: no change in forearm flow. (c) Legs and lower trunk raised: inset shows how central venous pressure rises, with no significant change in arterial pressure; large reflex vasodilatation. (d) Pneumatic cuff around neck inflated to 30 mmHg to reduce carotid sinus distension; very little reflex change in forearm blood flow. Analysis of oxygen content in deep and superficial veins established that the changes in forearm flow did not arise from the skin. (After Roddie, I. C., Shepherd, J. T. and Whelan, R. F. (1957) *Journal of Physiology*, **139**, 369, by permission)

fluid volume. The total diastolic volume of the heart rises by approximately 180 ml, stroke volume increases by about 30% via the Frank-Starling mechanism and arterial pressure rises approximately 10 mmHg. A substantial diuresis ensues owing to a reflex renal vasodilatation, a fall in plasma vasopressin concentration and a rise in plasma atrial natriuretic peptide. There is also a slower-acting fall in aldosterone level due to a fall in renin secretion. It is likely that these changes (aside from ANP secretion) are initiated by both the cardiopulmonary receptors and the arterial

baroreceptors. An analogous chain of events develops in astronauts subjected to zero gravity, leading to a diuresis and fall in extracellular fluid volume. The reduction of plasma volume, coupled with a weakened baroreflex, causes severe orthostatic intolerance on returning to earth.

13.3 Excitatory inputs: arterial chemoreceptors and muscle receptors

As well as the cardiopulmonary and baroreceptor depressor reflexes, which serve to stabilize the blood pressure, there are excitatory reflexes which help the cardiovascular system to respond positively to stresses such as exercise and hypoxia.

Arterial chemoreceptors and the chemoreflex

Arterial chemoreceptors are nerve terminals which respond to hypoxia, hypercapnia and acidosis. They are located mainly in the carotid and aortic bodies, which are small, highly vascularized nodules adjacent to the carotid sinus and aorta (see Figure 13.2). Their afferent fibres accompany with the baroreceptor afferents in the IXth and Xth cranial nerves. The chief role of the arterial chemoreceptors concerns the regulation of breathing, and their influence on the circulation is slight at normal gas tensions. When excitation is increased by hypoxia and hypercapnia, however, they elicit a sympathetically-mediated constriction of resistance vessels (except in the skin), and constriction of the splanchnic capacitance vessels. If breathing is held constant by artificial ventilation they also elicit a modest bradycardia but in spontaneously breathing animals the chemoreflex induces an increase in tidal volume, which in turn excites stretch receptors within the lungs. The pulmonary stretch receptors themselves elicit a 'lung inflation reflex', consisting of a modest vasodilatation and a marked tachycardia, and the latter opposes the direct bradycardial effect of the chemoreflex. The lung inflation reflex also contributes to sinus arrhythmia.

The cardiovascular elements of the chemoreflex become very important during asphyxia, producing a rise in blood pressure and thereby enhancing cerebral perfusion. The chemoreflex is also important during severe haemorrhage: severe hypotension impairs the perfusion of the chemoreceptor bodies and the resulting 'stagnant hypoxia' excites the chemoreceptors very strongly. Further stimulation is produced by the metabolic acidosis that develops during clinical hypotension (see Chapter 15), and the chemoreceptor-driven rise in sympathetic vasoconstrictor activity helps to support the blood pressure. This is particularly important in severe hypotension where the baroreflex has reached the limit of its range: the baroreceptor fibres fall silent below about 70 mmHg, whereas the chemoreceptors become progressively more excited the lower the perfusion falls. The support provided by the chemoreceptor input is proved by the sharp plunge in blood pressure if the chemoreceptor nerves are cut in a severely hypotensive animal. The chemoreflex also initiates the rapid breathing that is characteristic of hypotensive shock.

The role of chemoreceptors in the diving response is described in Chapter 14.

The work receptors of skeletal muscle

Exercise elicits a rise in heart rate, myocardial contractility and, in moderate to severe exercise, arterial blood pressure. The rise in pressure is especially marked during isometric exercise and is called the pressor response. These responses to exercise are initiated in part by the higher regions of the brain but they are also in part a reflex response initiated by receptors in the exercising muscle. In human subjects, local anaesthesia of the major limb nerves to

block selectively the sensory input from working muscle, impairs the tachycardia of exercise (see Figure 13.10). The afferent fibres carrying the excitatory input from muscle are small myelinated fibres (Group III) and small unmyelinated fibres (Group IV), and not the muscle spindle afferents (Group I). Their receptors are activated by chemicals released during exercise, notably K^+ ions and H^+ ions (due to lactic acid formation), and also by local pressure and active muscle tension. The importance of the chemical stimuli is revealed by inflating a pneumatic cuff around the human arm

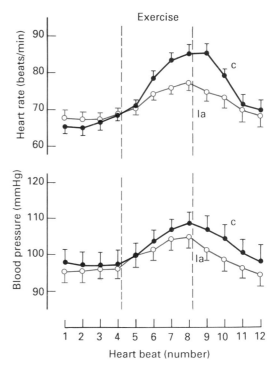

Figure 13.10 Evidence that afferent information from receptors in human working muscle contributes to the rise in heart rate during isometric exercise (4 s maximal voluntary handgrip). Local anaesthesia (la) of the axillary and radial nerves by lignocaine reduces the rate response compared with the control (c). The pressure response however is not significantly changed here. (From Lassen, A., Mitchell, J. H., Reeves, D. R. (1989) *Journal of Physiology*, **409**, 333–341, by permission)

just before forearm exercise is terminated, thereby trapping blood and chemical stimulants within the limb. The exercise pressor response is then partially maintained after the exercise is terminated, and only subsides fully when the cuff is released. The muscle receptor reflex (namely tachycardia, increased myocardial contractility and vasoconstriction in inactive tissues) serves to raise the pressure perfusing the active muscle. The reflex is strongest when the active muscle becomes ischaemic and produces lactic acid, so the reflex is probably best regarded as a system for sensing underperfusion of active muscle.

Influence of external receptors

Cardiovascular responses can also be evoked by receptors not concerned primarily with cardiovascular control. Somatic pain, for example, causes tachycardia and hypertension while severe visceral pain causes bradycardia, hypotension and even fainting. Ambient cold causes a rise in blood pressure, which increases left ventricular work and can trigger angina in susceptible patients. The special senses too influence the cardiovascular system: a sudden loud noise or the sight of a bus bearing down on one produces the alerting or defence response, involving a brisk tachycardia (Section 12.4). Sexual stimulation evokes a sharp tachycardia and hypertension (see Figure 7.8). Stimulation of facial receptors by cold water elicits a special 'diving reflex', which is described in Chapter 14.

13.4 Central pathways

In 1854, Claude Bernard showed that transection of the cervical spinal cord caused blood pressure to fall abruptly to around 40 mmHg, due to peripheral vasodilatation. This established that normal sympathetic vasoconstrictor activity depends on a tonic, net excitatory drive from the brain to the spinal sympathetic neurones. This tonic excitatory drive arises within the medulla oblongata, which is the most caudal (tail-end) part of the brainstem (see Figure 13.11). The medulla is by no means the only part of the brain involved in cardiovascular regulation; integration of the vast influx of sensory information relevant to the circulation requires the participation of the hypothalamus, cerebellum and the cortex. These central pathways are complex and are only partially characterized at present.

Role of the medulla: classic *versus* modern views

Traditionally, the medulla is described as possessing a 'cardiac centre' and a 'vasomotor centre', the latter being a diffuse scattering of cells within the dorsal reticular formation believed to regulate the sympathetic outflow to blood vessels. However, many recent observations have eroded confidence in this classical view. While it is true that cardiovascular changes can be elicited by electrical stimulation of the dorsal reticular region, so too can many non-vascular effects, and the same area has also been described as a 'respiratory centre', 'sleep-waking centre' and 'motor centre'. Its true role may in fact be to regulate the excitability of spinal neurones in general. For this and many other reasons, most workers would now agree with Hilton and Spyer, who in 1980 declared that the classic notion of a medullary vasomotor centre 'though reasonable when first proposed . . . has now become an impediment to research and is in any case untenable'. The modern view emphasizes longitudinal traffic up and down the brain between medulla, hypothalamus and cerebellum (see Figure 13.11) as much as transverse traffic within the medulla.

The roles of the medulla in circulatory control may be summarized as follows.

To receive the cardiovascular receptor traffic The dorsomedial medulla contains an elongated nucleus of cells, the nucleus

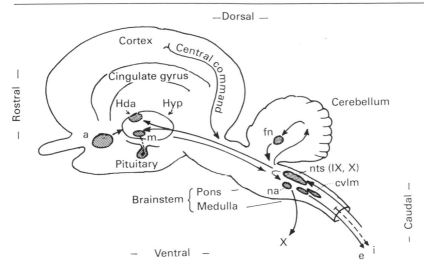

Figure 13.11 Longitudinal arrangement of central cardiovascular pathways in the cat brain. a, amygdala; cvlm, caudal ventrolateral medulla; e, excitatory drive to spinal sympathetic neurones from the rostal ventrolateral medulla; fn, fastigial nucleus; hda, hypothalamic depressor area; hyp, hypothalamus; i, descending inhibitory influence on spinal sympathetic neurones; m, magnocellular neurones in supraoptic and paraventricular nuclei of hypothalamus; na, nucleus ambiguus; nts, nucleus tractus solitarius

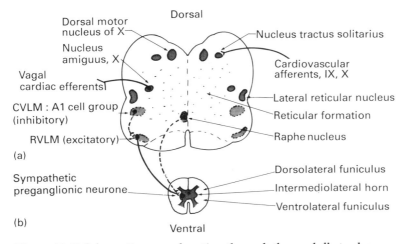

Figure 13.12 Schematic coronal section through the medulla to show relative positions of cardiovascular-related bodies in the dorso-ventral plane. RVLM, rostral ventrolateral medulla group; CVLM, caudal ventrolateral medulla group: the A1 cells are a group of noradrenaline-rich neurones. Dashed lines indicate inhibitory pathways. The structures occur at various rostrocaudal levels and would not in reality all be present in a single anatomical section. Thoracic spinal cord shown on a smaller scale at bottom

tractus solitarius (see Figures 13.11 and 13.12), which is the site of first synapse for virtually all the cardiovascular afferents – baroreceptors, cardiopulmonary afferents, arterial chemoreceptors, pulmonary stretch receptors and muscle work receptors. Destruction of the nucleus tractus solitarius causes a sustained hypertension. The muscle afferents also project to a lateral reticular nucleus (see Figure 13.12), destruction of which impairs the exercise pressor response.

To relay afferent information to other regions   The output of the nucleus tractus solitarius is relayed to various parts of the medulla, hypothalamus and cerebellum (see Figure 13.11). Within the medulla, a polysynaptic path relays a signal to the nucleus ambiguus, which contains the vagal cardiac motor neurones. Signals are also relayed to the caudal ventrolateral medulla, which influences sympathetic output by inhibiting the rostral ventrolateral medulla (see later). The nucleus tractus solitarius also relays information up to the hypothalamic cells that synthesize vasopressin (the magnocellular neurones of the supraoptic and paraventricular nuclei), and to the hypothalamic depressor area (see later).

To generate the vagal outflow to the heart   The cell bodies of the vagal preganglionic fibres controlling the cardiac pacemaker are located chiefly in the nucleus ambiguus (see Figures 13.12 and 13.13), and to a lesser extent in the dorsal motor nucleus. These vagal nuclei used to be called the 'cardioinhibitory centre'.

To regulate spinal sympathetic neurones   When the anaesthetic pentobarbitone is applied locally to the surface of the rostral ventrolateral medulla, a severe fall in blood pressure results. This led to the discovery of a rostral ventrolateral group of neurones that exert a tonic excitatory effect upon the sympathetic preganglionic neurones of the spinal intermediolateral columns. This tonic excitation is carried by bulbospinal fibres

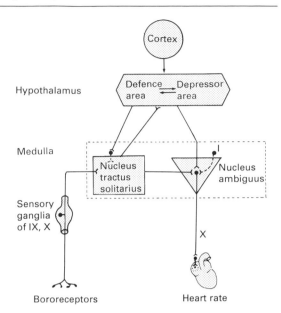

Figure 13.13 Central pathways governing vagal efferent activity to the heart. Synaptic complexities are not shown. I, inspiratory neurone input responsible for sinus arrhythmia. The inter-neurones mediating inhibition in the nucleus tractus solitarius are GABAergic; γ-aminobutyric acid is an inhibitory central neurotransmitter. IX, glossopharyngeal nerve; X, vagus. (After Spyer, K. M. (1984), see Further reading list, by permission)

that run down the dorsolateral funiculus of the spinal cord (see Figures 11.8 and 13.11). The tonic excitatory activity of the rostral ventrolateral medulla is modified by an inhibitory input from the caudal ventrolateral medulla. In addition there is a more direct, descending inhibitory influence on the spinal sympathetic neurones, arising from the raphe nuclei of the brainstem.

Role of the hypothalamus

The hypothalamus contains many regions involved in cardiovascular regulation, including the hypothalamic depressor area, the alerting or defence area, the temperature-regulating area and the magnocellular vasopressin-secreting nuclei.

Hypothalamic depressor area This is located in the dorsal part of the anterior hypothalamus and receives an input from the nucleus tractus solitarius (see Figure 13.11). When stimulated electrically the depressor area mimics the baroreflex, i.e. it activates the cardiac vagal fibres and inhibits sympathetic outflow. Although lesions here impair the baroreflex, they do not abolish it, so the region is evidently only one component in the central processing of the baroreflex.

Hypothalamic defence area The cardiovascular response in a cat or dog faced with sudden danger is both striking and stereotyped, consisting of tachycardia, acute hypertension, splanchnic and renal vasoconstriction and dilatation in skeletal muscle mediated by sympathetic cholinergic fibres (see Chapter 11). This pattern forms part of a reaction known variously as the defence, alerting, alarm or fear-fight-flight response. The response is generated by a discrete 'defence area' of the hypothalamus, the anterior perifornical region, and electrical stimulation here evokes not only the cardiovascular response but also, in a conscious cat, the behavioural manifestations of fear or rage (spitting, snarling and piloerection). The defence area is probably normally activated by the amygadala of the limbic system (see Figure 13.11), a system known to be involved in generating emotional behaviour patterns. Stimulation of the defence area indirectly inhibits those neurones in the nucleus tractus solitarius that are excited by baroreceptor traffic (see Figure 13.13). The hypothalamic defence area can also, by other pathways, influence the cardiac vagal motor neurones and the rostral ventrolateral medulla neurones that govern sympathetic outflow.

'Playing dead' reaction This is exhibited by the opossum and young creatures like the rabbit in the face of danger. It involves a profound bradycardia and hypotension, being in this respect the opposite of the defence response. The response originates

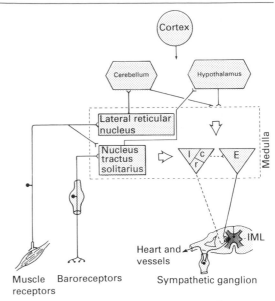

Figure 13.14 Central pathways governing sympathetic nerve activity. The open arrows indicate uncertainty as to exact pathways. Dashed lines denote inhibition. I, inhibitory influences, multifactorial; c, caudal ventrolateral medulla; r, raphe nuclei; E, excitatory neurones: notably those in rostral ventrolateral medulla; IML, intermediolateral horn containing preganglionic sympathetic neurones

in the cingulate gyrus, another part of the limbic system (see Figure 13.11). It has been suggested that human fainting in response to intolerable psychological stimuli ('swooning') is really a manifestation of the opossum response – the avoidance of a threatening situation by collapse.

Two other hypothalamic regions to note here are the temperature-regulating area in the anterior hypothalamus, which coordinates the output of the cutaneous vasomotor and sudomotor nerves, and the supraoptic and paraventricular nuclei, which contain the vasopressin-producing magnocellular neurones. The axons of the latter feed vasopressin to the capillary bed of the pituitary gland. Direct electrophysiological recordings show that their activity is controlled by inputs from local osmoreceptors and from the nucleus tractus solitarius.

Role of the cerebellum

The major role of the cerebellum is the coordination of muscular movement, and during exercise this region helps coordinate the cardiovascular response too. The areas involved are the fastigial nucleus and the associated vermal cortex, which receive projections from the medulla. Stimulation of the vermal cortex elicits renal vasoconstriction and muscle vasodilatation, i.e. the pattern seen in exercise, while destruction of the fastigial nucleus reduces both the tachycardia and pressor response of dogs to exercise.

Influence of the cerebral cortex: central command

In 1913, Krogh and Lindhard postulated the 'central command' hypothesis to explain the dramatic, rapid cardiovascular response to exercise. The hypothesis proposes that the cerebral cortex, which initiates muscular exercise, also initiates many of the cardiovascular responses by a direct action on the brainstem regions controlling vagal and sympathetic outflow. There are undoubtedly areas of cortex (sensorimotor, orbital and temporal areas) that initiate cardiovascular changes when stimulated electrically. A subthalamic region capable of evoking the exercise pattern of cardiovascular changes has also been identified. Biofeedback, the ability of some individuals to exert a degree of voluntary control over their heart rate or blood pressure, could be of cortical origin.

Overview of central pathways controlling the circulation

Figures 13.13 and 13.14 summarize, in an extremely simplified fashion, the central cardiovascular pathways.

Control of vagal outflow to the heart There are two main routes linking the barorecep-tor input with the vagal motor neurones controlling heart rate. One route, of short latency, remains within the medulla and passes from the nucleus tractus solitarius to the vagal motor nuclei, not necessarily directly. The other, of longer latency, passes from the nucleus tractus solitarius up to the hypothalamic depressor centre and from there to the vagal motor neurones.

Control of spinal preganglionic neurones The central pathways linking the nucleus tractus solitarius to the spinal sympathetic neurones are more complex and less well understood. Higher regions are again involved to some degree. Whatever the intermediate pathways, baroreceptor activation ultimately inhibits the descending excitatory drive from the rostral ventrolateral medulla to the spinal sympathetic preganglion neurones, and this inhibition is mediated partly by the caudal ventrolateral medulla. In addition there is a direct descending spinal inhibitory pathway from the brainstem raphe nuclei.

The spinal patient The activity of the spinal sympathetic preganglionic neurones depends partly on the activity of the descending bulbospinal fibres and partly on local inputs from within the spinal cord. As Claude Bernard showed, sectioning the cervical spinal cord cuts off the net excitatory influence of the brainstem and causes an abrupt hypotension. Sherrington and others soon pointed out, however, that over several weeks the blood pressure gradually recovers in patients with spinal transections, indicating that the sympathetic preganglionic neurones are capable of generating an output by local mechanisms. Some reflex modulation of this output can occur at a spinal level in these patients; for example, a full bladder or somatic pain cause a reflex rise in blood pressure. Nevertheless, spinal patients lack a baroreflex (except for the vagal control of heart rate) and they therefore suffer from a labile blood pressure and a proneness to postural hypotension.

Further reading

Bisset, G. W. and Chowdrey, H. S. (1988) Control of release of vasopressin by neuroendocrine reflexes. *Quarterly Journal of Experimental Physiology*, **73**, 811–872

Calaresu, F. R. and Yardley, C. P. (1988) Medullary basal sympathetic tone. *Annual Review of Physiology*, **50**, 511–524

Coleridge, H. M. and Coleridge, J. C. G. (1980) Cardiovascular afferents involved in regulation of peripheral vessels. *Annual Review of Physiology*, **42**, 413–427

Dorward, P. K. and Korner, P. I. (1987) Does the brain 'remember' the absolute blood pressure? *NIPS*, **2**, 10–13

Gebber, G. L. (1984) Brainstem systems involved in cardiovascular regulation. In *Nervous Control of Cardiovascular Function* (ed. W. C. Randall), Oxford University Press, New York, pp. 345–368

Linden, R. J. (1987) The function of atrial receptors. In *Cardiogenic Reflexes* (eds R. Hainsworth, P. N. McWilliam and D. A. S. G. Mary), Oxford University Press, Oxford, pp. 18–39

Mancia, G. and Mark, A. L. (1983) Arterial baroreflexes in humans. In *Handbook of Physiology, Cardiovascular System, Vol. 3* (eds J. T. Shepherd and F. M. Abboud), pp. 755–793. Also Cardiopulmonary baroreflexes in humans, ibid., pp. 794–813. American Physiological Society, Bethesda

Mitchell, J. H. and Schmidt, R. F. (1983) Cardiovascular reflex control by afferent fibres from skeletal muscle receptors. In *Handbook of Physiology, Cardiovascular System, Vol. 3* (eds J. T. Shepherd and F. M. Abboud), American Physiological Society, Bethesda, pp. 623–658

Needleman, P., Currie, M. G., Geller, D. M., Cole, B. R. and Adams, S. P. (1984) Atriopeptins; potential mediators of an endocrine relationship between heart and kidney. *Trends in Pharmaceutical Science*, **5**, 506–509

Rothe, C. F. (1983) Reflex control of veins and vascular capacitance. *Physiological Reviews*, **63**, 1281–1333

Shepherd, J. T. (1982) Reflex control of arterial pressure. *Cardiovascular Research*, **16**, 357–383

Spyer, K. M. (1984) Central control of the cardiovascular system. In *Recent Advances in Physiology, 10* (ed. P. F. Baker), Churchill Livingstone, London, pp. 163–200

Stone, H. L., Dormer, K. J., Foreman, R. D., Thies, R. and Blair, R. W. (1985) Neural regulation of the cardiovascular system during exercise. *Federal Proceedings*, **44**, 2271–2278

Wallin, B. G. and Fagius, J. (1988) Peripheral sympathetic neural activity in conscious humans. *Annual Review in Physiology*, **50**, 565–576

Chapter 14
Coordinated cardiovascular responses

All the individual elements of the circulation have been covered in the preceding chapters but, as with a jigsaw puzzle, it is not enough to view the individual pieces separately; what matters ultimately is how the pieces fit together to produce an effective whole. The purpose of this chapter is to illustrate how the components of the circulation respond in coordinated patterns to various challenges. One general principle will emerge, namely that *each major adaptation is achieved by the integration of several smaller responses*. To take a specific example, a 13-fold increase in the rate of oxygen absorption by the pulmonary circulation during strenuous exercise is not achieved by a 13-fold change in any one parameter but by the integration of, typically, a 1.5-fold rise in stroke volume, a threefold rise in heart rate, and a threefold increase in the arteriovenous difference in oxygen concentration across the lung. Other examples of integration will be found below.

14.1 Posture

The challenge Movement from a supine to a standing position (orthostasis) is a severe challenge to the human circulation owing to the effect of gravity on the distribution of venous blood. Gravity induces a tenfold rise in transmural pressure in the most dependent veins, increasing the dependent venous volume by approximately 500 ml. The redistribution of blood causes a 20% fall in intrathoracic blood volume over about 15 s (see Figure 7.18). Cardiac filling pressure falls several cmH$_2$O and the energy of myocardial contraction is reduced by the Frank–Starling mechanism. Stroke volume declines by 30–40% from about 70 ml to 45 ml, so pulse pressure falls substantially (see Figure 14.1). Mean pressure falls only transiently owing to reflex corrections (see later), but even so the transient hypotension can be severe enough to impair cerebral perfusion and cause dizziness and visual

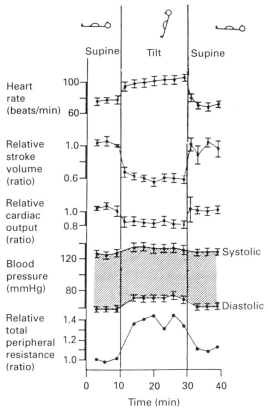

Heart rate (beats/min)
Relative stroke volume (ratio)
Relative cardiac output (ratio)
Blood pressure (mmHg)
Relative total peripheral resistance (ratio)

Supine   Tilt   Supine

Time (min)

Figure 14.1 Response of young adults to a 20-min head-up tilt. Points represent means, bars are standard errors. (From Smith, J. J., Bush, J. E., Weidmeier, V. T. and Tristani, F. E. (1970) *Journal of Applied Physiology,* **29**, 133, by permission)

fading. Even healthy individuals occasionally experience postural giddiness, especially when warm conditions cause cutaneous venodilatation, which further reduces central filling pressure. Postural hypotension is worse after prolonged bed rest or exposure to zero gravity but it does not usually progress to postural syncope (fainting) unless the compensatory reflexes are blocked by autonomic neuropathy or by pharmacological agents, such as the α-adrenergic receptor blockers.

The responses In healthy subjects, the cardiopulmonary and carotid sinus reflexes

quickly restore mean arterial pressure and prevent postural dizziness. Cardiopulmonary receptor traffic is reduced by the fall in cardiac blood volume. Carotid baroreceptor traffic is reduced by the fall in pulse pressure and by the fall in sinus pressure due to the direct effect of gravity; the sinus is close to the base of the skull and lies 25 cm or so above heart level in orthostasis. The reduced input to the nucleus tractus solitarius informs the brain of the gravity of the situation (!), eliciting a reflex reduction in vagal outflow to the heart and an increase in sympathetic outflow to the heart and vasculature. *Heart rate* increases by 15–20 beats/min, due chiefly to the carotid sinus reflex. Combined with a sympathetically-mediated rise in contractility this limits the fall in cardiac output to approximately 20%. Sympathetically-mediated *vasoconstriction* in the skeletal muscle, splanchnic and renal vascular beds raises the peripheral resistance by 30–40% which not only restores mean arterial pressure but even increases it to 10–14 mmHg above the supine value (see Figure 14.1). Splanchnic *venoconstriction* partially compensates for the dependent venous pooling but there is no sustained reflex venoconstriction in muscle or skin during orthostasis.

The above responses normally take less than a minute to complete. The effect of orthostasis on capillary filtration, leading to a 6–12% fall in plasma volume over about 40 min, was described in Section 9.9. Renal *salt and water excretion* is cut down by reflexly-induced increases in plasma vasopressin, renin, angiotensin and aldosterone, acting in combination with the reflex renal vasoconstriction.

The net result of this complex, integrated response is that arterial pressure and therefore cerebral perfusion pressure is safeguarded. It seems perverse that, despite all this physiological 'effort', the cerebral blood flow actually declines by 10–20% during orthostasis. The decline is caused by a rise in cerebral vascular resistance, which may be due partly to the increased ventilation that accompanies orthostasis and lowers the arterial P_{CO_2}, partly due to a sympathetically-induced constriction of cerebral vessels and partly to the collapse of the extracranial veins that drain cerebral blood.

14.2 Valsalva manoeuvre

Valsalva was an eighteenth-century Italian physiolgist. The eponymous manoeuvre is not, however, an obscure physiological rite but a natural event performed daily by most of us. It is a forced expiration against a

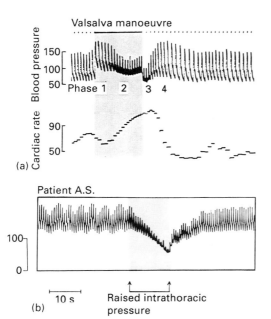

Figure 14.2 Effect of Valsalva manoeuvre on blood pressure and heart rate. (a) Normal subject. (b) Patient suffering from idiopathic orthostatic hypotension, caused by an autonomic defect. The patient's pressure failed to stabilize during phase 2, and there was no reflex bradycardia in phase 4. Blood pressure in mmHg; rate in beats/min; top time markers in seconds. ((a) From Bannister, Sir R. (1980) In *Arterial Blood Pressure and Hypertension* (ed. P. Sleight), Oxford University Press, Oxford, pp. 117–121 and (b) From Johnson, R. H. and Spalding, J. M. K. (1974) *Disorders of the Autonomic Nervous System*, Blackwell, London, by permission)

closed or narrowed glottis, and this is a normal accompaniment to defaecation, coughing, lifting heavy weights, singing a top A or playing the trumpet. The manoeuvre creates a high intrathoracic pressure which evokes a complex circulatory response with four phases (see Figure 14.2). Initially arterial pressure rises because the high intrathoracic pressure presses upon the thoracic aorta (*phase 1*). Mean arterial pressure and pulse pressure then begin to fall because the high intrathoracic pressure impedes venous return, reducing enddiastolic volume and impairing stroke volume by the Frank–Starling mechanism (*phase 2*). As pressure falls, cardiovascular receptors elicit a reflex tachycardia and peripheral vasoconstriction which halts the fall in pressure. When the Valsalva manoeuvre is stopped there is a sudden mechanical drop in blood pressure as periaortic pressure returns to normal (*phase 3*). The drop in intrathoracic pressure allows venous blood to surge into the thorax, distending the heart and increasing the stroke volume. As a result the pulse pressure and mean pressure rebound rapidly (*phase 4*), causing the baroreceptors to elicit a reflex bradycardia. The Valsalva response is therefore a useful test of the competence of the baroreflex in man. If the reflex is interrupted by a neurological disorder, the Valsalva test shows a continuing pressure fall in phase 2 and no pressure overshoot or bradycardia in phase 4. Individuals showing such a pattern are prone to postural hypotension.

14.3 Exercise

Overview From an evolutionary, survivalof-the-fittest point of view, perhaps the most important circulatory adjustment is the response to exercise. Muscular exercise imposes three tasks on the circulation: pulmonary blood flow must be increased to enhance gas exchange, blood flow through the working muscle must be raised, and a reasonably stable blood pressure must be maintained. The first requirement, a rise in pulmonary perfusion, is met by an increase in right ventricular output. The second requirement, an increase in muscle perfusion, is met primarily by local metabolic vasodilatation which reduces the resistance to blood flow through the working muscle; but in addition an increase in left ventricular output is necessary to supply the extra flow. The third requirement, arterial pressure stability in the face of huge changes in systemic vascular resistance and cardiac output, is achieved by a variable degree of vasoconstriction in non-active tissues. The net effect of these changes is the diversion of an increasing fraction of the raised left ventricular output into working muscle, as illustrated in Figure 14.3.

Cardiac output and oxygen uptake during exercise

During exercise, the heart shows a remarkable ability to increase its output in direct, almost linear proportion to whole-body oxygen consumption (see Figure 14.4). In an untrained adult, the maximum cardiac output is around 20 litres/min, representing roughly a fourfold increase from rest. Increased oxygen uptake in the lungs is due partly to this rise in cardiac output and partly to an increase in the oxygen uptake per litre of pulmonary blood; the latter can increase just over threefold, owing to the low oxygen content of venous blood entering the lungs. Applying the Fick principle to such data (Section 5.1), we find that the

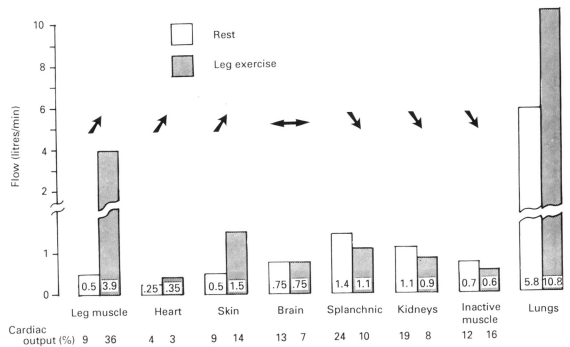

Figure 14.3 Redistribution of cardiac output during light exercise of the legs in the human adult. Cardiac output increased from 5.8 litres/min to 10.8 litres/min. Oxygen consumption during the exercise period was 0.4–0.6 litres/min per m². Room temperature. Number at base of each column is blood flow in litres/min. Arrows indicate direction of change. (After Wade, O. L. and Bishop, J. M. (1962) *Cardiac Output and Regional Blood Flow*, Blackwell, Oxford and Blair, D. A., Glover, W. E. and Roddie, I. C. (1961) *Circulation Research*, **9**, 264)

Table 14.1 Cardiovascular and pulmonary functional capacities determined during maximal exercise in college students and Olympic athletes[a]

	Exercising students			Olympic athletes
	Control	After bedrest	After training	
Maximal oxygen uptake, litres/min	3.30	2.43	3.91	5.38[b]
Maximal voluntary ventilation, litres/min	191.0	201.0	197.0	219.0
Transfer coefficient for O_2 (ml min^{-1} $mmHg^{-1}$)	96.0	83.0	86.0	95.0
Arterial O_2 capacity, vol %	21.9	20.5	20.8	22.4
Maximal cardiac output, litres/min	20.0	14.8	22.8	30.4[b]
Maximal stroke volume, ml	104.0	74.0	120.0	167.0[b]
Maximal heart rate, beats/min	192.0	197.0	190.0	182.0
Systemic arteriovenous O_2 difference, vol %	16.2	16.5	17.1	18.0

[a] Mean values, $n = 5$ and 6 respectively. Age, height and weight similar.
[b] Significantly different from college students after training, $P < 0.05$.
(From Blomqvist, C. G. and Saltin, B. (1983) *Annual Review of Physiology*, **45**, 169–189)

pulmonary oxygen uptake rate can increase by around 13-fold to over 3 litres/min in the untrained adult (see Table 14.1)*.

Heart rate rises linearly with exercise intensity, up to a maximum 180–200 beats/min in adults (see Figure 14.4). The tachycardia is due partly to withdrawal of vagal inhibition of the pacemaker and partly to sympathetic stimulation. The relative contributions of heart rate and stroke volume to the increased cardiac output depend partly on posture (see Section 6.9). In supine exercise, almost all of the increased output is due to tachycardia, stroke volume increasing at most by a mere 10–20%, whereas in the upright position stroke volume starts from a lower value and can increase by 50–100% (see Tables 14.2 and 14.3). The enhancement of stroke volume is

* The rate of oxygen consumption is as good measure of exercise intensity in the steady state. Resting consumption in an adult is approximately 0.25 litres/min. During light work, such as walking on the level at 3 km/h, oxygen consumption increases to 0.4–0.8 litres/min; moderate work, 0.8–1.6 litres/min; hard work, 1.6–2.4 litres/min; and severe work, such as running at 12 km/h, 2.4–3.0 litres/min. From Pugh, G. (1971) *Journal of Physiology*, **213**, 255.

achieved partly by a rise in filling pressure which increases the ventricular end-diastolic volume, and partly by a rise in ejection fraction which lowers the end-systolic volume (see the echocardiogram in Figure 6.20). Filling pressure rises by around 1 mmHg due to the skeletal muscle pump and sympathetically-mediated splanchnic venoconstriction. Ejection fraction and ejection velocity are raised by a sympathetically-mediated improvement in myocardial contractility. Patients with severe coronary disease cannot achieve this increase in ejection fraction, and partly for this reason their cardiac output during exercise is poor (see Table 14.3).

Changes in blood flow to active muscle

In a fit human male performing heavy dynamic exercise, the total flow to muscle can increase from 1 litres/min (rest) to around 19 litres/min (comprising >80% of cardiac output), and since many muscle groups are only lightly used even in heavy exercise (e.g. arm muscles during heavy running, cycling), it is probably that muscle

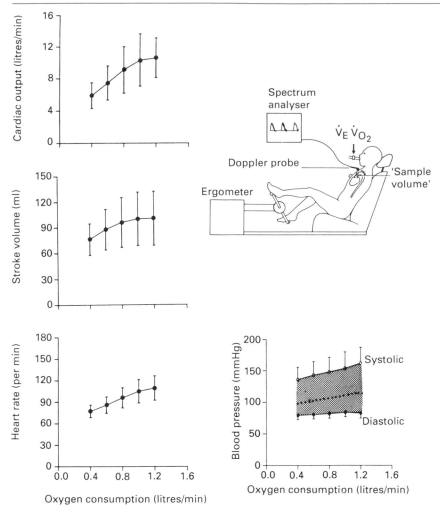

Figure 14.4 Cardiac response to exercise measured by pulsed Doppler method (see Chapter 5) in human adults of mean age 57 years. Bars indicate standard deviation of observations. Note the quantitative, almost linear relation between cardiac output and whole-body oxygen consumption. (From Innes, J. A., Simon, T. D., Murphy, K. and Guz, A. (1988) *Quarterly Journal of Physiology*, **73**, 323–341, by permission)

flow in maximally active groups increases as much as forty times. This hyperaemia is due chiefly to metabolic vasodilatation, aided in upright exercise by the muscle pump's amplification of the pressure gradient (Section 12.2). The fall in muscle vascular resistance also has a vital permissive effect on cardiac output: left ventricular output would otherwise be severely curtailed by a huge rise in arterial pressure which would oppose ejection.

Metabolic dilatation of terminal arterioles also causes capillary recruitment, shortening the diffusion distance to the contracting fibres (see Figure 8.15). The increased capillary blood flow, capillary recruitment

Table 14.2 Supine exercise *versus* upright exercise

		Stroke volume (ml)	Heart rate (beats/min)	Cardiac output (litres/min)
Supine	Rest	111	60	6.4
	Exercise	112	91	9.7
Upright	Rest	76	76	5.6
	Exercise	92	95	8.4

Mean cardiac response of 8 healthy males to pedalling at 30% of V_{O2} max. Stroke volume measured by the aortic Doppler flow technique
(From Loeppky, J. A., Greene, E. R., Hoekenga, D. E. *et al.* (1981) *Journal of Applied Physiology*, **50**, 1173–1182)

Table 14.3 Ventricular volume during upright submaximal exercise in normal subjects and patients with multiple coronary artery disease

	Normal		Coronary disease	
	Rest	Exercise	Rest	Exercise
Cardiac output (litres/min)	6.0	17.5	5.7	11.3
Heart rate (beats/min)	81	170	75	119
Stroke volume (ml)	76	102	76	96
End-diastolic volume (ml)	116	128	138	216
End-systolic volume (ml)	40	26	62	120
Ejection fraction	0.66	0.8	0.6	0.46

Means of 30 normal subjects and 20 patients. Upright submaximal bicycle exercise. Left ventricle dimensions determined by radionuclide angiocardiography. (After Rerych, S. K., Scholz, P. M., Newman, G. E. *et al.* (1978) *Annals of Surgery*, **187**, 449–458)

and decreased tissue concentration together raise the glucose transport rate into active fibres by an order of magnitude (see Table 8.2 and Figure 8.16). Arteriolar dilatation raises capillary pressure, and capillary filtration can reduce plasma volume by as much as 600 ml during prolonged heavy exercise (Section 9.9).

Changes in blood flow to other tissues

Blood flow to nearly every tissue in the body is altered during exercise, as shown in Figure 14.3. Coronary blood flow increases in proportion to cardiac work owing to metabolic vasodilatation. Skin is a battleground of conflicting demands: initially, cutaneous vessels may be constricted to support the blood pressure but if core temperature rises during the exercise the thermoregulatory role of skin becomes dominant and dilatation supervenes. This calls for a further rise in cardiac output, yet at the same time the cutaneous venodilatation is reducing the cardiac filling pressure. Consequently, the stroke volume tends to decline during prolonged heavy exercise while heart rate increases to compensate.

The fall in peripheral resistance occasioned by vasodilatation in skeletal muscle, myocardium and skin is so great during hard exercise that blood pressure would fall by 12–40 mmHg, despite the raised cardiac output, were it not for a *compensatory vasoconstriction* in the splanchnic and renal vascular beds and in non-exercising muscle. During leg exercise, for example, vascular resistance rises in the forearm. Although textbooks often stress that vasoconstriction in the resting tissues 'diverts' blood to working muscle, a simple tally of the changes in Figure 14.3 shows that the diverted flow is really rather small (0.6 litres) and only accounts for a very small part of the increased flow through the exercising muscle. The true importance of the vasoconstrictor response lies rather in supporting the arterial pressure.

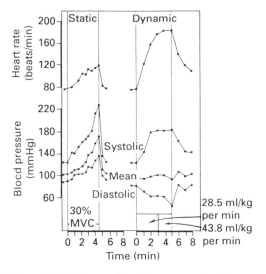

Figure 14.5 Effects of static compared with dynamic exercise. Static exercise caused a bigger rise in mean pressure. Dynamic exercise caused a bigger rise in pulse pressure and heart rate. MVC, maximal voluntary static contraction; arrowed numbers refer to oxygen consumption. (From Lind, R. A. and McNicol, G. W. (1967) *Canadian Medical Association Journal*, **96**, 706, by permission)

Blood pressure during static and dynamic exercise

Systemic arterial pressure during exercise depends very much on the severity, duration and nature of the exercise. In dynamic exercise (i.e. alternating contraction and relaxation), mean pressure rises by 20 mmHg or less. Systolic pressure and pulse pressure increase more than this, owing to the rise in stroke volume and ejection velocity, but diastolic pressure rises little (see Figure 14.4) or even falls (see Figure 14.5). In static exercise (e.g. a sustained handgrip), there is by contrast a large rise in diastolic pressure; simply supporting a 20 kg suitcase for 2–3 min can raise diastolic pressure by 30 mmHg (the pressor reflex, Section 13.3). This increases left ventricular work considerably, and isometric exercise is therefore best avoided by patients with ischaemic heart disease. The rise in pulmonary blood pressure during exercise was discussed in Section 12.5.

Denervated heart in exercise: role of circulating catecholamines

Normally, the rise in heart rate and ejection fraction are driven chiefly by the cardiac sympathetic nerves: yet patients with denervated, transplanted hearts can still undertake moderate levels of exercise. The cardiac response to exercise after chronic cardiac denervation has been investigated in racing greyhounds. Denervation of the greyhound heart reduces the animal's track speed by only 5% and there is still a substantial exercise tachycardia, though it is reduced and of sluggish onset (see Figure 14.6). This tachycardia is produced by a backup mechanism, namely the rise in plasma adrenaline and noradrenaline that occurs during exercise. If cardiac stimulation by the circulating catecholamines is

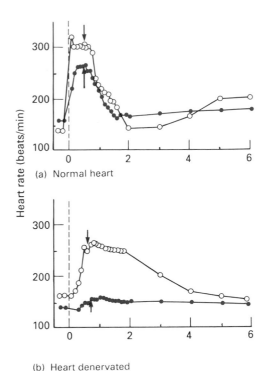

(a) Normal heart

(b) Heart denervated

Figure 14.6 Role of circulating catecholamines after cardiac denervation. The heart rate of a greyhound was monitored by telemetry (radio signals) on a race track. Arrows indicate time at which dog passed the 5/15th mile mark. Open symbols: normal heart (a) and denervated hearts (b), without other intervention; denervation impairs but does not abolish the tachycardia. Closed symbols: normal and denervated hearts after β-adrenoreceptor blockade by propanolol; β-blocked, denervated dogs were significantly slowed and became exhausted. Blockade was less effective against normal cardiac nerve activity than against circulating catecholamines. (From Donald, D. E. Ferguson, D. A. and Milburn, S. E. (1968) *Circulation Research*, **22**, 127–133, by permission)

blocked by a β-adrenoceptor antagonist, the exercise tachycardia is prevented, the denervated greyhound's track speed drops and the dog finishes the lap in a state of extreme exhaustion.

In strenuous human exercise, plasma noradrenaline rises from around 1 nM to 10–20 nM, due chiefly to local 'spillage' from sympathetic junctional gaps rather than adrenal gland secretion. Plasma adrenaline shows little change during light to moderate exercise in man (approximately 0.2 nM) but rises to 2–5 nM during maximal dynamic exercise owing to secretion by the adrenal medulla. The rise in catecholamine level during exercise benefits cardiac transplant patients considerably. A further backup mechanism of value to such patients is the skeletal muscle pump, which raises the cardiac filling pressure and hence stroke volume by the Frank–Starling mechanism. These backups illustrate a general principle enunciated by the respiratory physiologist, Julius H. Comroe: 'if a job is worth doing, the body has more than one way of doing it'.

What initiates the circulatory adjustments in exercise?

Aside from metabolic vasodilatation, the cardiovascular changes in exercise are caused by altered autonomic nerve activity. It is not entirely clear, however, what causes the brainstem to initiate these changes in autonomic activity. Two main hypotheses have been put forward: the central command hypothesis and the peripheral reflex hypothesis, and there is evidence that both play a part.

The *central command hypothesis* was advocated by Krogh and Linhard in 1913 and supposes that the cerebral cortex, or a closely related forebrain region, not only initiates the voluntary contraction of muscle but also directly 'commands' the autonomic and respiratory neurones of the brainstem. One observation compatible with central command is that the heart rate begins to increase at the first beat after the onset of exercise. Moreover, after partial neuromuscular blockade by tubocurarine, voluntary attempts at contracting the partially paralyzed muscle (requiring, presumably, a bigger central command signal) produce an enhanced rise in heart rate and blood pressure. Central command does not, however, easily explain a quintessential feature of the cardiac response, the near-linear relation between cardiac output and skeletal muscle oxygen consumption (see Figure 14.4).

The *peripheral reflex hypothesis* arose from work carried out by Alam and Smirk in 1938, who found that chemical excitation of receptors in working muscle contributes to the exercise pressor response (increased cardiac output and blood pressure, Section 13.3). Their key observation was that inflation of a pneumatic cuff around a limb, to retain the chemical stimulants within the working muscle, results in a partial maintenance of the pressor response even after exercise is terminated. An attractive feature of this hypothesis is that the chemosensitive endings could provide the necessary, quantitative link between muscle oxygen consumption and cardiac output. The progressive interstitial accumulation of chemical factors such as potassium and lactate could also explain the gradual increase in tachycardia over 1–2 min following the start of exercise (see Figure 14.5).

It seems likely therefore that both central command and muscle work receptors drive the cardiac response to exercise. Even so, the near-linear relation between cardiac output and whole-body oxygen consumption is a mystery only partially resolved.

14.4 Physical training

The circulatory adaptations that accompany fitness training are very important for endurance athletes, though less so for those involved in brief events like sprints and shot-putting, where straight muscle power is a key factor. In an endurance event such as a medium-distance race, the chief factor limiting performance appears to be the maximal rate of oxygen transport from lungs to active muscle (and not the lung diffusing capacity or the muscle oxidative capacity). The major effect of training on the vasculature of skeletal muscle is to stimulate the growth of new capillaries. This increases

the exchange area and prevents an increase in diffusion distance, which muscle fibre hypertrophy would otherwise cause. Dynamic training also affects cardiac structure and function. Structurally, the ventricular wall grows thicker, myocardial vascularity increases and the ventricular cavities enlarge. Ventricular end-diastolic volume increases from approximately 120 ml in the untrained resting adult to as much as 220 ml in the resting athlete, and the stroke volume is 100–125 ml at rest (cf. normal 70–80 ml). The resting cardiac index (output per unit body surface area) is the same in trained and untrained individuals but owing to a higher stroke volume the trained subject achieves his resting output at a lower heart rate (40–50 beats/min). The resting bradycardia is produced by vagal inhibition of the pacemaker. During exercise, the athlete achieves bigger stroke volumes than the untrained subject, some athletes achieving 170 ml during maximal exercise (see Table 14.1). The athlete's maximum heart rate is about the same as the untrained subject's, but since the athlete starts with a slower heart rate, he can achieve a proportionately greater change: a rise in heart rate from 40 beats/min to 180 beats/min is a 4.5-fold increase, in contrast to a rise from 70 beats/min to 180 beats/min, which is only a 2.6-fold increase. Along with the enhanced stroke volume this enables the athlete to increase his cardiac output up to 7-fold, outputs of up to 35 litres/min having been recorded in certain individuals.

14.5 Diving reflex

Diving animals like the duck, seal and whale show remarkable cardiovascular changes during a dive, and man shows the same responses too to a lesser degree. The diving reflex comprises three responses: apnoea, intense bradycardia and peripheral vasoconstriction. The circulatory changes conserve the limited oxygen store for the benefit of the heart and brain, permitting prolonged survival under water. The Wed-

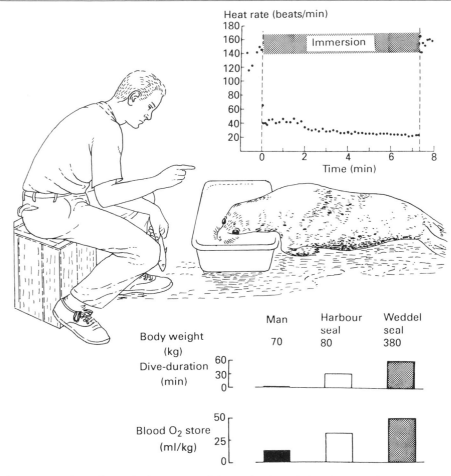

Figure 14.7 Physiological adaptation of the seal to breath-hold diving. Heart rate response of a seal trained to perform voluntary head immersion is shown at top. (Various sources)

dell seal can survive up to 70 min immersion and the whale 2 h, though feeding dives are usually shorter than this. By contrast the Amas, the Japanese and Korean women who dive for pearls, remain submerged for only 40–50 s. The vastly superior performance of diving animals is due partly to a larger store of oxygen in blood and in muscle myoglobin (see Figure 14.7) and partly to more extreme cardiovascular responses. In addition, diving animals are more tolerant of asphyxia than man; the arterial gas values in a harbour seal after a prolonged dive are 10 mmHg oxygen and 100 mmHg carbon dioxide, which not only vastly exceeds the human breath-hold breaking point but would probably be fatal in man.

The diving response is initiated by cold water touching the facial receptors of trigeminal nerve fibres, especially those around the eyes, nose and nasal mucosa. As the dive progresses, asphyxia develops and arterial chemoreceptor activity reinforces the cardiovascular responses.

Bradycardia The seal's heart rate can fall to 20 beats/min during a dive, owing to

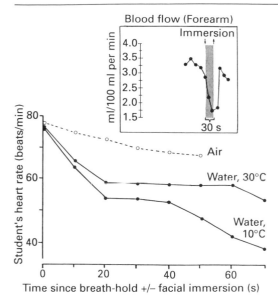

Figure 14.8 (a) Heart rate response of a medical
student to a voluntary breath-hold facial
immersion in water. Breath-hold in room air
(control) has only a small effect. Cold water
evokes a stronger reflex than warm water;
indeed, the experimenter stopped the
experiment in some alarm when the subject's
heart rate fell below 40 per min. Not all students
show such a pronounced diving reflex. Boxed
inset illustrates concomitant peripheral
vasoconstriciton in human forearm,
(Unpublished data of J. R. Henderson. Inset
from Heistad, D. D., Abboud, F. M. and
Eckstein, J. W. (1968) *Journal of Applied Physiology*,
25, 542–549, by permission)

vagal inhibition of the pacemaker potential.
Many human subjects display a pronounced bradycardia during facial immersion in cold water (see Figure 14.8), and
facial immersion in cold water has even
been used successfully to interrupt human
supraventricular tachycardias.

Peripheral vasoconstriction There is a
profound, sympathetically-mediated vasoconstriction in the splanchnic, renal and
skeletal muscle circulations, and this even
outweighs the metabolic vasodilator influence in active muscles. This peripheral
vasoconstriction maintains the blood press-

ure in the face of the extreme bradycardia,
and at the same time diverts most of the
greatly reduced cardiac output to the heart
and brain. Large arteries, which are safely
upstream of the metabolic vasodilator influence, vasoconstrict strongly. A large quantity of lactic acid accumulates in the swim-
ming muscles, causing a sharp vasodilatation when the animal resurfaces. Human
subjects too, respond to facial immersion
with a vasoconstriction in skin and skeletal
muscle (see Figure 14.8); forearm blood flow
falls by 25–50% and blood pressure rises by
approximately 20%. Immersion of the body
but not the face, or immersion of the face
wearing a breathing tube, fails to elicit this
fascinating reflex.

Further reading

Blix, A. S. and Folkow, B. (1983) Cardiovascular
adjustments to diving in mammals and birds.
*Handbook of Physiology, Cardiovascular System, Vol. 3,
Part 2, Peripheral Circulation* (eds J. T. Shepherd and
F. M. Abboud), American Physiological Society,
Bethesda, pp. 917–946

Blomqvist, C. G. and Saltin, B. (1983) Cardiovascular
adaptations to physical training. *Annual Review of
Physiology*, **45**, 169–189

Blomqvist, C. G. and Stone, H. L. (1983) Cardiovascu-
lar adjustments to gravitational stress. *Handbook of
Physiology, Cardiovascular System, Vol. 3, Part 2,
Peripheral Circulation* (eds J. T. Shepherd and F. M.
Abboud), American Physiological Society, Bethesda,
pp. 1025–1063

Christensen, N. J. and Galbo, H. (1983) Sympathetic
nervous activity during exercise. *Annual Review of
Physiology*, **45**, 139–153

de Burg Daly, M. (1984) Breath-hold diving: mechan-
isms of cardiovascular adjustments in the mammal.
In *Recent Advances in Physiology 10* (ed. P. F. Baker),
Churchill Livingstone, London, pp. 210–245

Lind, A. E. (1983) Cardiovascular adjustments to
isometric contractions: static effort. *Handbook of
Physiology, Cardiovascular System, Vol. 3, Part 2,
Peripheral Circulation* (eds J. T. Shepherd and F. M.
Abboud), American Physiological Society, Bethesda,
pp. 947–966

Mitchell, J. H., Reeves, D. R., Rogers, H. B., Secher, N.
H. and Victor, R. G. (1989) Autonomic blockade and
cardiovascular response to static exercise in partially
curarized man. *Journal of Physiology*, **413**, 433–455

Vatner, S. F. (1984) Neural control of the heart and
coronary circulation during exercise. In *Nervous
Control of Cardiovascular Function* (ed. W. C. Randall),
Oxford University Press, New York, pp. 414–424

Chapter 15
Cardiovascular responses in pathological situations

15.1 Shock and haemorrhage	**15.3 Essential hypertension**
15.2 Fainting (syncope)	**15.4 Chronic cardiac failure**

In the previous chapter, specific coordinated patterns of response were recognized to different physiological challenges. In this final chapter it seems appropriate to consider how the circulation reacts to pathological situations, drawing some examples from the field of human disease.

15.1 Shock and haemorrhage

Meaning of 'shock', and its causes

The term 'shock' is used by the medical profession and general public alike, but with quite different meanings. In general conversation the term refers to a withdrawn psychological state, often with physical manifestations such as a muscular tremor, but without an underlying organic cause. To the physician and surgeon, however, 'shock' is a serious, indeed potentially fatal, pathophysiological disorder characterized by an acute failure of the cardiovascular system to perfuse the tissues of the body adequately. The state is recognized by a characteristic pattern of physical signs. The skin is pale, cold and sweaty with constricted veins. The pulse is rapid and weak, due to a tachycardia and low stroke volume. Arterial pressure may be reduced or normal, but pulse pressure is always reduced. Breathing is rapid and shallow, urine output is impaired and the general condition is one of muscular weakness and reduced mental awareness or confusion.

The clinical causes of shock fall into three categories. *Hypovolaemic shock* is caused by a fall in blood or plasma volume, which may be due to external fluid loss (haemorrhage, diarrhoea and vomiting, dehydration) or to internal fluid loss (extensive burns, crushing injuries, pancreatitis). *Septicaemic shock* is caused by bacteraemia, often due to Gram-negative organisms that release a powerful cardiovascular toxin called endotoxin. *Cardiogenic shock* is caused by an acute organic impairment of cardiac function such as myocardial infarction, myocarditis or an arrhythmia. Although the details of the physiological response vary to some degree with the cause, there is a shared pattern, and this will be illustrated by considering the response to haemorrhage.

Haemorrhagic hypotension

A 10% blood loss (the volume withdrawn during a blood donation) elicits little change in mean blood pressure and does not produce the shock syndrome. A 20–30% blood loss lowers mean pressure to 60–80 mmHg and produces the shock syndrome, but does not usually threaten life. A 30–40% blood loss lowers pressure to 50–70 mmHg and causes severe, sometimes irreversible shock, with anuria and impaired cerebral and coronary perfusion. The arterial hypotension is mediated by the Frank–Starling mechanism: acute hypovolaemia lowers central blood volume and hence ventricular end-diastolic volume, which reduces the energy of contraction. Stroke volume therefore declines, reducing the arterial pressure. Arterial hypotension is thus an indirect rather than direct consequence of a bleed, being quite unlike the loss of pressure in a punctured tyre in this respect.

Hypovolaemia initiates a series of reflex responses that help preserve the perfusion of the brain and myocardium, and thus preserve life. Cardiopulmonary volume-receptor activity and arterial baroreceptor activity decline or cease, while arterial chemoreceptor activity increases, owing to metabolic acidosis (see Figure 15.1) and impaired chemoreceptor perfusion (stagnant hypoxia). The chemoreceptor input stimulates the rapid ventilation that characterizes shock. The altered inputs to the nucleus tractus solitarius evoke a reflex increase in sympathetic outflow and reflex secretion of adrenaline and vasopressin, and these reflexes play a vital role in the immediate defence against hypovolaemia. The defence can be divided into three phases: the responses which are effective within seconds, those that act more slowly (5–60 min) and those that act over longer periods (days–weeks).

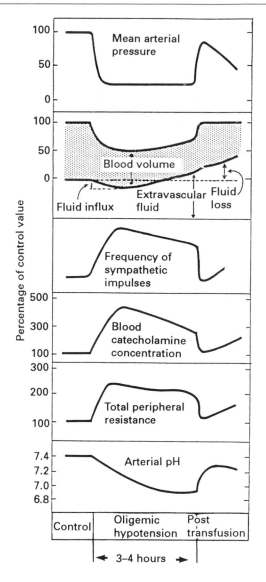

Figure 15.1 Responses to shock induced by blood withdrawal in the dog. The remaining blood volume is indicated by the shaded region in the second panel; the dip below the zero line indicates fluid transfer from the interstitial space into the bloodstream. In this very severe case of hypovolaemia, the fluid transfer begins to reverse again after approximately 2 hours as vasoconstriction fails, and blood volume declines further, leading to irreversible shock. (From Chien, S. (1967) *Physiological Reviews*, **47**, 214–288, by permission)

Immediately effective responses

The increased sympathetico-adrenal outflow constricts the resistance vessels in the cutaneous, skeletal muscle, splanchnic and renal vascular beds, raising the total peripheral resistance and supporting arterial pressure. The reduced perfusion of the tissues leads, however, to muscular weakness, lactic acidosis, oliguria (low urine flow) and pallor. Increased sympathetic cholinergic discharge to skin causes sweating and produces the 'clammy' skin characteristic of shock. A secondary factor, thought to impair tissue perfusion in severe shock, is the adhesion of white cells to the walls of microvessels.

The sympathetico-adrenal outflow helps to support cardiac output by stimulating a tachycardia, increased myocardial contractility, and active venoconstriction in the splanchnic and skin circulations: the peripheral venoconstriction partially restores thoracic blood volume and cardiac filling pressure. The attending physician often becomes aware of cutaneous venoconstriction when he attempts to cannulate a vein for intravenous fluid replacement.

The above neurally-mediated responses are reinforced by increased levels of circulating vasoconstrictor hormone, namely adrenaline, vasopressin and angiotensin II. Vasopressin is secreted by the hypothalamic magnocellular neurones as a reflex response to the fall in cardiac receptor and baroreceptor inputs; vasopressin enhances the peripheral vasoconstriction. Angiotensin II contributes to the peripheral vasoconstriction by both local and central actions (Section 11.6) and accounts for about 30% of the initial recovery in blood pressure in venesected dogs. The renin-angiotensin system is activated strongly by the combination of high renal sympathetic nerve activity, reduced renal artery pressure and reduced sodium load at the macula densa.

Because of the compensatory changes in peripheral resistance, venous capacitance and cardiac performance, the arterial pressure may not fall much after a moderate haemorrhage, making arterial pressure an unsafe guide to the severity of shock.

Intermediate term response: the 'internal transfusion'

The fall in blood pressure combines with the sympathetically-mediated increase in pre- to postcapillary resistance ratio to reduce capillary pressure substantially in shock. As a result, the osmotic pressure of the plasma proteins predominates across the capillary wall for a while, and causes the absorption of up to 500 ml of interstitial fluid into the vascular compartment in a human adult. The way in which the intracellular fluid compartment too is drawn upon is explained in Section 9.12. As a result of the plasma volume expansion, the haematocrit and plasma protein concentration fall soon after a haemorrhage.

Long-term responses

Although the above responses preserve cerebral and myocardial perfusion, the patient is left with a reduced perfusion of most major organs and a reduced body content of water, electrolytes, plasma protein and red cells. These deficiencies are corrected gradually over days and weeks. The water and salt deficits are corrected first by reduced renal excretion and increased fluid intake. Glomerular filtration rate is cut down by a sympathetically-mediated constriction of the afferent arterioles, while salt and water reabsorption is stimulated by a rise in plasma aldosterone and vasopressin (anti-diuretic hormone). The high plasma concentration of angiotensin II not only stimulates aldosterone secretion but also stimulates the subfornicular organ of the brain, producing the intense thirst that patients experience after a haemorrhage. The resulting increase in water intake, in combination with the oliguria, quickly replenishes the body water content. Salt retention by the renal tubules, combined with a normal dietary intake of salt (2–10 g/ day), replenishes the extracellular salt mass within a few days.

The synthesis of albumin by the liver gradually restores plasma protein mass over the course of a week. Red cell production by the bone marrow is stimulated by an erythropoietic factor secreted by the kidney, restoring the haematocrit to normal over a period of some weeks, provided that iron intake is adequate.

The outcome of shock: irreversible shock and other complications

The above sequence of events occurs in reversible shock, such as might be produced by a 25% blood loss. If, however, the loss exceeds 30% and has lasted over 3–4 h before fluid replacement begins (as in Figure 15.1), shock is often irreversible, even if the whole loss is subsequently made good by transfusion. In such cases, blood pressure may be maintained for a while by the high sympathetic-adrenal outflow, but pressure then begins to fall, leading to myocardial hypoperfusion and possibly death. The cause of the delayed, fatal fall in pressure is controversial: some workers attribute it to depression of myocardial contractility while others attribute it to failure of peripheral vasoconstriction caused by peripheral neurotransmitter depletion, accumulation of competing metabolic vasodilator substances and changes in a central opioid pathway within the brain.

Several other serious complications can arise during severe, prolonged hypotension. Probably the commonest of these is acute tubular necrosis, a form of acute renal failure caused by hypoxic damage to the renal tubules; this is heralded by failure of the urine output to improve after a day or so. For this reason the urine output of a patient in shock is closely monitored. Another serious complication in patients with pre-existing ischaemic heart disease is myocardial infarction, triggered by the fall in perfusion pressure.

15.2 Fainting (syncope)

A faint is a sudden, transient loss of consciousness that occurs when cerebral blood flow falls to less than half normal owing to an abrupt fall in arterial pressure. The critical cerebral artery pressure is approximately 40 mmHg, which corresponds in an upright subject to approximately 70 mmHg mean pressure at heart level. The initiating factor may be a pathophysiological stress such as hypovolaemia or orthostasis, or it may be a psychological stress such as fear, pain or horror (see Figure 15.2). For example, the sight of blood, especially one's own, often induces emotional fainting in young adults. In such

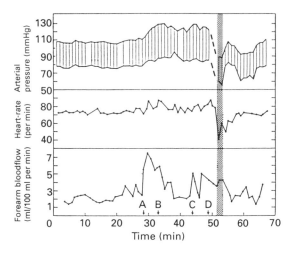

Figure 15.2 Circulatory changes in a male student during an emotional faint. The student showed forearm vasodilatation (alarm response) while watching the preparation for venepuncture (A) and venepuncture of a colleague (B), but did not faint as he had on an earlier occasion (and as the experimenter intended). Insertion of a needle into his arm (C) again produced vasodilatation, but no faint. The subject was therefore asked to drink some of the blood taken from his colleague (D). He became pale, yawned, said 'I'm going' and fainted (stippled area). No heart beat was detected by ECG for 11 s and heart rate then averaged 37 per min. Forearm blood flow remained above resting level despite the slump in blood pressure, showing the vasodilatation had occurred. Consciousness was regained after 2 min. (From Greenfield, A. D. M. (1951) *The Lancet*, p. 1303, by permission)

psychogenic fainting, the circulation initially evinces a normal alarm response, namely tachycardia, muscle vasodilatation, cutaneous vasoconstriction and sweating. During this pre-faint period the subject looks pale and sweaty, hyperventilates and, very characteristically, yawns. Then a sudden increase in vagal outflow causes a profound bradycardia, and at the same time the peripheral resistance vessels dilate, due probably to a fall in sympathetic vasoconstrictor drive. As a result, blood pressure falls precipitously, and reduced cerebral perfusion is followed within seconds by loss of consciousness. This sequence is sometimes called a 'vaso-vagal attack'. The cause of the sudden changes in vagal and vasomotor activity is not certain; in the case of psychogenic fainting, the response could be related to the 'playing dead' response of small animals, which emanates from the cingulate gyrus (Section 13.4). In the case of post-haemorrhagic syncope, the response may be initiated by activation of left ventricular mechanoreceptors in the near-empty heart at end-systole.

The supine position resulting from a faint raises the intrathoracic blood volume and filling pressure, and cardiac output and arterial pressure are quickly restored. Consciousness is recovered in about 2 min. It is a mistake, albeit a well-intentioned one, to prop up the patient during a faint, since this deprives him of the benefit of the Frank–Starling mechanism.

15.3 Essential hypertension
Definition and classification

The medical condition 'hypertension' can be defined as a chronic, usually progressive, raised arterial pressure. To some extent this definition begs the question, for the dividing line between normal and raised pressure is a rather difficult issue, the distribution of pressure in the population being unimodal. Most physicians would diagnose hypertension if repeated measurement of resting pressure exceeded 140/90 mmHg in a patient under 50 years old, or 160/95 mmHg in an older patient. The disease is in itself almost symptomless, and it may come to light only as a result of a routine medical examination; or it may present later with one of its many sequellae, namely heart failure, coronary artery disease, a cerebrovascular accident (stroke), retinopathy or chronic renal failure. Discrete identifiable causes are relatively uncommon but they include hyperaldosteronism (Conn's syndrome), renal artery stenosis and phaeochromocytoma (a catecholamine-secreting tumour); such cases are termed 'secondary hypertension'. In 90% or more cases, however, no such cause is found and the disease is classed as primary or essential hypertension.

Pathophysiology of essential hypertension

Since blood pressure depends on the balance between cardiac output and peripheral resistance, hypertension must be regarded as an imbalance in cardiovascular regulation. In the early stages of the disorder, when hypertension is both marginal and labile, the cardiac output is raised while the peripheral resistance is only slightly above normal. When the disease is well established, however, the cardiac output is normal or slightly reduced and the hypertension is therefore due to an increase in peripheral resistance (see Table 15.1). The increase in vascular resistance affects virtually every organ, including the kidney, and is caused partly by a narrowing of the arterioles and partly by rarefaction. *Rarefaction* is a reduction in the number of vessels present in unit volume of tissue, and the occurrence of rarefaction has recently been confirmed in both the retina and the intestine of hypertensive patients. *Narrowing* of the arterioles is due to increased vascular tone in the early stages, and is fully reversible by vasodilator drugs at this stage. As time pases, however, the smooth muscle of the tunica media responds to the chronic-

Table 15.1 Change in vascular resistance of the hand in hypertension. (Percentage change from normal in parenthesis)

	Normal	Hypertensive
Mean arterial pressure (mmHg)	82 (100%)	116 (141%)
Peripheral resistance $(PRU_{100})^*$	3.8 (100%)	5.9 (156%)
Peripheral resistance at maximal vasodilatation $(PRU_{100})^*$	1.6 (100%)	2.5 (163%)
Sensitivity of resistance to noradrenaline (slope of response curve)	100%	178%

(Hypertensive vessels can dilate and contract but even at maximal dilatation their resistance is higher than normal maximally dilated vessels.)
*PRU_{100} is a peripheral resistance unit, mmHg/(ml/min)/100g.
(After Sivertsson, R. and Olander, B. 1968, *Life Science*, **7**, 1291–1297)

ally raised pressure load by hypertrophying, and this leads to organic narrowing of the lumen. The elevated resistance can then no longer be fully abolished during maximal vasodilatation (see Table 15.1). Medial hypertrophy takes only a few weeks to develop in rats subjected to experimental hypertension by clipping one of the renal arteries (which stimulates the renin-angiotensin-aldosterone system).

Stiffening of the carotid artery wall reduces the sensitivity of the baroreflex in hypertensive patients, but this is an effect rather than a cause of the hypertension.

What initiates the hypertension?

Despite enormous research efforts, the answer to this question remains uncertain. There is a strong familial tendency to hypertension and also a rather controversial epidemiological association with a high salt intake. These observations, coupled with direct experimental work, have led to a number of causative theories.

Neurogenic hypothesis This proposes that an excessive sympathetic outflow in response to stress initiates bouts of reversible hypertension, which gradually induce a structural reaction (medial hypertrophy) that perpetuates the condition. Experimental lesions of the nucleus tractus solitarius raises the sympathetic outflow and leads to chronic hypertension in experimental animals. Moreover, repeated exposure of laboratory animals to psychogenic stress can cause chronic hypertension.

Salt imbalance hypothesis Here the proposition is that a small but sustained discrepancy between renal salt excretion and dietary intake leads to an increase in extracellular salt mass and, owing to the osmotic effect of the salt, an increase in extracellular water. This leads to a rise in plasma volume, filling pressure, stroke volume and blood pressure, which in turn evokes medial hypertrophy and the non-reversible rise in peripheral resistance. The controversial association of hypertension with a high-salt diet would fit this hypothesis. Also, a genetic strain of rat exists in which a salt diet, harmless to ordinary rats, causes hypertension. The renin-angiotensin-aldosterone system, which stimulates salt retention by renal tubules, is very active in many (but not all) human hypertensives. Extracellular fluid volume is found to be increased or normal in hypertension, but plasma volume is reduced in the established disease.

Abnormal transport of salt by cell membranes It has become clear in recent years that the active Na^+-K^+ transport system is abnormal in the membranes of blood cells taken from hypertensive patients, and this appears to be related to a defect in calcium binding by the cell membrane. It is postulated that the resulting membrane abnormalities are widespread and multiple, and that similar defects in vascular smooth muscle cells might increase their myogenic activity, leading to hypertension.

Multifactorial hypothesis The long-term control of blood pressure involves neural, endocrine and renal mechanisms and many workers suspect that hypertension develops only if more than one regulatory process is abnormal. Whatever the initial cause, the process is thought to become self-perpetuating once medial hypertrophy develops, since a rise in pressure evokes further hypertrophy.

The treatment of hypertension is based on diuretic drugs to lower extracellular fluid volume, captopril to block the angiotensin-converting enzyme and therefore reduce angiotensin and aldosterone levels, peripheral vasodilators such as calcium-channel blockers (verapamil, nifedipine) and α_1-adrenoreceptor blockers (prazosin) to lower peripheral resistance, and β-adrenoreceptor blockers like propanolol to reduce cardiac output.

15.4 Chronic cardiac failure

Definition and causes

Chronic or congestive cardiac failure (CCF) may be defined as an intrinsic inability of the heart to maintain an adequate perfusion of the tissues at a normal filling pressure.

This contrasts with the disorder 'shock', where the low output state is secondary to a low filling pressure. Starling, working with the isolated heart preparation, noted long ago that when a heart begins to fail, it requires a higher filling pressure and higher end-diastolic volume to maintain its stroke volume; at a normal filling pressure the stroke volume became subnormal. Thus the immediate cause of cardiac failure is a fall in the energy of contraction at any given end-diastolic volume, in other words, a reduction in contractility.

In many patients, a recognizable pathological condition initiates the chronic failure; for example, diffuse coronary artery disease, reduction in functional muscle mass after a myocardial infarct, or chronic work overload caused by hypertension. In other cases there is no obvious precipitating pathology and the cause is related to myocyte biochemistry. Studies of myocardium from failing hearts indicate that although contractility is impaired, energy production is normal, judging by the normal levels of ATP and creatine phosphate. Energy utilization, however, is impaired: both myofibrillar ATPase activity and myofibril content per gram of myocardium are low. The most serious abnormality may involve intracellular calcium, the key factor in excitation-contraction coupling (Section 3.5); calcium transport into the sarcoplasmic reticulum is impaired, so the internal calcium store may be low. Possible impairment of the affinity of troponin for calcium is also being investigated.

Impairment of cardiac performance

Owing to the reduction in contractility, the ventricular function curve of the failing ventricle(s) is depressed and its slope is reduced (see Figure 15.3). The pump function curve (stroke volume against arterial pressure, Figure 6.13) is depressed too. The rate of tension development is slow, and the ejection fraction falls from the normal 66% to as little as 10–20%. In severe failure, this reduces the stroke volume, but in mild

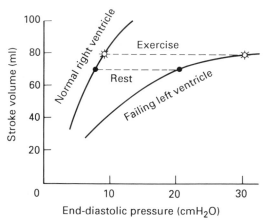

Figure 15.3 Diagram illustrating the operation of the Frank–Starling mechanism in a patient with a failing left ventricle but relatively healthy right ventricle. The left ventricular function curve is depressed and its slope is reduced. Under resting conditions (closed circles) the left ventricle requires an elevated filling pressure to match the right ventricle's stroke volume. In exercise (open symbols) the disparity in filling pressure becomes extreme owing to the near-plateau on the function curve of the failing left ventricle

failure the stroke volume may be almost normal owing to a compensatory increase in end-diastolic volume (see later). Consequently, the cardiac ouput at rest may be either within the normal range (compensated failure) or subnormal (decompensated failure; mean 3.8 litres/min in one series). The impaired cardiac performance becomes much more obvious during an exercise test, because the failing heart cannot increase its output to a normal extent, as shown by the data in Table 14.3: the patient's exercise tolerance is poor and he/she complains of excessive fatigue. The poor response to exercise is rather interesting, for it is not only due to impairment of the stroke volume response but also to impairment of the heart rate response. The inability of *stroke volume* to rise to a normal extent is caused by the decreased sensitivity to filling pressure (i.e. the reduced slope of the ventricular function curve), by the decreased ability to cope with a rise in arterial

pressure, and by a decrease in the responsiveness of contractility to catecholamines (see later). The impaired *heart rate* response (see Table 14.3) is caused partly by a depletion of noradrenaline from the cardiac sympathetic nerve terminals due to a fall in tyrosine hydroxylase activity, and partly by 'down regulation' of the myocyte β_1-adrenoreceptors – that is to say, a decrease in the number and affinity of the receptors.

Pathophysiological responses to heart failure

The responses of the circulation and other systems to heart failure include compensatory influences on the heart, the redistribution of cardiac output, renal retention of salt and water, and oedema.

Compensatory influences on the heart

The output of the failing heart is supported by two compensatory mechanisms, an increase in filling pressure and an increased level of circulating catecholamine.

Raised ventricular filling pressure: good and bad aspects Filling pressure rises to well over $12\,cmH_2O$, distending the failing ventricle and, in mild failure, improving its contractile energy by the Frank–Starling mechanism. The rise in filling pressure is due to a combination of increased plasma volume and peripheral venoconstriction (see later). The resulting cardiac dilatation can be gross, and is readily detected in chest radiograms (see Figure 15.4). Although the increase in end-diastolic volume shifts the ventricle along the ventricular function curve, this is of little benefit beyond a certain point because the curve reaches a virtual plateau (see Figure 15.3). Moreover, excessive cardiac dilatation can be harmful because contraction becomes mechanically inefficient: the active tension required to generate systolic pressure increases with ventricular diameter (see 'Laplace effect', Figure 6.9), raising the energy cost of

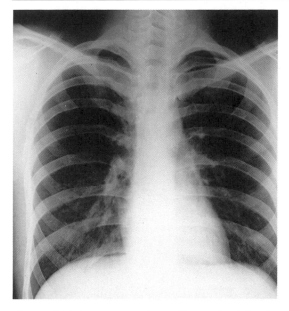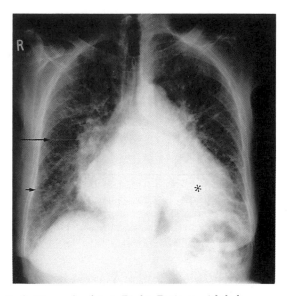

Figure 15.4 Anteroposterior radiograph of the chest. Left: Normal subject. Right: Patient with left ventricular failure. Asterisk marks grossly dilated left ventricle. Upper, long arrow: line of oedema fluid in fissure between upper and middle lobes. Lower, short arrow: septal line caused by interstitial oedema (Kerley B line). Radiating opacities due to pulmonary interstitial oedema are also present in the left lung field (Kerley A lines). (Courtesy of Dr A. Wilson, St. George's Hospital, London)

systole in a ventricle that can ill afford extra energy costs. In addition, gross dilatation can widen the atrioventricular orifice to such an extent that the atrioventricular valve becomes functionally incompetent, further reducing the ventricular ejection fraction. A further ill-effect of a high filling pressure is the generation of oedema (see later). There are thus several reasons for trying to reduce the filling pressure in severe cardiac failure, even though this does shift the ventricle back along the Starling curve.

Stimulation by circulating catecholamines
The cardiac nerves themselves become depleted of catecholamine, as mentioned earlier, but there is a marked rise in plasma catecholamine level in severe failure which helps to support the inotropic state. The support is somewhat mitigated, however, by a downregulation of myocardial β_1-adrenoceptors as the disease progresses.

Changes in peripheral vascular beds

The limited cardiac output is preferentially distributed to the coronary, cerebral and skeletal muscle circulations at the expense of other peripheral tissues (see Figure 15.5). The perfusion of the renal, splanchnic and cutaneous vascular beds is severely reduced, owing to sympathetic vasoconstrictor nerve activity coupled with a rise in plasma angiotensin II. This peripheral vasoconstriction maintains the arterial pressure, which would otherwise be threatened by the low cardiac output. The increases in sympathetic outflow and circulating angiotensin II also induce cutaneous and splanchnic venoconstriction, which contributes to the rise in cardiac filling pressure.

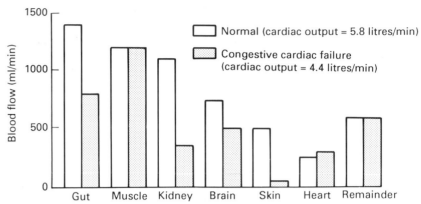

Figure 15.5 Redistribution of cardiac output in a resting patient with chronic cardiac failure and an output of 4.4 litres/min (filled columns). Note the poor perfusion of kidney, gut and skin. (From Wade, O. L. and Bishop, J. M. (1962) *Cardiac Output and Regional Bloodflow*, Blackwell, Oxford, by permission)

Although these changes may be beneficial in mild failure, they cause problems in severe failure, due partly to the harmful effects of excessive cardiac dilatation and partly to the curtailment of stroke volume when a failing ventricle has to eject against a normal arterial pressure.

Renal retention of salt and water

The kidneys retain salt and water in isotonic proportion in cardiac failure, expanding the extracellular fluid compartment by up to 30% and contributing to cardiac dilatation and oedema formation. The mechanisms underlying the salt and water retention are only partially understood but they include altered renal haemodynamics and stimulation of the renin-angiotensin-aldosterone system. Plasma aldosterone is further elevated by a reduced degradation rate in the congested, underperfused liver.

Peripheral and pulmonary oedema in cardiac failure

Oedema of the lungs and/or periphery is a prominent clinical feature in cardiac failure. The oedema is caused primarily by a rise in capillary pressure following the rise in venous pressure; pressure in the venous limbs of human finger capillaries reaches 20–40 mmHg in right ventricular failure. Another contributory factor is the fall in plasma colloid osmotic pressure by approximately 7 mmHg due to plasma volume expansion. These changes tip the balance of Starling forces across the venous capillary wall in favour of an excessive filtration rate, leading to oedema (Section 9.10). The oedema may be worse in the periphery or in the lungs depending on whether the right side or left side filling pressure is more severely affected.

Pulmonary oedema If the left ventricle is weaker than the right (as is common in ischaemic heart disease), the pressure in the pulmonary veins is raised. This is due to the operation of the Frank–Starling mechanism which ensures that left output equals right output, even in failure. If the left side transiently pumps out less blood than the right, more blood enters the left side, raising left ventricular filling pressure until by the Frank–Starling mechanism, the left ventricle output achieves parity with the right (see Figure 15.3). Because pulmonary venous pressure is raised, oedema develops in the lungs and such patients display pulmonary vein congestion (see Figure 15.4), reduced lung compliance, pulmonary interstitial oedema and dyspnoea (difficulty in breathing). This form of dyspnoea is especially marked during the night (paroxysmal nocturnal dyspnoea) because the supine position increases pulmonary congestion and pulmonary capillary filtration pressure: such patients find it more comfortable to sleep propped up by pillows. In moderate pulmonary oedema, the excess fluid collects mainly in the pulmonary interstitium around the bronchi and larger vessels but in severe pulmonary oedema the fluid floods into the alveolar spaces too, impairing oxygen transport with potentially fatal results.

Peripheral oedema If the right ventricle fails (for example secondary to pulmonary hypertension caused by lung disease), the combined effects of the Frank–Starling mechanism and renal salt-and-water retention is to raise the systemic venous pressure. This gives rise to peripheral oedema in the dependent tissues, namely the ankles in ambulant patients and over the sacrum in bed-ridden patients. Such patients show a combination of distended jugular veins and pitting ankle or sacral oedema. Not infrequently both ventricles fail and oedema occurs both in the periphery and the lungs.

Principles of treatment

The treatment of cardiac failure merits a mention here because it is an exercise in applied physiology. The aims of treatment from a physiological point of view are (1) to reduce cardiac work, (2) to reduce the excessive plasma volume and cardiac dilatation, and (3) to improve myocardial contractility if possible.

Cardiac work can be reduced by rest, by reducing arterial pressure and by reducing filling pressure. To this end, peripheral vasodilator drugs are used, such as the α_1-adrenoreceptor blocker prazosin and the calcium-channel blocker nifedipine. By reducing the arterial pressure opposing ejection such drugs improve the cardiac ejection fraction. Peripheral venodilators like nitroglycerine and nitroprusside lower filling pressure and relieve pulmonary congestion. Captopril and enalapril (angiotensin-converting enzyme inhibitors) are beneficial too.

Cardiac dilatation, plasma volume and oedema can be reduced by diuretic drugs like frusemide and the thiazides, or by captopril, which blocks angiotensin-converting enzyme and so reduces aldosterone levels. The advantage of reducing gross cardiac dilatation is that the heart can then operate at a better mechanical advantage (Laplace's law), which more than makes up for the concomitant movement down the ventricular function curve. (It is interesting to note that both vasodilator and diuretic drugs are partially reversing the natural compensatory responses to cardiac failure; one can view the natural compensations as 'overdone' in cardiac failure.)

The third line of attack is to enhance myocardial contractility by the inotropic drug digoxin. The discovery of this agent by William Withering in 1785 makes an interesting story. Dr. Withering was journeying through Shropshire when he was asked to see a woman suffering from severe 'dropsy' (cardiac failure). He could do little for her and on his return journey was astonished to find her not only alive but much improved. On enquiring, he discovered that she had been taking a local folklore remedy, an infusion of the leaves of the foxglove, *Digitalis purpurea*. The efficacy of a digitalis infusion is illustrated in Figure 3.12, and its mechanism of action (enhancing the intracellular store of calcium) is explained there. However, digoxin is also rather cardiotoxic and difficult to control therapeutically, and its use has fallen out of favour recently. An exception is made in cases of failure associated with atrial fibrillation (a common association), where digoxin also helps to slow and regularize the heart beat. In other cases, however, good responses are often obtained simply by a combination of rest, vasodilator and diuretic therapy.

The end but not the end

'Begin at the beginning', said the King, very gravely, to the White Rabbit, 'and go on till you come to the end: then stop'. Heart failure seems to offer a natural ending to this text, but the scientific investigation of the circulation is far from at an end – too many mysteries remain unsolved. The discerning reader will have recognized the superabundance of unresolved problems from the frequent use of 'perhaps', 'may be', 'probably', 'is thought to' and so on. Our subject began essentially with the work of William Harvey nearly three centuries ago, yet a comment by Harvey still makes an apt conclusion to today's textbook: 'I see a field of such vast extent . . . that my whole life perchance would not suffice for its completion'. (A. Mallock (1929) *William Harvey*, Hoeber, New York.)

Further reading

Bohr, D. F. (1989) Cell membrane in hypertension. *News in Physiological Science,* **4**, 85–88

Cowley, A. W., Barber, W. J., Lombard, J. H., Osborn, J. L. and Liard, J. F. (1986) Relationship between body fluid volume and arterial pressure. *Federal Proceedings,* **45**, 2864–2870

Geltman, E. M. (1987) Congestive heart failure. In *Cardiovascular Pathophysiology* (ed. G. G. Ahumada), Oxford University Press, New York, pp. 74–95

Janssen, H. F. and Barnes, C. D. (eds) (1985) *Circulatory Shock: Basic and Clinical Implications,* Academic Press, London

Ludbroke, J. and Evans, R. (1989) Posthemorrhagic syncope. *News in Physiological Science,* **4**, 120–133

Prewitt, R. L., Stacy, D. L. and Ono, Z. (1987) The microcirculation in hypertension: which are the resistance vessels? *NIPS,* **2**, 139–141

Sleight, P. and Freis, E. (eds) (1982) *Cardiology 1. Hypertension,* Butterworths, London

Technical appendix

The units employed below are in the main 'standard international units', based on metres, kilograms and seconds (the SI system). Some of the literature uses the older system of units based on centimetres, grams and seconds (c.g.s. system).

Arterial input impedance. Resistance is the ratio of a mean pressure drop to mean flow: but arterial pressure and flow oscillate, and do so out of phase owing to the distensibility of the arterial tree (see Figure 7.9). Consequently, the ratio of pressure to flow alters from moment to moment. To take account of this, the concept of 'impedance' has been adapted from the theory of alternating electrical currents. Arterial input impedance is a measure of the opposition of the circulation to an oscillating input (i.e. stroke volume). The input impedance depends not only on peripheral vascular resistance but also on arterial viscoelastic compliance and the frequency of oscillation (i.e. heart rate).

Avogadro's number (N, N_A). This is the number of molecules in 1 mole (1 gram-molecule) of a substance: 6.0×10^{23}/mole.

Brownian motion. Tiny random movements of supramolecular particles suspended in a fluid. First observed in a pollen suspension by the Scottish botanist Robert Brown in 1828.

Density (ρ, rho). Density is mass per unit volume. Important values in physiology are water 1.00 g/ml; blood 1.06 g/ml; mercury 13.55 g/ml; these values are at 20°C.

Electrical conductance of a membrane permeable to ions. The relation between the electrical conductance of a membrane (G) and its permeability to an ion (P) is given by:

$$G = \frac{P(V_m C_o F)}{(RT/F)^2 . (1 - e^{-V_m F/RT})}$$

where V_m is the membrane potential, R the gas constant, F the Faraday constant, T the absolute temperature and C_o is ion concentration.

Equilibrium. A system is said to be in equilibrium when its components have the same free energy level, for example, two solutions containing solute at the same concentration or, more accurately, at the same chemical potential. Equilibrium should not be confused with steady state (see later).

Faraday's constant (F). This is the charge carried by one mole of monovalent ion; 96 484 coulombs/mole.

Flux. Flux is the rate of movement of a material (e.g. a diffusing solute) across unit area of surface. In physiology the word is sometimes used loosely, omitting the 'per unit area of surface' aspect.

Force, work, energy and power. A force of one newton (N) is one that accelerates 1 kg mass at $1 \, m/s^2$; $1 N = 1 \, kg \, m/s^2$. The c.g.s unit of force, the dyne ($1 \, g \, cm/s^2$), equals $10^{-5} N$. *Work* is defined as force times distance moved by the point of application of the force. Energy is defined as the capacity to do work and has the same units as work. One joule of work or energy (J) equals a force of 1 newton displaced over 1 metre (1 Nm). It equals 10^7 ergs (the c.g.s unit, 1 dyne cm). *Power* is defined as rate of work or rate of change of energy; its unit, the watt (W), equals 1 J/s.

Gas constant (R). This quantifies the relation between energy level and absolute temperature for one mole of substance; 8.316 joules K^{-1} $mole^{-1}$.

Gravity (g). The force of gravity varies at different points on the earth's surface, depending on latitude and altitude. It is $9.81 \, m/s^2$ at latitude 50°N (e.g. Land's End, Cornwall).

Osmole. An osmole is defined by analogy with the ideal gas law; it is the mass of a substance which when distributed in 22.4 litres of solvent at 0°C exerts an osmotic pressure of 1 atmosphere. This definition stems from van't Hoff law's, osmotic pressure π = RTC, where C is molal concentration (moles/kg solvent), T is absolute temperature and R is the gas constant. A one *osmolar* solution contains 1 osmole of solute per litre of solution. A one *osmolal* solution contains 1 osmole per kilogram of solvent. Mammalian body fluids contain approximately 0.3 osmoles/kg water and have a potential osmotic pressure of 5800 mmHg at body temperature ($22.4 \times 0.3 \times 310/273 = 7.6$ atmospheres). Human plasma protein (60–80 g/l) by contrast exerts an osmotic pressure of only approximately 25 mmHg.

Pressure. Pressure is the force exerted by a gas or liquid upon unit area of surface. Pressure acts equally in all directions, unlike stress. The SI unit of pressure, the pascal (Pa), equals 1 newton per square metre (N/m^2). One atmosphere of pressure is 100 100 Pa or 100.1 kPa (kilopascals). This pressure will support a column of mercury 760 mm high so an atmosphere is commonly quoted as 760 mmHg pressure. Body fluid pressures are conventionally expressed relative to atmospheric pressure; a venous pressure of '0 mmHg' really means an absolute pressure of 760 mmHg, and an interstitial pressure of '−5 mmHg' means an absolute pressure of 755 mmHg. The pressure exerted by a 1 mm-tall column of mercury (1 mmHg) is 133 Pa or 1.36 cmH$_2$O at 20°C, being fluid height × density × gravity. The pressure exerted by a 1 cm-tall column of water (1 cmH$_2$O) is 98.1 Pa (981 dynes/cm^2) at 20°C.

Quantity of ions exchanged during an action potential. A cylindrical myocyte of length 10^{-2} cm and radius 10^{-3} cm has a volume of 3.14×10^{-8} cm^3 and a surface area of 6.28×10^{-5} cm^2. Each cm^2 of cell membrane requires 1 microcoulomb of charge to alter its potential by 1 volt, i.e. its capacitance is 1 V/cm^2 (1 milliFarad). An action potential of 0.1 V (from −80 mV to +20 mV) requires a net transfer of 6.28×10^{-6} microcoulombs per cell. One mole of monovalent ion carries 96 500 coulombs (the Faraday constant), so the quantity of ions transferred works out to be 6.51×10^{-17} moles or, applying Avogadro's number, 3.9×10^7 ions. From the cell volume and intracellular concentrations in Table 3.1, the cell actually contains 1.9×10^{11} sodium ions and 2.6×10^{12} potassium ions. The fractional change in ion concentration after a single action potential is thus miniscule.

Second messengers. The 'first messenger' is a hormone such as noradrenaline, which binds to a cell surface receptor. The receptor is an integral membrane protein and undergoes a conformational change upon binding to the ligand (first messenger). This conformational change can have two main effects. (1) It may open up a nearby ionic channel across the membrane, which either produces a small ionic current leading to a change in membrane potential (e.g. an excitatory junction potential) or else allows a major influx of ions (e.g. receptor-operated calcium channels in vascular smooth muscle). (2) It may activate one of a class of membrane proteins called GTP-binding protein (guanosine triphosphate-binding proteins or G proteins), which in turn activate membrane-bound enzymes (adenylate cyclase, guanylate cyclase or phospholipase C). The latter catalyse production of intracellular 'second messenger' at the cytoplasmic boundary. The main second messengers are cyclic adenosine monophosphate (cAMP), cyclic guanosine monophosphate (cGMP) and inositol triphosphate (IP3). The second messenger activates enzymes called protein kinases which acts on ion channels in the surface membrane or on intracellular organelles to achieve the response associated with the ligand. (Robinshaw, D. and Foster, K. A. (1989) Role of G proteins in the regulation of the cardiovascular system. *Annual Review of Physiology*, **51**, 229–244.)

Steady state. If two components at different energy levels are brought into contact, material or energy will flow from the higher level to the lower. If this transfer is occurring at a steady rate and without any change in the energy level at either end, the system is said to be in a steady state. This should not be confused with an equilibrium state (see earlier). If there is a very slow, almost negligible change in the energy levels, the system is said to be in a quasi steady state (L. 'quasi' = as if).

Stokes–Einstein radius (a, r_{se}). The hydrodynamic resistance which a solute particle experiences as it diffuses through a solvent depends partly on the size of the solute particle. The hydrodynamic resistance encountered by a sphere of known radius was worked out by Stokes and Einstein, and this enables the average dimension of a solute (whatever its true shape) to be represented by a sphere of equal hydrodynamic resistance (the Stokes–Einstein radius). The equation relating the free diffusion coefficient of the solute, D (a measure of its hydrodynamic drag), to the Stokes–Einstein radius is:

$$a = RT/D\,N_A\,6\pi\,\eta$$

where η (eta) is the solvent viscosity and T is absolute temperature.

Strain and stress. When a solid body is subjected to a force, the force per unit cross-sectional area of material is called the stress (N/m^2). The change in size, divided by the original size, is called the strain (dimensionless). The ratio, stress/strain, is Young's modulus of elasticity.

Viscoelasticity. When a solid, perfectly elastic body is subjected to a deforming stress (e.g. a spring with a suspended weight) the strain is linearly proportional to stress and is independent of time; and on removing the stress the body reverts exactly to its original conformation, without any dissipation of energy. Virtually all biological materials, e.g. artery wall, behave differently. After the application of a stress, there is an initial rapid deformation, but this is followed by a slower, ever-decreasing deformation as time passes ('creep'). On removal of the stress, the recovery of shape follows a non-symmetrical pathway (hysteresis) owing to a dissipation of energy. This form of mechanical behaviour can be mimicked by an arrangement of elastic elements (springs) and viscous flowing elements (dashpots), and is called viscoelasticity.

Index